# The Power of Technology in the Bronze Age Eastern Mediterranean

*Monographs in Mediterranean Archaeology*
Series Editor: A. Bernard Knapp, University of Glasgow and Cyprus American Archaeological Research Institute



Aimed at the international archaeological community, *Monographs in Mediterranean Archaeology (MMA)* seeks significant new contributions from the multicultural world of Mediterranean archaeology, publishing problem-oriented studies that present a solid, extensive corpus of archaeological data within a sound theoretical and/or methodological framework. *MMA* volumes deal with major archaeological issues related to the islands and lands or regions that border (or impact on) the Mediterranean Sea. No constraints are placed on the period of focus, from Palaeolithic through early Modern. We encourage contributions that treat the social, politico-economic and ideological aspects of local or regional production and development; issues related to social interaction and change or exchange; or more specific and contemporary issues such as gender, agency, identity, representation, phenomenology and landscape.

Published volumes:

*Mortuary Ritual and Society in Bronze Age Cyprus*
(Volume 9)
Priscilla Keswani

*Debating Orientalization: Multidisciplinary Approaches to Change in the Ancient Mediterranean*
(Volume 10, Volume Editor: John F. Cherry))
Edited by Corinna Riva and Nicholas C. Vella

*Rural Landscapes of the Punic World*
(Volume 11)
Peter van Dommelen and Carlos Gómez Bellard

*The Power of Technology in the Bronze Age Eastern Mediterranean: The Case of the Painted Plaster*
(Volume 12)
Ann Brysbaert

*Citadel and Cemetery in Early Bronze Age Anatolia*
(Volume 13)
Christoph Bachhuber

*The Archaeology of Nuragic Sardinia*
(Volume 14)
Gary Webster

# The Power of Technology in the Bronze Age Eastern Mediterranean

## The Case of the Painted Plaster

**Ann Brysbaert**

**Equinox Publishing Ltd**

SHEFFIELD UK BRISTOL CT

Published by Equinox Publishing Ltd.

UK: Office 415, The Workstation, 15 Paternoster Row, Sheffield, South Yorkshire, S1 2BX
USA: ISD, 70 Enterprise Drive, Bristol, CT 06010

www.equinoxpub.com

First published 2009. This paperback edition published 2015.



British Library Cataloguing-in-Publication Data

A catalogue record for this book is available from the British Library.

ISBN-13 978 1 84553 433 2 (hardback)
978 1 78179 253 7 (paperback)

Library of Congress Cataloging-in-Publication Data

Brysbaert, Ann.
The power of technology in the Bronze Age eastern Mediterranean : the case of the painted plaster / Ann Brysbaert.
p. cm. -- (Monographs in Mediterranean archaeology ; 12)
Includes bibliographical references and index.
ISBN 978-1-84553-433-2 (hb)
ISBN 978-1-78179-253-7 (pb)
1. Bronze age--Mediterranean Region. 2. Industries, Prehistoric--Mediterranean Region. 3. Painting, Prehistoric--Mediterranean Region. 4. Plastering--Mediterranean Region--History. 5. Mediterranean Region--Antiquities. I. Title. GN778.25B79 2008
930.1'509822--dc22
2008023918

Typeset by ISB Typesetting, Sheffield
www.sheffieldtypesetting.com
Printed and bound by Lightning Source Inc. (La Vergne, TN), Lighting Source UK Ltd. (Milton Keynes), Lightning Source AU Pty. (Scoresby, Victoria)

# Contents

# List of Figures

In colour: indicated with (*) before caption

# List of Tables

# List of Abbreviations

## Used in text and tables

| | |
|---|---|
| EC | = Early Cycladic |
| EH | = Early Helladic |
| EM | = Early Minoan |
| LC | = Late Cycladic |
| LH | = Late Helladic |
| LM | = Late Minoan |
| MC | = Middle Cycladic |
| MH | = Middle Helladic |
| MM | = Middle Minoan |

## Used in tables only

| | |
|---|---|
| AL | = Alalakh |
| DB | = Tell el-Dab'a |
| GL | = Gla |
| TK | = Tel Kabri |
| KN | = Knossos |
| MI | = Miletus |
| MO | = Monastiraki |
| MP | = Myrtos-Pyrgos |
| MY | = Mycenae |
| OR, ORA | = Orchomenos |
| PH | = Phylakopi |
| PK | = Palaikastro |
| QA | = Qatna |
| SA | = Tel Sakka |
| TH | = Thebes |
| TI | = Tiryns |

## Dedication

For my parents
Who taught me to follow my dreams
Wherever they would take me

# Acknowledgements

My most sincere expression of gratitude goes to Professor A.B. Knapp. As one of my former supervisors and as editor of MMA he voluntarily and generously gave of his expertise and support throughout the time I worked on this book. Dr R. Jones was my main supervisor for my PhD thesis at the University of Glasgow and he encouraged my work throughout. Many thanks go to him for his support. I want to express very warm thanks to Professor V. Perdikatsis at the University of Chania, Crete, who taught me many aspects of instrumental analysis on painted plaster and who has been a great collaborator for almost a decade now. Dr S. Fox, director of the Wiener Laboratory-ASCSA, is gratefully acknowledged for her practical and moral support during long periods in the laboratory between 2001 and 2004, and on several occasions after that. My colleagues in the Department of Museum Studies, University of Leicester, are warmly thanked for allowing early study leave so that this book could move forward. I also want to thank Professor M. van der Veen for her advice in writing the book proposal and for useful discussions about the book structure. Jeanette Atkinson and Heidi Robbins were the best proofreaders I could imagine for this book—thank you very much.

The XRD and some of the SEM work were carried out at the laboratories of the Institute for Geology and Mineral Exploration, Athens, and I would like to thank the staff of the Institute, especially Dr G. Economou, Dr D. Tarenidis, Dr E. Dimou, M. Sakalis, V. Spiropoulos and P. Salaounis for all their kind and often last-minute help. Many thanks go to Professor L. Moens and Dr P. Vandenabeele at the University of Ghent, Belgium, for their time and preparedness to share their knowledge on MRS. At the Foundation for Research and Technology, Hellas – Institute of Electronic Structure and Laser on Crete, Dr D. Anglos and Ms K. Melessanaki invited me to employ and test out the LIBS instrument on a set of my samples. Thank you very much for a great collaboration.

Many field directors and Ephoreias have expressed their trust in my project by granting permission to study and sample material. My deep gratitude is expressed to all of them, especially the following:

Dr V. Aravantinos, Ephor of Boeotia, Dr A. Papadimitriou, Epimelete at the Ephoreias of the Argolid, Dr A. Karetsou, Ephor of Central Crete in 2000, Dr N. Papadaki, Ephor of east Crete in 2000, Dr M. Marthari, Ephor of the Cyclades, Dr M. Vlassaki, Ephor of

Chania, Dr E. Kountouri at Thebes, Professor S. Iakovidis at Gla, Professor K. Fitsschen at Orchomenos, Professor J. Maran at Tiryns, Dr E. French at Mycenae, Dr N. Momigliano and Professor D. Wilson at Knossos, the BSA at Knossos, Dr A. MacGillivray and Mr H. Sackett at Palaikastro, Professor G. Cadogan at Myrtos-Pyrgos, Professor A. Kanta at Monastiraki, Professor C. Renfrew at Phylakopi, Professor W.-D. Niemeier at Miletus and Tel Kabri, Dr R. Moorey at the Ashmolean Museum for Alalakh, Professor M. Bietak at Tell el-Dab'a, Professor P. Pfaelzner at Qatna and Dr A. Taraqji at Tel Sakka, Dr J. Seeher at Hattusha, and Dr N. Sheftelowitz and Dr A. Yassur-Landau for Tel Kabri.

This book, including analytical work and travel, has been funded over the years by the following organizations: University of Glasgow, The Flemish and Greek Governments, Wiener Laboratory-ASCSA, COST G8, Bryn Mawr College (Mary Hamilton Swindler Memorial Prize for Aegean Painting and Art), SCIEM 2000, INSTAP, M. Aylwin Cotton Foundation, BSA, University of Leicester, Department of Museum Studies, British Academy.

Long periods of time were spent at the BSA. I would like to thank Mr D. Blackman and Ms H. Clark for their invaluable help with permit applications. At the BSA at Knossos, my warm thanks go to Dr E. Hatzaki and Dr Don Evely for facilitating my periods of study.

It was in beautiful Tuscany where I ended up learning to mix and apply plaster, and to paint *al fresco*. I very much want to thank Mrs S. Hocombe for her patience with an untalented student.

Writing a book is not an easy job and I realized that the learning curve is steep and not just in the academic sense. More than once these past years have seemed like a rite of passage for me, and without the constant and unconditional love and support of my parents, M. Brysbaert and N. Claeys, I would never have come to a conclusion.

My special gratitude goes to Roxana Docsan, who constantly offered advice on illustrations and painstakingly prepared the fantastic maps for this book, and who has been a close friend for over a decade.

I want to thank a real host of friends in Belgium, Greece, the UK, Germany and the Netherlands, all of whom have kept me away from my desk and computer in order to preserve my sanity. In the company of Eleni, Iota and Manoli, Melissa, Marina, Clare, Katherina, Lisanne and Rob, Virginie and Manoli, Don, Rob, Alain, Roxana, Astrid, Diana and Jeff, Anne and Yanni, Heinrich, Rana and James, Jeanette, Kosta and Anna, Maria and Kosta, Alexandros, Sheila, Alida and Giorgo, Areti, Ingrid, Pelagia, and many others, these years have been some of the best of my life.

I would like to thank the two anonymous reviewers who provided me with many useful and constructive comments about this manuscript in order to improve it to this level. Any remaining erors are my own.

# 1 A Tale of 'Frescoes'

## Introduction

This book is concerned with understanding the relationship between power and technology during the eastern Mediterranean Bronze Age. More specifically, it explores issues about the power of technology and technological knowledge, and about craft specialization and its role in the relationships pre-industrial people had with each other. Its content, therefore, explores the role of material culture and technology in expressing dynamic social identities and in forging complex interwoven human relationships. In other words, this book is about past people's actions and interactions, what they thought and felt and expressed through their material world. To contextualize these concepts, the book analyzes and assesses the intriguing case study of Bronze Age Aegean and eastern Mediterranean painted plaster (Figure 1.1) and the current question about the direction of technological and iconographic transfer therein. Briefly, 'transfer' is understood here as the act of passing on, or transmitting from one person or group to another, a set of materials, concepts, ideas and possibly also knowledge and designs. This transmission can take place on a technological level or an iconographic level or both. The book also discusses how and why technological transfer developed and what broader impact this had on the wider social dynamics of the late Middle Bronze Age (ca. 2050–1700 BC) and Late Bronze Age (ca. 1700–1200/1180 BC, Table 1.1) in the eastern Mediterranean.

This study offers new insights into the social relationships and power structures that existed among a variety of peoples involved with several aspects of technology. It also confirms one of the existing hypotheses about travelling artisans in which they travelled from the Aegean to the east; their direction and the data set presented here form a much firmer basis upon which such a statement can be constructed. As a result, the book encourages the archaeologist, art historian or general reader to continue, or start thinking about, approaching material culture from different angles and combining the data sets generated. The book suggests that a single approach to material culture, such as for example iconographic *or* technological *or* stylistic studies of a given group of objects, results necessarily in a single data set, saying one thing about the material under study. Such singular approaches, grounded in the traditional divide between 'arts and sciences' (Dobres 2000), can no longer be sustained when artifacts/objects are to be compared to each other and are to be placed in their economic and socio-political contexts where their

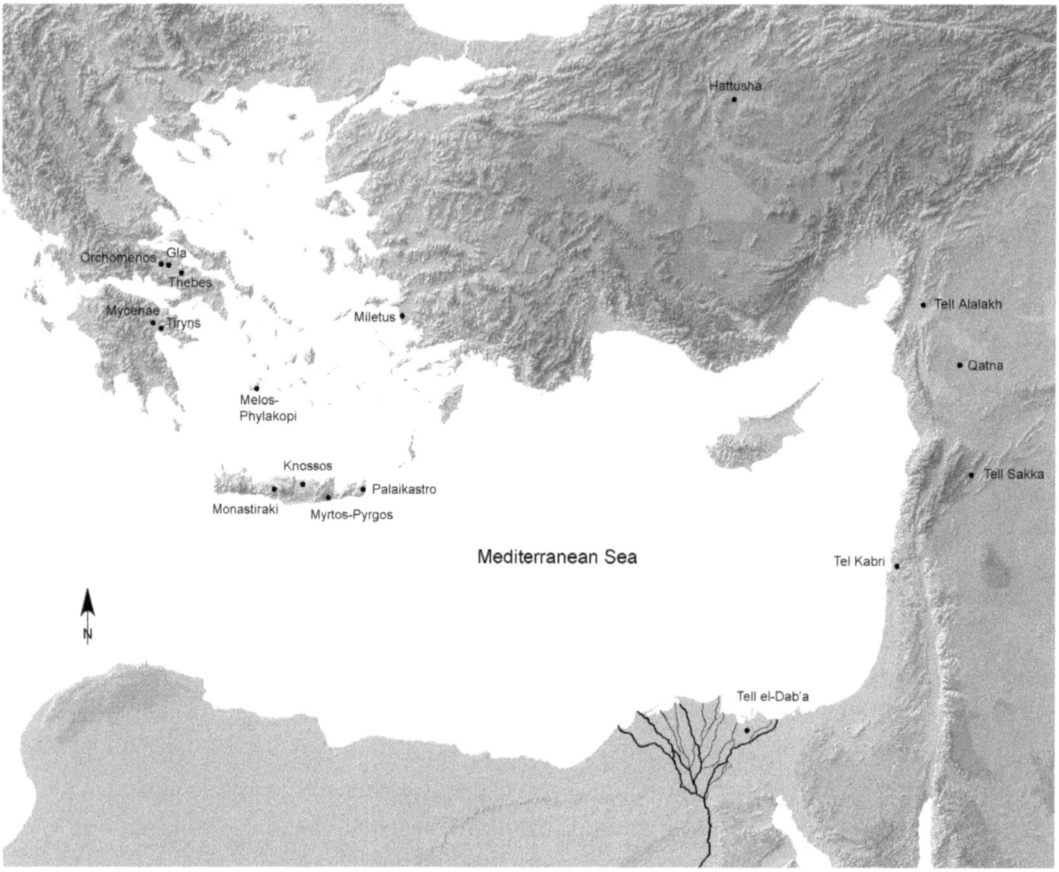


Figure 1.1 Map of the eastern Mediterranean indicating the sites investigated.

| | *Crete and Cyclades* | *Greek Mainland* | *Eastern Mediterranean* |
|---|---|---|---|
| *Chronology* | MM IB-II (20–/19th–18th centuries BC) | LH II/IIIA1 (1400 BC) | 1) Middle Bronze Age (17th century BC) |
| | Until | Until | 2) Late Bronze Age (16–14th centuries BC) |
| | LM/LH IIIB (1200 BC) | LH IIIB (1200 BC) | 3) LM IA (17th century BC) |
| | | | 4) 18th Dynasty |
| *Sites* | Knossos | Mycenae | 1) Alalakh (Turkey) |
| | Palaikastro | Tiryns | 2) Hattusha (Turkey) |
| | Monastiraki | Thebes | 3) Miletus (Turkey) |
| | Myrtos-Pyrgos | Orchomenos | 2) Qatna (Syria) |
| | | Gla | 1) Tell Sakka (Syria) |
| | Phylakopi (Melos) | | 1) Tel Kabri (Palestine) |
| | | | 4) Tell el-Dab'a (Egypt) |

Table 1.1 Spatial and chronological outline of the sites studied in the book.

meaning becomes overt. Since an artifact can no longer be considered purely aesthetic *or* functional but actually may have both attributes as well as several social attributes, this should subsequently guide our way of looking at it, thinking about it and analyzing it in a much more holistic way.

It has long been clear that painted plaster played a significant role in the Aegean and eastern Mediterranean, not only in the existing architecture but also in the broader scale of themes and motives, when brought into contact with other artifacts (see for instance the older publications by Evans 1921; Woolley 1955), and in their technological make-up (Heaton 1910; 1912; Woolley 1955). These paintings have been extensively studied for over a century now resulting in numerous publications, some of which are discussed below (e.g. Immerwahr 1990; Nunn 1988). Apart from the few times that the artisans were mentioned in this context, these paintings were 'orphans' for a very long period and were never really looked upon as the product of real people (but see Cameron 1975). This book, therefore, also emphasizes the importance of looking at material culture from a range of angles and allowing people to play their role.

Another reason for writing this book is to try to offer a broader perspective on painted plaster studies in the eastern Mediterranean, and more specifically on technological studies. Although such studies were often carried out alongside iconographic studies, they concentrated on the well-known, largest, 'most significant' palatial sites in the Aegean only. Consequently no comparative study employing a holistic approach has ever been undertaken in this context, encompassing a broader range of sites in the Aegean and applying it to the larger area of the eastern Mediterranean. This book thus forms an update in our knowledge of the technology of painted plaster in this region during the late Middle Bronze Age and Late Bronze Age. It furthermore demonstrates that technology, iconography and style are intricately interwoven with each other. This interwovenness developed over time and culminates in a full-on 'technological style' from the final phase of the Middle Bronze Age onwards into the Late Bronze Age.

By looking at painted plaster from a variety of angles, the book tries to answer certain questions which touch upon two main issues. These issues are, first, the pure technological aspects of the plaster, and, second, social structures underlying technological transfer, resulting in the following questions:

- Can evidence be found for technological transfer within the Aegean painted plaster tradition, and between this tradition and that in the eastern Mediterranean?
- If so, is this technological transfer in support of the well-studied corresponding case of iconographic transfer?
- What are the forces behind such a technological transfer?
- Based on the data available, can we investigate human agency and social identities within the sphere of craft specialization in the eastern Mediterranean Late Bronze Age?

By studying these paintings in an interdisciplinary manner, multiple strands of information about them were uncovered. Experimental replication of some of the paintings, for instance, provided, on the one hand, clear insights into technical aspects and, on the other hand, illuminated specific human aspects of the craft, its labour division and complexity. When people and their interactions take centre stage, it can be demonstrated exactly how transfer was facilitated and how people's social identities and their interrelationships were expressed through what they made and through the ways they organized themselves.

Similar studies have been published in Costin and Wright (1998), but none dealt with the unique situation of the Aegean or eastern Mediterranean Late Bronze Age. The smaller states of the Aegean and eastern Mediterranean were influenced by the larger states and empires of ancient western Asia (some of which are discussed in Costin and Wright 1998), but equally they were also in a position to interact intensively with these larger states. The unique situation of the Aegean area was the result of such Near Eastern influences combined with their own local environment, geography, material and other riches. As a result, they created their own unique style, visible, for instance, in their painted plaster. Exactly how this uniqueness came about is an issue treated in detail in this study.

As a result of the data produced via an interdisciplinary research project, this book suggests and demonstrates that the technological transfer most likely moved from west to east and that this art/craft was very much under the control of a social elite. It also shows how the smaller regions of the Aegean were capable of staying within the complex trade and exchange networks that comprised the large powers of the eastern Mediterranean and Egypt. The people of the Minoan, Cycladic and Mycenaean cultural regions had their own resources to offer in Late Bronze Age trade, gift-giving and exchange networks, and these were clearly appreciated at several large centres. We see this, for example, in the presence of oil and wine vessels in many centres in both the eastern and western Mediterranean (Buxeda I Garrigós *et al.* 2003), and it has been suggested that Aegean textiles were also widely appreciated (Barber 1991: 340; Stevenson Smith 1998: 138-39).

Despite its often fragmentary state, painted plaster is very visible in the archaeological record. Since this plaster cannot be 'transported' without its artisans, I therefore suggest specific interactions and mobility of several groups of people. At all of these larger eastern centres (Alalakh, Miletus, Hattusha, Qatna, Tel Kabri, Tell el-Dab'a) other material remains of Aegean origin were found. These finds reinforce the notion of contact, but they do not necessarily suggest the physical presence of Aegean people on a permanent basis. The physical production of paintings on the walls in these locations required the temporary presence at least of a small team of painters and plasterers. Precisely this need for a temporary presence of the artisans forms an argument in support of travelling artisans. The relationships of these artisans to their society and the elites they worked for in turn bring to light broader aspects of contact, trade and exchange mechanisms during the late Middle Bronze Age and Late Bronze Age.

Beyond this tale of 'frescoes', however, fundamental patterns of power relationships form the basis for understanding how human interaction via and through technology are crucial to any given pre-industrial society supported by, or founded upon, groups of craft specialists. In pursuit of such insights, this book, therefore, offers different ways of looking at the role that technology and its associated aspects (aesthetics, functionality, style and many more) play within pre-industrial societies, and how power and technology, as a result, are intimately interlinked. Considering the complexity of the relationships between people and their material culture, an interdisciplinary approach by combining instrumental analysis, experimental replication and social theory offers a promising methodology for acquiring insights into this remarkable labyrinth of human relationships that produced the material culture during the Late Bronze Age.

Structurally, the book is divided into two parts. Part One (Chapters Two to Four) offers the theoretical, methodological and practical foundation delineating its interdisciplinary character, while Part Two (Chapters Five to Seven) provides concepts and ideas developed previously. Hence it concentrates on the case study of Aegean and eastern Mediterranean painted plaster and offers contextualized results and meanings.

After a summary overview, presented within this chapter, about research done in the wide field of painted plaster, Chapter Two formulates and discusses existing ideas and theories about technology and how it is interwoven with human action. This is demonstrated, for instance, by means of the concept of *chaîne opératoire*. Furthermore, the chapter explains who the artisans were, how they were organized, and what their social identities may have been. Traditionally, pre-industrial societies and their artisans have been seen as subordinate to the elite upon whom they 'depended' and for whom they worked (e.g. Zaccagnini 1983). Chapter Two, therefore, formulates the different facets of 'power' and how each relates to the context of the actor(s) involved. These contextualized relationships are then developed further through an approach whereby people, their changing and shifting identities and their social organization are central to considerations of technology and technological activities.

Chapter Three deals with issues of iconography, technology and style in visual art expressions. It explains what each of these is, how they were studied separately in the past and why they can and should be viewed together when investigating visual art forms. Technology is seen as an integral part of the aesthetic forms of any aspect of material culture as much as iconography and style are, and their interrelationship is indicated with the concept of 'technological style' (Lechtman and Steinberg 1979). An interdisciplinary approach reveals multi-faceted identities, and the study shows and explains how experimental work and scientific approaches can contribute to discussing social issues in Chapter Four.

Having formulated and outlined the theoretical, methodological and practical framework in Part One, the chapters in Part Two present the case study in its various aspects. Chapter Five starts off by giving a summary outline of some specific trends that seem to have emerged from cross-cultural contacts and crafts in the east Mediterranean. The chapter then goes on by explaining the role of painted plaster in the region. It subsequently presents a geographical and chronological overview of the selected sites (see Table 1.2), outlining where painted plaster was found and what

| *Region/country* | *Site* | *Date* |
|---|---|---|
| *Crete, Greece* | Knossos: South East Houses, MUM† | EM III, LM I |
| | Palaikastro, Building 6 | MM-LM |
| | Monastiraki: Palatial structure | MM I-II |
| | Myrtos-Pyrgos: Villa | LM IA |
| *Melos, Greece* | Phylakopi | LM IA |
| *Argolid, Greece* | Mycenae: Early palace, below HOM‡ | LH IIA, IIIA2 |
| | Tiryns: Old palace, Epichosis, (Lower Citadel) | LH IIIA, IIIB, (IIIC) |
| *Boeotia, Greece* | Thebes: Old Kadmeion, Treasury, O. Spourli | LH IIIA1, LH IIIB1, LH IIIB |
| | Orchomenos: Palatial structure, area T | LH IIIB |
| | Gla: Melathron, Room N1, Room H | LH IIIB1 |
| *East Delta, Egypt* | Tell el-Dab'a: H I, H II, H III, H IV | 18th Dynasty |
| *Antakya, Turkey* | Alalakh: Palace, level VII | MB* IIB |
| *Asia Minor, Turkey* | Miletus: Building A | LM IA |
| *Hattusha, Turkey* | Hattusha Temple 5, 9, palace Büyükkale | LB** |
| *Israel-Palestine* | Tel Kabri: Palace | MB* IIB |
| *Syria* | Qatna: Palace: Room N | LB** |
| | Tel Sakka: Palace: central court | MB IIB |

**Table 1.2** Overview of the sites studied for this book, and their dates (* Middle Bronze Age, **Late Bronze Age, †MUM = Minoan Unexplored Mansion, ‡HOM = House of the Oil Merchant).

its main features are. The importance of the painted plaster found on those sites is discussed within the context of its overall architecture and socio-political function.

Chapter Six combines the data from instrumental analysis with experimental replication to discuss the material culture remains of painted plaster and demonstrates the existing and inherent technological and iconographical interrelationships. Through the elucidation of patterns of material transfer and mobility, the study shows that the triptych of technology, iconography and style is clearly interwoven and forms a 'technological style' in its own right in the Aegean painting craft.

Combining the above meaningful data sets with human input and impact, Chapter Seven demonstrates how and why the painted plaster remains also an aid in shaping the complex social relationships and shifting social identities of people. The chapter suggests that the human actor plays a central role as the connector between all relationships and any significant aspect of painted plaster technology, irrespective of the latter's level or scale.

Chapter Eight draws together Parts One and Two by synthesizing all the arguments. It demonstrates the importance of studying human agency in painted plaster production in order to illuminate the importance and power of technology and trade in the eastern Mediterranean Late Bronze Age's socio-political and economic context as a whole. The study finally shows how these socio-political and economic changes in pre-industrial societies have had a direct impact upon the changes in material culture as a way of expressing, confirming and manipulating social identities and relationships.

## Analytical, Technological and Conservation Studies

While there is no doubt that iconographic studies of painted plaster are crucial to answering many archaeological questions, reviewing all of them is far beyond the scope of this book. This summary overview, therefore, focuses specifically and only on technological studies, which is a broad field in its own right. I prefer to see these technological studies divided into three groups, namely, (1) identifying materials and painting techniques, (2) presenting new techniques in the field and (3) conservation studies. Only the first group deals explicitly with the region and period of interest. The other two groups are discussed below only when they refer specifically to the material under study here. Mora *et al*'s book, *Conservation of Wall Paintings* (1984), however, which refers to the material from the Bronze Age Aegean and Alalakh as the oldest 'frescoes' in the world, should be mentioned because they indirectly make a statement about the direction of transfer of the paintings under study here.

An overview and discussion of technological studies in the Aegean has been recently published (Jones 2004; see also Jones 1999), which focuses mainly on materials identification, the state of knowledge on tools and installations, and on experimental replication studies carried out until 1998. My summary overview focuses on conservation issues and on the technological studies that have appeared since 1998, and on those studies that deal with social aspects of the craft of painted plaster.

### The Aegean

Conservation and preservation studies include a range of factors, of which the most important are:

- Condition survey
- Passive conservation treatment
- Active conservation treatment
- Documentation of every step undertaken

Depending on the preservation state of painted plaster and subsequent conservation treatments, each of the above four factors has its impact on the condition of excavated painted plaster, thus also on the level at which painted plaster material can be studied.

Bosanquet, who excavated Palaikastro and Phylakopi, carried out the first chemical tests on specific pigments at the National Museum in Athens. Most interesting was his worrying observation on the blue pigments at Phylakopi, which he described as 'changing over time' (Atkinson *et al.* 1904: 71, 79), possibly referring to organic-based pigments. Brysbaert and Vandenabeele (2004; 2006) discuss in this context the analysis of a specific organic pigment from Thebes, and Brysbaert (2006) reports on the consequences of analyzing another organic pigment from Gla.

Heaton (1910: 211-12; see also Lang 1969: 7), Evans' scientific adviser at Knossos, referred to the difference between wear and tear during the life of painted plaster as a functional mural and the decay during burial. Decay and loss *upon* excavation, however, were never mentioned by Heaton, but they were taken into account when interpreting data for this study.

Evans, too, was concerned with the conservation of the excavated remains at Knossos (Brown 1994: 12). More specifically, his great interest in the painted plaster made him decide to:

> restore *in situ* the painted stucco coating of the walls…This has been attempted by means of permanent copies, in the case of … the Priest-King Relief' in the South-North Corridor (Evans 1928: 353, 710, 798; and 1935: 5).

This specific case is very interesting because Evans knew that these fragments did not come from one scene only. So, in addition to setting the figure in a wooden frame, an erroneous replica was ordered for the Palace (see also Niemeier 1988), based on a few fragments from different deposits (see early notebooks at the Ashmolean museum). Evans was not too concerned about the original location of these paintings (but cf. Evans 1928: 710, 798), as can be seen in the room above the Throne Room at Knossos (Brysbaert 1996: 33, citing Cameron 1993: 103, 109).

In a conservation report published for the Ayia Triada sarcophagus, Levi (1956: 193-95) deplored bad conservation treatments, their deleterious effects upon painted plaster and the fact that no scientific studies had been conducted upon these materials. His correct methodological approach to fragile material contrasted starkly to what happened elsewhere (e.g. flattening of painted plaster at Amnissos, use of shellac as adhesive and consolidant).

M. Cameron was very interested in an interdisciplinary approach as is proven both by his scientific collaborations (Cameron and Jones 1976; Cameron *et al.* 1977), and by his experimental replicas (Cameron *et al.* 1977; see also Jones 1999; 2004: 221-22). He also cared for preservation and conservation issues (Cameron 1964: 48, 51; 1967: 46, 50, 58). His chapter on 'media of restoration' and treatment recommendations, however, (Cameron 1976: 25-26) is now out of date. Another work on Minoan and Mycenaean floors by Hirsch (1975: 261-66; 1977) provided interesting results employing an innovative approach with non-destructive infra-red equipment. Lastly, the exceptional preservation situation at Akrotiri has led to now-standardized methods of recording fragmentary paintings during their excavation (Patrianakou-Iliaki 1978).

Moving on to technological studies, Forbes' massive work on ancient technology should briefly be mentioned here because he locates the origin for *al fresco* painting (and *fresco-secco*) in Crete and not in Egypt or the Near East (Forbes 1966a: 249-50). For Forbes, it must have been obvious that the techniques used in Egypt were not *al fresco* or even *fresco-secco*. Forbes' writing has influenced later discussions about technological and iconographic transfer of the painting tradition from east to west or west to east.

The most thorough studies of Bronze Age painted plaster were conducted at Knossos (Cameron *et al.* 1977) and at Akrotiri (Perdikatsis *et al.* 2000). The only missing information from the Knossos study is a microscopic cross-section study, especially on fragments of the pre-second palace phase (MM IB-MM IIB or ±1900–±1740 BC) which, in my opinion, is crucial for understanding the formation of *al fresco* painting method on Crete.

On the one hand, the most recent technological studies are few and, as before, they each focus on one site (Perdikatsis 1998; Perdikatsis *et al.* 2000; Photos-Jones *et al.* 2003). Others are based on very few samples for comparative work and sometimes use just one parameter to make statements about painting techniques (Dandrau 1999; 2000; 2001). On the other hand, they show a new generation of work, including different analytical instrumentation such as the Scanning Electron Microscope and microprobe, good use of the reflected light microscope on cross-sections and, most recently, digital imaging in combination with image analysis software (Brysbaert and Perdikatsis 2008: Brysbaert 2004b). The main downside of the entire discipline is still the predominant focus on material from palatial sites or on very well preserved examples, and even this book is guilty of that to some extent. E. Chryssikopoulou's PhD in progress may shed more light on non-palatial sites because she has sampled widely both on Crete (Zakros palace and houses, Petras, Galatas, Prasa, Amnissos) and outside Crete (Akrotiri on Santorini and Trianda on Rhodes). The comparative work between Zakros town and its palace material will be of particular interest. To my knowledge, none of this work has been published apart from the technical study at Akrotiri (Perdikatsis *et al.* 2000), while the Zakros work has been presented in a recent Cretological Conference (Chania, October 2006). Similar comparative projects at Mycenae and Tiryns are under way on palatial and non-palatial plaster by the author. Since 2000 an interdisciplinary technological study of the painted plaster at Pylos is being undertaken under the general direction of J. Davis and S. Stocker. Of interest is their discovery of organic binders, indicating *al secco* additions to certain *al fresco* paintings (see Lang 1969: 10-11, 16, 23). The publication of this work is keenly awaited.

Another encouraging trend is the macroscopic work presented at the First International Wall Paintings Symposium on Thera in 1997. Apart from producing interesting results with regard to the use of technical painting aids (Birtacha and Zacharioudakis 2000), such work is totally non-destructive. Another purely macroscopic study was carried out on the Palaikastro painted plaster (Brysbaert 2000).

Finally, two pilot studies involving non-destructive analytical techniques have been published recently (Brysbaert and Vandenabeele 2004; Brysbaert *et al.* 2006). The use particularly of micro-Raman Spectroscopy seems to be extremely promising because the technique can also analyze organic pigments, while actual sampling for analysis with Laser Induced Breakdown Microscopy is virtually unnecessary. Several recent studies provide the first evidence for organic pigments on Bronze Age Aegean painted plaster (Chryssikopoulou and Sotiropoulou 2003; Brysbaert and Perdikatsis 2008; Brysbaert and Vandenabeele 2004; 2006; Brysbaert 2006).

## Egypt

Early site reports on Egyptian material offered interesting conclusions on the transfer of materials: red ochre was considered to be imported from Crete into Egypt (Petrie 1903: 38-39). Material identification and conservation reports continue to be published up to the present day (Rassart-Debergh 1997; Preusser 1991; Seeber 2000;

several papers in Davies 2001), but very few deal with the plaster itself. Robins (1994; 2001), however, provides indirect clues about the plaster and painting techniques in her research on the use and application of the grid system.

In their overview of work done in the Aegean, Duell and Gettens (1942: 203-204) mentioned that the *al secco* painting technique was applied in Egypt whereby pigments were mixed with a medium and painted upon a dry support. This was confirmed by Lucas and Harris (1962: 74-79, 339-54) and by Lee and Quirke (2000: 104-120); the change from gypsum to lime plaster only occurred during the Ptolemaic Period (Lucas and Harris 1962: 73-74; but see Lee and Quirke 2000: 118; and Gourdin and Kingery 1975: 150). Robins (1994: 26, 119-21; 2001: 60) pointed out that the grid system was applied to a dry wall with red paint, but there were technical changes over time, especially during the Amarna phase (see Lucas and Harris 1962: 78; Weatherhead 2001: 58). Lucas and Harris, Forbes (1966a) and Lee and Quirke (2000) provided overviews of pigments, binding media and plaster and their use, even beyond murals.

Preliminary analytical results from Tell el-Dab'a were recently published (Seeber 2000). I too analyzed material from this site, and the results are discussed in Chapter Six in (see also Brysbaert 2002), and in a recent publication (Brysbaert 2007c). An interesting ongoing project is under way at the Max Planck Institute, where El Goresy (2000) managed to use the composition of Egyptian blue as a dating method for Egyptian painted plaster. Therefore, the material from Tell el-Dab'a plays not only an important role in Egypt and its relation to the Aegean. A different picture, however, emerged from Aegean Egyptian blue, which I describe in Chapters Six and Seven.

Specifically in relation to wall paintings, Swindler (1929: 40) mentioned finds such as colour, brushes, mortars and pestles (see also Polz 1997: 34-35), mixing jars and palettes from Egypt. These tools, combined with wall paintings depicting painters at work, inform us about this craft in Egypt. Unfortunately for the Aegean, many of these organic tools were not preserved (but see Blitzer 1995: 456-58; Evely 2000).

### The Near East and the Levant

In comparison to Egypt and the Aegean, few studies have dealt with iconography and the technology of painted plaster in the Near Eastern Bronze Age (see Frankfort 1996; Nunn 1988). Apart from Woolley's early technical account (1955: 228-31), Moortgat's studies (1959; 1969: 17, 70, 84, 99, 118) on Near Eastern painting were the earliest dealing with technical aspects. The most recent overview was provided by A. Nunn (1988), who presented the state of knowledge on pigments, plaster and painting techniques. Painted lime plaster was known by 8500 BC at Ain Mallaha III, in the southern Levant (Nunn 1988: 10, 14). The idea of using lime plaster, and the knowledge of its pyrotechnological process (Bar-Yosef 1983; Gourdin and Kingery 1975: 133-34; Kingery *et al.* 1988: 226) might have been shared and transferred between the Near East and northern Greece by groups of people leaving the Anatolian and/or Levantine areas (Perlès 2001: 54-56, 63, 96).

One incomplete report for pigment use and painting techniques at Tel Kabri in Israel has been published recently (Cornale *et al.* 2002: 287-88), while Moorey (1985: 188-93; 1994) provided an in-depth study of materials and techniques used in the Near East in which the origin and appearance of Egyptian blue is given some attention (see also Lee and Quirke 2000: 110). Moorey also mentions that the precise painting technique of western Asiatic murals has not yet been analyzed (1994: 328-29). Chapter Six of this book has now closed this gap partially, because it deals with most known painted plaster from this area and period, apart from the paintings from Kar-Tukulti-Ninurta at Ashur (Assyria), those from Mari (Syria), the paintings from Yorgan Tepe (Nuzi), from Dur-Kurigalzu (Babylonia, Kassite art) (all discussed in Moortgat 1969), and the recently found paintings from Tell Burak (Lebanon, work is ongoing).

The most recent studies on technological aspects—including Near Eastern examples from Miletus and Alalakh—that have emerged from my own research (Brysbaert 2002b; 2003b) present non-analytical comparative results between Near Eastern, Egyptian and Aegean painted plaster based on non-destructive macro- and microscopic observations. Analytical work from Near Eastern sites (including Qatna and Tell Sakka) and its results have been presented in full detail in Brysbaert (2004b). This book also integrates the study of the paintings found at Hattusha and Tel Kabri. Finally, Feldman's recent publication (2006) makes no reference to the painted plaster, and while she erroneously states that applying scientific materials analysis leads to inconclusive results (Feldman 2006: 4), this is contradicted in several cases in my own work here (see Chapters Six and Seven).

## Social Aspects to the Interface between Technology, Iconography and Style

A small number of studies have touched upon the social context of the artisans (for Egypt: I. Shaw 2004), and some have tried to explain the apparent transfer of themes and motifs both geographically and chronologically (Crowley 1989), which resulted in a wide-spread artistic *koine* (Boulotis 2000). These explanations are based on: (1) the idea of travelling craftspeople executing paintings wherever they were called upon by the royal courts or local ruler; (2) the organization of these artisans among themselves, including possible training schools; and (3) the interrelation between the artisans and the people they worked for.

Certain of these references used, firstly, analogies of travelling craftspeople known from cuneiform evidence such as that from Mari (Zaccagnini 1983; Caubet 1999; Durand 1999; Guichard 1999; Niemeier and Niemeier 2000: 790), Amarna (Haider 1996) and Ugarit (Walls 1996), where particular rulers were asking for artisans to carry out specific tasks (Egypt: Matthews 1995; Mesopotamia: Drenkhahn 1995; Feldman 2006: 12). While analogies with the artisans' situations in the Near East and Egypt seem logical, no positive written evidence has ever been found in the Aegean, apart from references in the Linear B tablets to foreign female slave use (in textile production) in Mycenaean palaces (see Killen 2001: 172; Warren 2000: 24) or to the purchase of slaves at Knossos (see Shelmerdine 2001: 359 referring to Killen 1985; Olivier 1987: 497). Muhly (2005) and Hitchcock (2005) wrote some of the most recent papers about craftspeople and the specific aspect of 'travelling'. From their papers it is clear that craftspeople were travelling only under specific conditions within elite contexts. Moreover, no written evidence seems to exist on travelling painters anywhere in the eastern Mediterranean. Documentary evidence from Egypt and the Near East, however, does refer to the Aegean (Amenophis III at Kom el-Hetan, see Warren 2000: 25; Mari letters referring to Kaptara involved in trade contacts, see Cline 1994: 27; Portugali and Knapp 1985: 66). Mythology has been used to explain the phenomenon of travelling artisans, more specifically referring to a mythological palace builder and decorator from Crete (Morris 1989; 1993: 73-101; Niemeier 1991: 199; Niemeier and Niemeier 2000: 790-93). Zaccagnini (1983: 245-64) brought together many social issues about artisans, several of which may potentially be applicable by analogy to the Aegean Bronze Age, if applied cautiously (see Chapter Seven).

Egyptian iconographic evidence has been used to illustrate that Cretans, or possibly 'Keftiu', were in contact with Egypt (e.g. Koehl 2000: 97, fig. 3). Their typical hairstyle, clothing and tributes shown in the tombs of Senmut, Menkheperra-Seneb and Rekhmire at Thebes (all 18th Dynasty) are thought to represent Minoans (Koehl 2000; Stevenson Smith 1998). Moreover, a British Museum papyrus shows Egyptians fighting together with men in 'Aegean-style' clothes and helmets (Parkinson and Schofield 1995: 125-26). Recently, a similar iconographic connection has been made between the Aegean and the Mari paintings (Kanta 2003: 24), but this cannot be supported by material evidence (see Chapter Seven).

The organization of the artisans themselves may have provided them with either a freelance status (Negbi 1994: 87; Boulotis 2000: 846), or their complete dependency on the elite (Televantou 1992: 146; Boulotis 2000: 846; indirectly: Niemeier and Niemeier 2000: 792-93). Other scholars have hypothesized that the painters may have accompanied the supposed Aegean-Knossian princess who formed part of a diplomatic marriage in Egypt at Tell el-Dab'a (Hankey 1993: 13-14; Bietak 1992: 28; 1995: 26-28). There is also the notion of Minoan painters trained in the Knossian tradition who came to Tell el-Dab'a to paint its walls (see Bietak 1992: 27). Schools of painters (Cameron 1975: 360, 466, 569-70), and even their individual styles, have been recognized, and were held responsible for similarities between paintings in different regions (Cameron 1978: 588; Militello 1998: 274-75; Shaw 1996). While iconography may be the best way to approach this issue, technological studies at a micro-level have certainly added to this debate (see Chapters Six and Seven). Finally, professional interaction between artisans from different crafts has been observed and studied iconographically (Boulotis 2000; Blakolmer 1999; Marthari 2000 and many more). We still lack a certain technological support and a social framework into which to place these data for most of these interactions. Feldman (2007) has made an attempt to place a small selection of these paintings into a social framework, but had no proper technological data to work with, hence her quickly produced and incomplete picture. In Chapter Seven I have sought to bring together the social theory and analytical data in order to provide insights into the social aspects of cross-craft interaction visible in the painted plaster craft (Brysbaert 2007a).

Lastly, M.C. Shaw's paper (1997) studied the relationships between patrons and painters. The evidence suggests the presence of these paintings in elite contexts, but that does not explain the actual relationship(s) between the parties involved. One of the missing aspects is any consideration of the actual logistics involved in the transport of such artisans. Warren (2000) has touched upon this in a general fashion, while the session on 'social dimensions' in the First International Symposium on Theran wall paintings was revealing (Sherratt 2000, vol. II). Even though this session focused almost entirely upon the Theran situation, the papers by Davis, Marthari, Televantou and Boulotis show a shift in approach from studying the pure material as a self-contained entity towards 'putting the people back in the picture'. Without a technological and socially informed framework, however, these have created, in some cases, completely opposite views (see Davis versus Televantou about the *fresco-secco* issue).

Many of the questions posed here may never receive satisfactory answers, but that does not stop us from trying to find possible explanations. Based upon the material evidence and approaching it from an interdisciplinary perspective, including the wider literature and ideas stemming from technology, sociology and anthropology, these questions may be answered more meaningfully. A good example of what can be achieved is reflected in Shortland's edited volume on Egypt and the Near East (2001) and in the publication of a second workshop on the same theme of technological choices (Bourriau and Phillips 2004). These efforts resulted in meaningful discussions about the role and influence of technology within society and *vice versa*, while several papers show an agency-centred approach.

Some deficiencies in Bronze Age painted plaster studies in the Aegean and eastern Mediterranean are unavoidable since no work is complete. This study, therefore, builds on existing achievements with the aim of filling in more of the lacunae which form at different levels. First, the technological understanding itself seems problematic and often arises from the lack of a clear definition of certain technological terms (*al fresco* versus *al secco*, use of *fresco* as a generic term for paintings on the wall, etc., see glossary for correct use of terminology). Moreover, the interdisciplinary relationship between technology, iconography and social dimensions has not yet been explored in the field of painted plaster, even though some earlier hints were made by Cameron. Subsequently, where researchers do not take into account preservation issues, for instance, erroneous data collection may occur and may be followed by improper technological and iconographic interpretations. Examples are known and published. Finally, there is still some way to go, methodologically, and that comprises both detailed and focused technological studies, but also approaches painted plaster as a craft in its own right. Lacunae in the research and in the literature published have been found on many levels and are summarized below.

- Poorly understood interconnection between pigments and plaster (i.e. painting technique(s)), and between plaster and the wall structure
- No theoretical framework or study of the connection between technology, style, iconography and social aspects resulting in poor understanding of the complexity of transfer
- Not enough consideration of the preservation problems in both technological and iconographic studies
- Lack of systematic and problem-focused experiments linked to a lack of usable methodologies for the study of painting techniques
- Lack of comparative work geographically, chronologically and socially (e.g. palace versus non-palace)

Combining multiple approaches will produce the best methodology to obtain meaningful answers to any archaeological question. Since iconography, technology, art, use and many other issues are forms of a society's expression, all have to be taken together to gain the clearest picture of just what constitutes that social expression. Due to space limitations, the scope of this research cannot investigate all possible angles of approaching painted plaster, but it seeks to bring them somewhat closer together.

# 2 The Power of Technology, Knowledge and Social Agency

## Introduction

Technological studies have existed since archaeology became a discipline in its own right. Such studies, which adopt an interdisciplinary approach and use scientific analysis and experimental replication, have been conducted in other domains as well. In the field of Aegean and eastern Mediterranean Bronze Age painted plaster, however, no such research combines a theoretical-methodological approach together with both analytical work and experimental replication. The reasons for such an interdisciplinary approach are stated in Chapter Four.

The aim of the following discussion is to zoom in on human action and many of its fascinating aspects, here viewed as active forces that may be responsible for technological features such as innovation, continuity and change. Looking at human action may also shed light on technological transfer within a given craft. The concept of technological transfer is explained below.

Ultimately, the reason for studying human action is to attempt to answer questions about *why* people acted the way they did. People are social beings and (inter)act within their social and physical environment with other people and things. Therefore the point is to explain the social phenomena that are expressed in the material culture under study (see Table 2.1). By studying and conducting archaeology, we are trying to make

*In the framework of elite exchange:*

↓

| *Phenomena* | *Chronological/spatial* | *Reasons* |
|---|---|---|
| Mobility of people, material, knowledge | Middle Bronze Age-Late Bronze Age | Demonstration of power/status/ knowledge for elite and craftspeople |
| Technological sharing and transfer, possibly supporting iconographic sharing and transfer | Eastern Mediterranean and Aegean | Performance exhibition for both elite and craftspeople |
| Technological tradition and innovation | Palatial versus non-palatial art | Technology was initially 'independent from' Style, developed later into 'in function of' Style |

**Table 2.1** Schematic overview of the research: its questions, context, and human input.

useful and meaningful statements about human behaviour in the past, both within a general and also a more specific social context.

This chapter discusses existing ideas and theories about what technology is, how it is interwoven with human action and how the concept of the *chaîne opératoire* aids in demonstrating this relation between technology and human action. It discusses and looks into who artisans were, how they were organized in pre-industrial societies, and what their social identities may have been. It also touches upon the advantages and disadvantages of applying modern concepts to past societies, concepts such as 'part-time' versus 'full-time' labour, and being 'attached' or 'independent' as a worker.

Traditionally, pre-industrial societies and their artisans have been seen as subordinate to the elite upon whom they 'depended' and for whom they worked. This chapter focuses in on this situation and formulates the different facets of 'power' and how each relates to the context of the actor(s) involved. Through an approach in which people, their identities and social organization are central to considerations of technology and technological activities, the essence of such activities can be seen in a framework of complex relationships whereby social identities change and shift according to the relationship under discussion.

Since the main focus of this book is on painted plaster, and technology is another focus, here is the appropriate place to familiarize ourselves with some basic relevant terminology relating to the technology of painted plaster (see also the glossary). Furthermore, all consecutive steps of paintings on plaster are sketched here. Finally, some definitions relating to a social approach to technology will elucidate both social and technological aspects encountered when focusing on ancient technologies as an all-encompassing concept (on why I consider definitions methodologically necessary here, see Costin 2005: 1078-79).

Apart from the definitions, this chapter is divided into several more sections for clarity. The way that these sections overlap and link is meant to reflect how people are really central to this discussion and how they form the connector throughout. Human agency, in fact, weaves this chapter together.

## Painted Plaster Definitions

Renfrew (2003) has argued that, for about five centuries now, our Western thinking about art has been subject to the consequences of the Renaissance which referred to the Classical Greek prototypes of sculptures and paintings (some transmitted through Roman art). He has called this phenomenon 'the Tyranny of the Renaissance' (Renfrew 2003: 58), and explains that this Renaissance view in which everything that is not inspired by Classical Greek art is bad art is responsible for the fact that many other art forms (earlier, from other regions) have been considered 'barbaric and primitive' (Renfrew 2003: 59). In this case, the 'Tyranny of the Renaissance' also had an effect on the technological understanding of the specific art form under discussion. Literature on Renaissance knowledge of the *al fresco* painting technique has fostered incorrect use of some of the practical terminology relating to this topic (see Cennini's '*Il Libro dell'Arte*': Thompson 1960; Gettens and Stout 1966). Since then, many scholars have transferred this knowledge to the study of painted plaster, irrespective of the context of these paintings, as if *al fresco* technology manifested itself materially in just one specific way, the 'Renaissance way'. The main issue is the correct definition of *al fresco* and *al secco*. Both terms should only be used to describe painting techniques, not actual (floor, ceiling, wall) paintings. Moreover, before a painting can be determined as a 'fresco' it has to satisfy a series of technological and material criteria without which no *al fresco* is possible. The erroneous way of defining any painted plaster surface as 'frescoes' has caused confusion, and consequently has resulted in erroneous (usually unstudied) identification of painting techniques.

*Al fresco* (or *buon fresco*): painting technique whereby pigments, suspended in water, are painted onto a damp lime plaster surface (Hocombe 1999: 6; Kakoulli 2002: 58).

*Al secco* (also called *secco fresco*, Jones 1999: 148): painting technique whereby pigments, mixed with a binder (casein, egg white or yolk, oil, honey), are applied onto a dry support of any material (lime, gypsum, mud plaster, stone, wood, gesso, canvas) (Kakoulli 2002: 57). Some people equate *al secco* with tempera, but tempera should only refer to *al secco* painting where pigments are mixed with egg or size (Croft *et al.* 1994: 841; Hocombe 1999: 88, 91).

*Fresco-secco*: (1) painting technique whereby part of the painting is done *al fresco* and part *al secco*, or (2) the technique whereby the pigments, mixed with lime water or slaked lime (Cameron *et al.* 1977: 161, citing M. Lang 1969), are painted onto a dry surface. This can be confused with the term *al secco*, which may also be replaced by *secco fresco*.

To avoid confusion, this book refers only to *al fresco* and *al secco* and tempera if the specific painting technique, described above, has been clearly identified by the necessary analysis. When a combination of *al fresco* and *al secco* has been observed, this is stated as such.

Now that we have distinguished *al fresco* and *al secco*, the steps involved in making these paintings follow logically below (Table 2.2).

## Technology Definitions

### Technology and Techniques

Both of these terms come from different traditions, each with their own meaning (Ingold 1999), but for this book, they are defined as one and the same. Chilton (1999: 3) defines both as 'the way that things are made' (my quotation marks) in its broadest sense. Technology

| *Step* | *Description* |
|---|---|
| 1 | Limestone is quarried (or collected) and calcined in a limekiln during which the limestone breaks down chemically into quicklime.<br>**$CaCO_3$ (e.g. limestone) + heat (860-920 °C) → CaO (quicklime) + $CO_2$** |
| 2 | The very reactive quicklime powder is mixed with water to produce lime putty, called slaked lime. Ideally, this putty is allowed to mature for about four years in oxygen-sealed slaking pits.<br>**CaO + $H_2O$ (slaking) → $Ca(OH)_2$ (slaked lime)** |
| 3 | When ready, the putty is mixed with the desired amount of filler and applied onto the surface. This surface is prepared to receive this putty by being keyed or made rough so the plaster adheres well. |
| 4 | The putty reacts with the carbon dioxide in the air to cure, set and dry out.<br>**$Ca(OH)_2$ + $CO_2$ (drying, and water evaporates) → $CaCO_3$ (plaster)** |
| 5 | When *al fresco* paintings are planned, the plaster is painted before it has dried out. Before the surface is ready to be painted, it has been floated and pressed in order to obtain a very fine textured and flat surface and to draw the lime to the surface in this process to better 'lock in' the pigments upon carbonization of that surface. This higher lime concentration at the surface will allow a better quality top surface after painting.<br>When *al secco* is planned, the pigments are applied when the plaster is dry, a binder is added to make the colour adhere to the surface. |
| 6 | For *al fresco* work the surface may be prepared by the painters by means of string line impressions, incision lines and other aids before any painting is applied. When the plaster is still damp and cold to the touch, but no longer soft, painting can start.<br>For *al secco* paintings, all preparation work is drawn or painted onto the dry wall, often with red, yellow or black lines. |
| 7 | The pigments for *al fresco* painting, most often mineral-based and alkaline-resistant, are then mixed with water and applied onto the damp wall.<br>Almost any pigment may be used for *al secco*, as long as they are compatible with the binder. |
| 8 | When the wall of an *al fresco* painting dries out, the top surface forms a 'crust' of calcium carbonate, locking in the applied pigments as part of that crust. This way, the painted surface is very durable and cannot be removed from the plaster without removing and damaging that top surface. In *al secco* painting, the binder keeps the colour in place as long as the binder does not deteriorate. |

Table 2.2 Consecutive steps in making painted plaster.

is a phenomenon that brings together the material, the social and the symbolic in a complex web of associations (Pfaffenberger 1988: 236, 240). To Lechtman and Steinberg (1979: 136-37), technology is one of the subsystems of culture while all such subsystems (e.g. religion, art, mythology, values, social structure) are functionally interdependent to varying degrees. For example, in the city-state of Lagash (in Sumer), technology or, more specifically, craft production stood in clear relation to religion where the god Ea/Enki played an active role in the processes of crafting (Winter 2003: 405). In order to understand technology (with its own life and style) Lechtman and Steinberg suggest that technology is thus integrated within the cultural matrix in which it manifests itself.

Goodenough (2003: 7) sees a clear link between activities and culture connected through tradition (see also e.g. Childs 1999), and both technology and tradition involve activities. Since both technology and tradition are structured concepts, DeMarais (2004: 13) suggests that a dialectic between structure and agency exists: 'people engage with material things as part of the unfolding of the activities of daily life' and technology may be the way in which this engagement takes place. Ingold (1999: ix) describes technology as the mutual relationship between people and things through skilled performance, and, with this definition, human agents have regained their crucial place in the processes of technology. What technology means for any society can only be understood when we accept that people, through their minds *in* their bodies (rather than separated from them, after Malafouris 2004: 55) interact with their material world here and now. Technology, when embedded in human behaviour, takes an active part in creating, maintaining and transforming social relationships (Gero 1989: 92). Schiffer (2004: 579) also places people centrally: 'Virtually every activity consists of interactions among people and one or more technologies'. For Dobres (1999: 17), agency, technology and gender are completely inseparable, while a necessary dialectic is going on between these fields in the everyday production of social collectivities.

This is why technology can never be defined by *equipment and tools to make things,* an essentially positivist view that excludes humans completely. As such, I agree with Killick (2004: 575), who believes that both a functionalist and a social constructivist approach to technology are complementary rather than conflicting. The way technology is understood in the context of this research puts the human actors—the people who stand in relation with others while working, acting, being, thinking, speaking, believing and observing—right in the centre of all stages and processes that make up technology.

## Technological Transfer

Derived from iconographic transfer (Crowley 1989 among many others), I use the term 'technological transfer' to describe the process whereby technological aspects or elements are moved or adopted or carried from one place to another (Brysbaert 2002b: 95), from one period to another, and from one material medium to another. In the context of this book, the main aspect of transfer is one of geographic directionality, i.e. transfer from one place to another. Following this definition, I believe that technological transfer is somewhat easier and clearer to detect than iconographic transfer, because the elements transferred in iconography can move across the medium on which they appear. For instance, flowers can be depicted on textiles, on pottery and on metal. Certain technical aspects, however, cannot travel across media because they are inherent in one specific medium (warp and weft patterns, repoussée technique, matrix-filler ratios). Consequently, they are easier to track down and to follow when transferred spatially. The concept of technological transfer can best be illustrated with contemporary everyday examples of eating Chinese food with chopsticks in any Chinatown in the west, or driving right-hand drive cars on the left side of the road in Cyprus, Australia, India and many other countries that were formerly part of the British Empire.

### Craft Specialization

While Stein (1996: 25) argues that craft specialization formed the key to the political economy of complex societies, craft specialization has traditionally been seen only in association with highly ranked and stratified societies, and often mainly in 'urban' settings, but this is no longer the case (Kenoyer *et al.* 1991: 46; Dobres 1999; Rosen 1997: 82; Patterson 2005: 314). To craft is to create actively with a specific form, objective or goal in mind, so such activities can take place in non-urban settings and in less stratified societies too. Considered mainly a human activity (as opposed to primate and bird activities), crafting is seen as social labour because it serves to explore the importance of activities such as creating personhood, defining social categories and generating social relationships (Costin 1998a: 4). Crafting may add value to common materials (Costin 1998a: 7). An important case study in this context has been published by Feldman (2006) who mentions the importance of crafting (with luxurious and exotic materials) in creating and maintaining diplomatic relationships between rulers.

Craft specialization and its organization may be closely interconnected with the use and manipulation of social relationships, rituals and cultural values considered essential for artisans to be successful in their work (Childs 1998: 113; see also Wailes 1996: 4). Thus, by actively crafting, people create, recreate, confirm and reconfirm their social identities. Finally, no craft stands on its own and the notion of cross-craft interaction has been discussed to some degree in various technological studies (e.g. McGovern 1989; Knapp and Cherry 1994: 9, 30; Brysbaert 2007a). We will come back to this concept in Chapters Three and Seven.

### Specialization

Specialization can be seen as any form of culture-specific activity that includes some of the features below in combination:

- High investment of labour
- Artisans produce goods for others (Costin 2005: 1065)
- A particular way of organizing 'production' (my quotation marks) (Costin 2005: 1038)
- Giving specific status to the producers, hence providing a proportion of the subsistence as remuneration for the work (Costin 1991: 3) and offering higher status to consumers too (Reents-Budet 1998; but not always: see Wright 1998: 57)
- Use of rare and special raw materials
- High level of skills and knowledge
- Specific access to all or a selection of the above

- Form of production where artisans may be partially or totally freed from subsistence production
- Certain level of standardization (but see Costin 1991: 33-34; Costin 2005: 1067 for other occurrences of standardization: Berg 2004), pointing to a specific level of social organization, of specific labour division (Patterson 2005: 321 argues that different sorts of craft specialization stand in relation to different bases upon which labour division is constructed or understood)
- Serving the elite(s) in society or a selective group within society (but not necessarily), who in return may offer higher or different remuneration. Thus specialization may be seen as both an economic and social relationship between producers and consumers (Costin 2005: 1062)

## Social Identity

Hall (1997: 30) has defined social identity as the knowledge, value and significance attached to membership of a social group. Social identity is a dynamic, changing, shifting and diverse phenomenon for everyone, and it is very much linked to its temporal and spatial context and, in some cases, affected by the concentration of production. Everyone who interacts with others and the material world also has, as a result, multiple social identities that may be altered repeatedly in changing social situations (see for instance Rowlands 1994: 132; Brumfiel 1998: 145). People, therefore, maintain multiple identities as a result of belonging to various groups: class, religious, linguistic, occupational or other, and they affiliate themselves with those identities that suit them best for the situation they are in. Social agency, therefore, can be seen as someone or something that

> 'initiate[s] "actions" which are "caused" by themselves, by their intentions… An agent is the source, the origin, of causal events, independently of the state of the physical universe' (Gell 1998: 16, original quotation marks).

According to Costin (1998a: 7), social identity comprises age, gender, experience, skills, social, political and legal status, economic and religious status, kinship, possession of specific esoteric knowledge (related to religion, craft production, distance, power, culture-specific symbolism, etc.), possession of material wealth (quantitative and/or qualitative) and, finally, craft production itself. Any of these culture-specific identifiers may shift in importance depending on the social relationships established.

## Labour Division

Most work (i.e. the entire *chaîne opératoire*) consists of a range of tasks and, hence, needs to be carried out by a range of people. In this, each person is doing that for which they are responsible, whether this responsibility is taken by the individual actors' initiative or whether it was forced upon them. Labour, and hence craft, is thus social production. Division of labour is then a fundamental integrating social force existing at all levels of a society (see Patterson 2005: 321), even in the (modern!) household (Costin 1998a: 10).

So being or becoming part of the labour force is socially and economically crucial for constructing people's social identity and inculcating feelings of belonging to social networks (Costin 1998a: 10). Remuneration for each task is of equal importance for the social identity and recognition of the person in question.

Within the context of this study, the social relationship between the producer and the patron must be considered. Issues such as 'independent' versus 'attached' need to be investigated in view of who holds the power and control.

### Chaîne Opératoire

Nowadays this notion is defined as the technical chain of sequential material operations by which natural resources are acquired and physically transformed into cultural commodities. Dobres (2000: 154-56), however, also suggests that people are located within their social context as well, and that the *chaîne opératoire* really should link people with the production of artifacts since it highlights the sequential nature of both material and social production. Furthermore, this link between people and their artifact production also recognizes that technical gestures are performed in the 'public domain' (=social context). In this sense, the *chaîne opératoire* is a unifier of technology, agents, spatial analysis and social organization (after Given and Knapp 2003: 302), forming a potent methodological tool to study technology and social agency in relation to each other (see below). Most recently, the concept has been connected with evolutionary theory (Riede 2006), and it has been contextualized with aspects of time and space in discussing past landscapes (Conneler 2006). The concept of multiple *chaînes opératoires* as a unifier has also been explored and employed in order to explain cross-craft interactions in an Aegean and eastern Mediterranean context (see Chapter Seven and Brysbaert 2007a).

When human action is chosen to become the central focus of a technological study, certain key questions emerge:

- *Human action in relation to technology*: who were the craftspeople, how were they organized and how were they looked upon by their society? Which technological choices and restraints did they live with, and how did they pass on their knowledge and skills?
- *Mobility and transfer in relation to technology*: what (materials, knowledge, craftspeople) was transferred and how (elite interaction during the Late Bronze Age, gift exchange, cross-craft interaction, craftspeople) and why (power display, changing aesthetic taste/style, status)?

Such questions will help us to understand the social interactions that people had with each other (elite ruler versus other elites and versus craftspeople) and with their material world (power symbols, gift exchange, etc.) during the late Middle and Late Bronze Age.

In order to elucidate the role that technology plays in the everyday lives of people, we need to look at who the people involved actually are, how they interact through technology, how and why they make things and distribute them. More specifically, this study deals with technological transfer and its direction within the Aegean and eastern Mediterranean Bronze Age painting traditions (see key questions above).

The first question about the artisans is basically concerned with their social identity (see section 4). This question is directly relevant to my research because I attempt to determine the direction of the technological transfer in relation to the iconographic transfer. As explained above, technological transfer is a much more stable phenomenon through which to study directions of transfer than iconography. This, however, does not imply a simple one-way and easy-to-follow thread as far as transfer is concerned because it does not mean single-linear.

Secondly, the question about who the artisans were and how they were socially organized within their craft (see section 5) is also a question about their *multiple* social identities (see definition above). Furthermore, this question also relates to how they interact socially with others while they are doing business (Dobres 1999: 14). Identities are constructed by people in relation to other people, within or beyond 'other' groups. So the opposition that marks differences or similarities aids in setting up categories defined by absence or by mutual exclusivity. Such views reflect how *we* may construct the thoughts and notions of others, but also how people in the past may have done this with respect to things they did not know (see for instance Morris 1989: 529, on the idealized Egyptian landscape of the Theran wall paintings).

Going back to the question about the directionality of transfer, we see two opposite opinions in the literature (see section 4). But is it really that simple? The material in question shows a more complex picture, which will become clearer in the following chapters.

## Who Were the Artisans?

Material culture has been, and still is, used to map out specific claims to ethnicity, identity and power (Hall 2001: 50). In the past, ethnic groups were too often fused with specific races, for instance in Nazi Germany (Jones 1997: 3) or in Israel, where direct genealogical methods are used to trace present groups back into prehistory (as in 'biblical archaeology', for which see Jones 1997: 8). Present-day archaeology often still refers to 'archaeological cultures' (Gordon Childe's term), which seems to be parading as an ideology-free substitute for the term 'ethnic unit' (Jones 1997: 3 citing Veit 1989: 42), and implying a specific group of people versus other groups.

Many scholars are opposed to labelling archaeological cultures in order to explain the ethnic origin of specific objects (see for instance Knapp 1998: 201-202), but it remains deeply ingrained within the discipline to use terms such as 'Minoan' and 'Mycenaean'. What is defined in this book by each 'label' such as Minoan, Mycenaean and Cycladic is a set of cultural phenomena (material, religious, social, physical, political) that is to some degree different from another such set. This may be based on, but is not limited to, a specific, well-defined locale. In fact, using labels for archaeological cultures may not be that bad *as long as* these are understood not to imply ethnic groups.

As indicated earlier, social identity produces an image of diverse, dynamic, competing configurations stretching from the local to the global, and engaging in multiple regimes of power (Jones 1997: 8, citing Clifford 1992: 101, 108). I believe that we can see past cultural groups in this light, independently from any (modern) state boundaries. Subsequently, discussing social identity, rather than ethnic identity, becomes relevant. Seeing social identity in this way, involving active maintenance of cultural boundaries while socially interacting, becomes an aspect of human processes. Moreover, it also becomes another component in the social system, alongside subsistence, economics, religion etc. (Jones 1997: 28).

A discussion of social identity in this study is necessary because it refers indirectly to one of my research questions, that is the direction of technological transfer within the painted plaster tradition. In the current debate on the direction of transfer, two groups have profiled themselves. One group suggests a transfer from west to east (most Aegean archaeologists, Bietak, Novák and Pfälzner), while the other suggests a possible transfer from east to west (Woolley, Sherratt, Knapp). Even without labelling the people responsible for the transfer, a transfer from west to east or from east to west is still a question worth answering for a variety of reasons:

- Technology and iconography, forming together this painting tradition, undergo a series of changes over time. It is meaningful to find out what these changes were, why and how they happened and who was responsible for them.
- Changes and transfer in iconography have been discussed at length, but in isolation only. Therefore, it is important to see how the changes and transfer in iconography relate to those in technology, how they influence each other, and how both approaches can only be studied holistically if we want to understand the craft of painted plaster as a whole.
- Since transfer and changes within this craft are part of the same entity, I want to investigate who was behind these, why that happened and how. In other words, I attempt to determine which people were responsible for the choices made, why *these* choices were made and what was expressed in this way.

In this specific case study, each of the (competing?) social or cultural groups of the broader eastern Mediterranean will have had their input (whether this is technical, iconographic or otherwise) in what eventually became an 'artistic *koine*'. This artistic *koine* was a hybrid mixture, so well known from all areas of the Aegean and eastern Mediterranean during the Late Bronze Age (Feldman 2006: 8-10: on definitions of the international artistic *koine*) and of which painted plaster, in my view, was an integral part, even though the craft objects were not portable (Feldman 2006: 10). This phenomenon may be called 'cultural pluralism', lying at the core of such an artistic *and* technological *koine*. An artistic *koine* should not be seen as the homogeneous result of a mixture of elements, as if they were blended together. Such a normative view potentially takes the risk of seeing all Mediterranean peoples as part of the same homogeneous cultural horizon and of masking the essential differences and variability between people's behaviour, expressed in their material culture. This is clearly not the case when archaeological studies compare local assemblages, studies at the intra- and inter-site level (see Dobres 1999 for Magdalenian material culture). What we have in the eastern Mediterranean and Aegean is an apparent sharing and exchanging of an 'assemblage' of items, arguably confined to the elite, coexisting with very localized traditions in every region and period.

Cultural pluralism can be seen in several different ways, but most clearly in the fact that each individual possesses many different social identities. For example Frida Kahlo was Mexican (identity [ID] related to her country and her mother's origin as Mestiza) and European (ID relating to her education and her father's origin as a German immigrant), and a painter (ID relating to her profession) (www.tate.org.uk/modern/exhibitions/kahlo/). This example illustrates the importance of geographical rootedness or belonging in defining people's social identity (see also Bhattal 2003: 2). When American Indians came to be seen as just one large homogeneous group in relation to the Whites, they lost their individual group identity. Once spaceless, their specific local identities were denied (Preucel and Hodder 1996: 602). Textual evidence from the Bronze Age also mentions a wide range of specific locations and it is often assumed that these names might also describe the people coming from these places, Kaptara from Crete (Mari tablets: Zaccagnini 1983; Niemeier and Niemeier 2000: 790), Ahhiyawa from the Greek mainland or Mycenae more specifically (Hittite tablets from Hattusha: Niemeier 1999: 143-44; but see also Güterbock 1983: 137-38; 1984), and Alashiyans from Cyprus (Amarna tablets: Moran 1992). Consequently, I believe that geographical locations, combined with the feeling of being rooted or belonging as part of self-perception, play a crucial role in issues of social identity.

*If* artisans had travelled (see Chapter Six), their social identity would have been important since they would have arrived, lived and stayed in places which were not 'home'. On the one hand, the social status of people who returned from 'other' places where they had lived within local but foreign communities may have been very different from their social status in relationships within their own community. In a new cultural and geographic environment, material things and ideas become redefined (Pydyn 1998: 100), but so too do social relationships. People's social identity may have played to their advantage when going to other places (Helms 1993: 36). This appears to have been the case especially for the Egyptian physicians and craftspeople who were invited to royal courts for their knowledge and power to cure people or execute specific tasks (Drenkhahn 1995). Another example can be seen in the Aztec civilization, where groups of artisans kept their ethnic origin (Toltec) very much alive when going to other regions, because they were seen as the heirs to the craftsmen of the Tollan, the mythical seat of central Mexican civilization (Brumfiel 1998: 147; see also Helms 1993: 32-33). In some cases, however, social identity worked against people's wellbeing (for examples see Costin 1998b: 124).

In each case, social status clearly changes with its context (Bender 2001: 81). People arriving in a new locale may decide to express themselves to others by what they are known for, including their social status, through their interaction with material and its transformation. The question then arises, how would their interaction with materials influence their social relations, if it did? Moreover, vice versa, how would their social interaction with others in the new place influence their interaction with materials?

Material culture within the archaeological record has often been employed to assign identities to people (see for instance Niemeier and Niemeier 1999). The main question is whether social identity is *really* visible in the archaeological record. As argued in other chapters (Chapters Three and Five), only multiple items, factors and assemblages can aid in correlating material remains to the presence (permanent or temporary) of a specific social group. For example, Childe argued that an artifact *assemblage* was more important than individual artifact types for identifying social groups (see Jones 1997: 18, citing Childe 1956). Moreover, Shennan (1989: 6, 11-14) distinguishes between archaeological cultures and the identity of social groups and he suggests that we should not treat cultures as entities, because it stands in the way of seeing social identity as an evanescent situational construct instead of a fixed fact through which we can trace people's destinations.

In this case, the presence of 'Aegean-style' painted plaster at Qatna, Syria, would not be convincing enough to conclude that Qatna was a Minoan settlement. In contrast, 'Aegean-style' painted plaster, Linear A on roundels, a possible ashlar building 'in Cretan style', weaving equipment, stacks of conical cups, Minoan pottery, Minoan religious paraphernalia and certain subsistence evidence (Niemeier and Niemeier 1999: 547-49)—all present at Miletus—make a much stronger case for a settlement that was at least strongly influenced by Minoans. Nevertheless, is such evidence enough to suggest a Minoan group of people residing at Miletus? When do we have qualitatively and quantitatively enough evidence to assign a social identity to material culture (see now also Broodbank's article on Minoanisation [2004] in this context)? In tandem with the 'Minoan settlement' example above, one can investigate technically (and/or iconographically) a specific group of materials in great empirical detail in an attempt to identify more rigorously and convincingly similarities, changes and differences over time and space. It will be exactly these distinctions and similarities that can contribute to discussions about the nature and intensity of the social interactions that lie at the base of the formation and alteration of social identities.

Cultural continuity or diversity may not necessarily indicate social continuity or diversity, as we see, for instance, in the debate over the use and users of hieroglyphic script and Linear A on Crete, or even in discussions about the relationship between Linear A and B (for discussion on both linear scripts see Tomas 2003). Very much the same will also become clear from the present study when we scrutinize the iconographic and technological evidence present at the different localities investigated. As will be shown, many more aspects of social identity come to play a role in explaining the 'how' and 'why' of technological features within the painting tradition discussed.

From this viewpoint, we see that human actors are the driving force behind similarities, changes and differences in material culture. If we then understand that material culture and cultural continuity may be insufficient to define people's social identity with specific reference to their geographic rootedness or belonging, it may become difficult to assign such social identity to the makers of specific groups of artifacts. What we can do, however, is to inform ourselves more broadly about the multiple social identities of which the geographic factor only forms one segment. All aspects of one's social identity are informed by each other and hence may shed light upon each other. By investigating the material record in such a way that it allows consideration of other aspects of such identities, and by seeing social identity in relation to a certain limited geographical place (which can be narrowed down by a process of negative selection), we may be able to make significant statements about who was responsible for the technological transfer, how it happened and why.

## Social Identities and Social Organization of Artisans

Crafts intersect with all cultural domains—economic, ritual, political, social (Lechtman and Steinberg 1979)—because everything made and used by pre-industrial people is the object or outcome of crafting (Costin 1998a: 3). Crafting involves relationships between producer and consumer, between producer and product (see Gell 1998: 17-18), and entails certain patterns of labour division that may be related to a variety of forces. Table 2.3 illustrates these forces, the possible reasons behind them, and provides a few examples. The table demonstrates that many of the labour-dividing forces overlap in meaning and in their social and natural outcome. In any given culture or society, there always will be a combination of these forces that are culture-specific. A large range of possible labour divisions indicates an equally large range of social actors in craft production that, therefore, cannot reflect a single social identity. Labour division also brings about differences in social status, clearly illustrated in Egyptian art. The example given by Baines and Málek (1988: 194) reflects very much the manner in which the ruler is depicted in relation to his subjects. Differences in status may exist between people from different crafts but also within one craft (Wattenmaker 1998: 53). As Costin (1998a: 6) states, artisans differ in experience (see examples by Laffineur 1995: 192), background, goals and desires, so they need to get 'a face' in discussions on social and political change. Peregrine (1991: 1, 8) goes one step further by seeing craft specialization as being as much a political and economical activity as it is artistic, and he calls artisans 'political personnel' for the role they play in confirming and reconfirming the power of elite groups through their manipulation. In Minoan Crete during MM I-II we can already see such a pattern emerging, whereby Burke (1999) has argued that the development of the *murex* dye industry may have played a role in the rise of the second palaces (MM IIIA-LM IB). Therefore, craft specialization may have furthered the establishment of political powers.

When we want to investigate the social organization of artisans, we see specific traditional issues addressed in the literature:

| *Forces* | *Reason* | *Example* |
|---|---|---|
| Task specific | Differentiation in skills, knowledge | Painters and plasterers |
| Gender specific | Physical ability, culturally restrained | Mining, breast-feeding |
| Age specific | Physical effort, curiosity | Children grinding pigments |
| Status specific | Confirms status, control by employer | Knossian painters |
| Ethnicity specific | Group known for their abilities | Pharaonic Egyptian physicians |
| Knowledge specific | To meet highest standards-quality | 'Master' versus 'apprentice' |
| Cross-craft specific | Technique and materials overlap | Amarna's joint glass/metal workshop |
| Religious specific | Taboo, sacrilege | Kosher cooking |
| Location specific | Production concentration | Village versus city |

**Table 2.3** Forces affecting labour division, with reasons and examples.

- Did they work full-time or part-time?
- Were they attached or independent workers?
- How was knowledge passed on over time?

In relation to the above, a few new questions arise from the present research:

- Are such oppositions always applicable and relevant?
- If or when artisans were travelling, and if it can be demonstrated that they were travelling, how was that practically organized and who travelled?
- What was the (legal, economic, social, political) status of artisans and how was this constructed and reconstructed?
- What was the relationship between the artisan (producer) and consumer (elite, ruler)?
- In passing on knowledge, who was teaching whom and why?
- What were all the different steps in the craft of painting plaster, where were they executed, and how many people were involved and why?
- How many of these issues can be discerned in the archaeological record?

I now focus on some of the issues raised through these questions. The advantages and disadvantages of talking about full-time versus part-time are discussed in section 1. The relationship between 'attached' and 'independent' is investigated in section 2. Finally, the concept of the *chaîne opératoire*, dealt with in section 3, is contextualized in this study and relates, as such, to issues involved with passing on knowledge and skills, among other unifying issues.

## 1. Full-Time versus Part-Time

Many scholars think and see evidence for the fact that in most pre-industrial societies, people engaged in full-time subsistence work combined with part-time craft activities (Patterson 2005: 315 gives published examples where there does not seem to be a disparity between the two types of activities: physical [agricultural] and mental [crafts] labour. I do not agree with the division between mental and physical labour since one cannot exclude the other). When the agricultural season was over, other work was done instead. This still seems to be the case when we consider task divisions within an agrarian community, for instance, many villages on Crete in the present century. After the olive collection and processing in the winter period, two months pass by before new sowing can start. During this period, people paint houses, repair and renew broken fences and roofs, repair curtains and clothes and do a range of activities around the house, local taverna or community. But when the agricultural season starts again, people still carry out certain other tasks; thus traditional crafts do not stop where subsistence work starts (for another example of combining multiple tasks *and* being mobile in order to earn a living, see the ethnographic work by Kardulias and team 2000: 42-43). The division between part-time and full-time becomes vague and irrelevant and may largely be regarded as a modern concept, usable in specific socio-economic contexts only.

The crucial point here, however, is whether the archaeological record is sufficiently robust to discern such differences. Kenoyer *et al.* (1991: 48) hold a rather pessimistic view for both 'full-time/part-time' and 'attached/independent' (see below) oppositions; Roux (1990: 142) likewise does not see full-time/part-time divisions in the archaeological record. Clark and Houston (1998: 38) also suggest that the craft-subsistence division is not always helpful in understanding the social organization of people's activities (see also Wright 1998: 59), let alone the distinction between full-time and part-time labour input.

By contrast, Kim (2001: 461-62) discusses how a different sort of wealth is achieved through a subsistence economy versus a prestige economy, the latter embedding specialized craft production. Brumfiel (1998: 146-47), for instance, has shown that part-time and full-time craft work may depend on where this work took place, while there also seems to be a link between full-time or part-time work and the production of primary and/or secondary commodities (after Blake and Knapp 2005: 12). This difference in location, and in what people produced and for whom, affected their social status and hence their social identity, and how they expressed this in their relationships with other people within the community (see also Childs 1998: 116-17 on the Toro iron workers and the seasonal factor of part-time work). In some cases remuneration was also different depending on whether work was carried out part-time or full-time (Wright 1998: 62; Wattenmaker 1998: 49-50). Full-time crafting may also have been induced by a change in technique, while this technical innovation went hand-in-hand with several other innovations during the same period, as illustrated by Knappett (1999: 101).

So which other factors affect whether people worked part-time or full-time? For instance, is the craft feasible every day, all day long, all year round? Are there any aspects of its technology that limit this work to specific seasons or make specific seasons favourable? Certain activities of the *chaîne opératoire* may need specific (short-long) concentrated periods of part-time or full-time labour for the end-product to be successful, while others need periods of rest. This can be inherent in the technology itself, but also in cultural (religious) restrictions. Certain activities may not be allowed during specific periods of time. This could mean that people may *have* to carry out something different during these periods.

Another important factor is the possible difference between people who carry out their craft at home versus another location. When at home, people can choose to carry out other tasks; this may not be the case when working in a foreign context as illustrated by examples from the Mari Tablets (Zaccagnini 1983: 249, 253). Such examples (see also Wattenmaker 1998 for Ebla) seem to show that there may be a relation between part-time/full-time and the employer for whom the artisans worked, possibly also affecting their legal status of being attached versus independent.

All these different factors show that working part-time or full-time depends on so many dictates from within the technology itself and from within the cultural, social and political context whence the technology derives. Each case is unique and difficult to assess without written sources to support it. It seems, however, that full understanding of the entire production system (Costin 1991: 18) or *chaîne opératoire* of the craft is crucial, in order to be able to see whether the involved tasks actually *can and should* be seen within such a temporal framework, and whether this yields information about the organization of the craft in general. Subsequently, if it is possible to discern for whom the artisans worked and under what conditions, another valuable contribution can be made towards discussing part-time and full-time labour input.

## 2. Attached versus Independent

Some time ago attached specialization was presented as a developing part of political processes versus independent specialization that emerged to meet utilitarian and economic needs (Brumfiel and Earle 1987; Costin 1998a: 11). Nowadays, views on this theme are somewhat expanded, and evidence in the Near East during the third millennium BC, for instance, shows that the division is altogether not that straightforward any more (Zettler 1996: 20), because many artisans appear to have been attached, while they also played a role in wider economic systems that included private sectors. Costin (2005: 1072-73) sees the dichotomy between 'attached' and 'independent' in relation to different levels of control, and, as a result, each forms a heuristically useful device to describe different relations in production.

In discussing artisans and their social organization within an Aegean context, two connections have been made. The first is the connection between travelling craftspeople and their independence (Negbi 1994: 87; Boulotis 2000: 846). The second suggests that travelling painters were sent from royal court to royal court and were seen as being completely dependent on the elite (Televantou 1992: 146; Boulotis 2000: 846; indirectly: Niemeier and Niemeier 2000: 792-93; for Zaccagnini 1983: 253, sending artisans or diplomatic gifts is seen as the same thing).

Is it that easy, however? Is independence a condition for travelling or vice versa? We certainly cannot say, as a result or consequence of the former equation, that artisans working locally were all attached. Does 'attached' mean being a slave, chained up in the property of one's owner (Zaccagnini 1983: 247), or living and working in the immediate vicinity of the employer (Costin 1991: 25; Watrous 2001: 200) and being strictly controlled by him or her? Does it mean working only for one person or group and *also* being dependent on them for subsistence (Wattenmaker 1998: 49-50, 52-53; Watrous 2001: 199-200)? Does it mean not having the choice to leave or not wanting to leave an elite group? Horden and Purcell (2000: 343) argue that no Mediterranean history can be complete if it does not include the buying and selling of people. Linear B texts also mention selling slaves at Knossos (Olivier 1987: 497), and Killen (2001: 171-76) discusses the range of different statuses that some of the Mycenaean artisans had in relation to their craft. What about artisans who are indeed in captivity but express their resistance through interaction with their material or with others, and how permanent is this status (see Clark and Houston 1998: 37)? Moreover, to how many crafts within a society does such a status refer?

Can such sharp opposition be applied (for difficulties with this, see Brumfiel 1998: 147; Wright 1998: 67)? The question about 'attached' versus 'independent' is very much a question about the status of the artisan. Since status, whether legal, economic, political, or religious, is part of the make-up of someone's social identity, it cannot be seen in isolation (see also Wright 1998: 67). This impacts on how we should see 'attached' versus 'independent', integrated within a social framework of interrelations between many people. This is simply illustrated by the questions, attached to whom, how and why? The status of an artisan may depend on a range of influences, such as their gender, age, kinship, legal, economic or political status, or whom they work for. It may be influenced by their social status, in itself formed purely by what they do (for artisans in Classical Greece: Vickers and Gill 1994: 95), by their level of skills, their practical and esoteric knowledge, who their employer is, what the value of the goods is that they produce, what remuneration they get for their efforts, whom they have been trained by, and with what possible divine, ancestral, religious or elite lineage they identify themselves (Brumfiel 1998: 147-48; Reents-Budet 1998: 77).

It is also important to see artisans as 'designers', not only as transformers of raw material into finished products, but also how they organize themselves and plan everything. Kingery (2001: 123-24) situates design at the core of technological change because design is required at every stage of the production process and use. For him, the actual idea of designing something implies technological change and any design consists of a set of technical choices (Skibo and Schiffer 2001: 143). We can see this, for instance, through the *chaîne opératoire* as a dialogue between people and material (Schlanger 1994), and through time management because craftspeople never carried out an isolated single activity (Dobres 1999: 21; see also Lechtman and Steinberg 1979: 144). In that sense, one could argue that people who worked part-time in a craft while also doing other things were certainly equally skilled as full-time workers, but at a different level. An example of how different technological activities were combined is illustrated by Vickers and Gill (1994: 90), and can be seen as 'embedded technologies' (see also Brysbaert 2007a: 334-35). Artisans as designers may not have known in advance if they were going to be successful because they may not have known the outcome of their design (see also Schiffer and Skibo's 'technoscience' [1992: 48]). Equally, they may not *want* to succeed in order to show resistance to their employer (Dobres 2000: 156).

Similarly, one might ask whether artisans made technological choices within the craft they performed. It has been shown that ineffective technological behaviours in certain societies in the past and present are not due to lack of knowledge about better alternatives (Lemonnier 1993: 2). Technological choices represent the processes of selection of technological features borrowed from outside or invented locally. An example of this is offered by the sudden and one-off use of lime plaster at Tell el-Dab'a, Egypt, while the rest of Egypt used gypsum plaster or painted straight onto the wall. This is where the independent mind, of both artisan and patron, plays a very important role. Even though the mind of an artisan may be limited by technical, cultural and social constraints, s/he may still choose to follow these rules or go against them.

### 3. *Chaîne Opératoire* as a Unifier

About 70 years ago Mauss (Dobres 2000: 153) stated that technology is a total social act, and he tried to understand how bodily behaviour reflects and is conditioned by social tradition. He argued that both the cultural significance of objects and those who make and use them are revealed in how technical acts and gestures unfold materially in a social milieu. Dobres (2000: 153-56) acknowledges that social identities, based upon a broad understanding of gender, are equally sequenced as the production of artifacts. This becomes important when we discuss knowledge transfer within a context of apprenticeship and the relationships involved (on apprenticeship and children's role in this: Grimm 2000: 62-67; Greenfield 2000; Baxter 2005: 50-52, 63-66). Knapp and Cherry (1994: 29) state that 'as "sedimentation" of experience, technology is an active ingredient of personal and social identity'. That 'sedimentation of experience' may be achieved, for example, through different levels of apprenticeships, and the subsequent layers of experience through learning may influence the level of status and, as such, social identity, over time.

In reply to my earlier question of 'who was teaching whom and why', a process of deliberate choices in these relationships may have been made. As such, apprenticeship can be seen as one of the crucial steps in the entire *chaîne opératoire* and hence in the process of social identity construction (Clark and Houston 1998: 42). In the Aztec civilization, independent workers trained their own children to become skilful in producing both utilitarian and elite goods, while attached workers also taught the children of the elite certain crafts (Brumfiel 1998: 147). This differentiation within the same society must have had implications for the social identities of all: the 'teachers', the 'pupils' and the elite themselves. We also know that Canaanite elite youths were trained in military and other technologies at the Egyptian court during the Empire period (1400–1200 BC) in order to keep ties of obligations with the Canaanite elite in general (Thyrza-Sparks 2004; Moran 1992: 23-29 on Amarna letter 296). Small-size fingerprints also suggested the use of children in clay tablet production at Knossos (Sjöquist and Åström 1991, cited in Laffineur 1995: 210 by T. Palaima in the discussion). Finally, Matson came across very revealing contemporary examples of apprenticeship training in the Middle East, whereby artisans said that 'the young are soft as clay—you can form them as you wish' (Matson 1989: 14). These activities of passing on technological knowledge are seen by Killick (2004) as verbal and non-verbal communication forms. In that sense, apprenticeships can be seen as guided imitation by means of instructions and demonstrations. I suggest that apprentice training and interaction is present across the entire production process, from developing ideas and sharing knowledge to learning specific aspects of the craft, into learning how, when and where to trade it.

It is not my goal here to treat the applicability of the *chaîne opératoire* in technological studies as it has been covered extensively (Pfaffenberger 1992; Lemonnier 1993; Schlanger 1994; Dobres 2000). More recently, Kuhn (2004: 563-64) has critiqued the use of the *chaîne opératoire*. The most important point is that the purely technical part of the *chaîne opératoire*, reconstructed from the data, can be used to infer aspects of social organization that may not be obvious from archaeological remains (Given and Knapp 2003: 302). A second point is that a thorough study and an understanding of technological features of material culture remain crucial if we wish to observe site-specific material variability and patterns. Dobres (2000: 160; 1999: 20), for instance, noted in her detailed study of Magdalenian lithic assemblages that an overall package of similar technical activities was present in the entire region but that specific intra-site differences were observed as well (see also Lechtman and Steinberg 1979: 140). She explained this phenomenon by suggesting that social interactions, made visible in the material, were unfolding differently at each site. Therefore, low-level variability between two sites or within one specific site may refer to individual agents, interacting with each other.

The present discussion has not yet explained why and how social interactions between people are different. Technical activities need to be seen in a wider framework placed within the society where gender, age, kinship relations, hierarchy, power relations and several other factors determine (and are determined by) and shape (and are shaped by) human interaction. Technical activities, therefore, have as their goal the construction of one's own social identities within (or without resistance to) existing social boundaries.

On a methodological level, being able to reconstruct the entire *chaîne opératoire* depends to some degree on the preservation of the archaeological remains. In this case study there are several quite distinct technologies that form together, often in a specific sequence, the end product of painted plaster (for technical details, see Table 3.2). Next to the training aspect, these are:

1. Procurement of raw materials (limestone, (in)organic fillers, coloured minerals, everything needed for brushes, paint palettes and containers, scaffolding)
2. Preparation installations to transform raw material into usable products (kiln for calcining limestone, pits for slaking quicklime, grinders, pestle and mortars for pigment crushing, etc.)
3. Preparation of the surfaces to be plastered
4. Plastering the surfaces

5. Preparations to facilitate the painting
6. The painting itself.

Each step needs different tools, instruments, installations, creativeness, technical anticipation and experimentation, human energy, labour investment, knowledge, skills, spatial arrangements, time management, practical organization and co-ordination, experience and technological expertise, social interaction and approval, and so on. If *each* step employs the above, then flexibility at each interface of all these steps is required. Clearly, the end result of painted plaster is far from a simple, everyday technical achievement. It shows how human involvement through mind in the body within an embodied and contextualized arena of social interaction is *the* most important aspect of all. In this technological context, therefore, we see material and mind (in the body) unified. This unifying relationship of the dualistic concepts of mind and matter wherein both exist in mutual co-dependency has been termed 'engagement' (Renfrew 2004: 23).

As stated previously, however, no human action can be discussed within the context of social interaction through material culture without scrutinizing all archaeological remains and material attributes, spatial arrangements, and evidence of variability and patterns. Yet, this is exactly where the problem lies as far as the Aegean and Near Eastern areas are concerned. No remains have been securely identified for step 1, for several parts of step 2, and only negative evidence for 5 and 6 has been recognized. We are, nevertheless, not completely at a loss because useful insights may be found from later (or earlier) periods and from other areas via ethnoarchaeological research and via experimental replication (Keller 1994: 60). Such information may cover at least the material part of the gap in the *chaîne opératoire*. All these approaches intersect with each other and provide thorough step-by-step insights into each aspect of the technology *and* into the role of human action (Dobres 2000: 220) and interaction within and with that technology (see also Brysbaert 2003a: 11).

## Technology within an Elite Context

We have already touched upon the subject of travelling artisans when discussing the importance of social identity (see earlier in this chapter). This section discusses three things, namely (1) whether artisans themselves decided to travel, or not, and why, (2) what role highly skilled artisans played while working in an elite context and how they performed that role, and (3) what role non-human factors may have played in relation to human actors in a complex set of relationships and expressions of power.

First, I will introduce some possible reasons why artisans travelled, that is, *if* they could decide for themselves (see most recently Muhly 2005: 689; Hitchcock 2005: 699). This issue refers to the discussion above about attached versus independent artisans (in the physical sense), but does not have to be linked to whether they worked in an elite context or not. If they did not make the decision, we can assume that they may have been forced to do so by others. As already discussed above, documentary information from Near Eastern cultures has demonstrated that artisans were frequently requested by royal courts in order to execute specific tasks (see Zaccagnini 1983 amongst others).

*If* they could decide for themselves, several reasons can be suggested. First, practicalities, relating to the location of raw materials, may 'force' artisans to travel because it facilitated their work or the task could only be executed in a specific place (e.g. builders of specific features required in distant places). Second, artisans may have to travel for economic reasons, for instance, to broaden the market scope for their produce or to satisfy the demand for their produce away from 'home'. It may also be the nature of their work that requires them to travel, or they may travel because they enjoy the prestige it may bring (Pydyn 1998: 101). Third, non-practical reasons to travel exist, although usually for highly skilled artisans only (Helms 1993: 33-34, 36). These are: (1) prestige associated with acquiring services from a famous master craftsman, and (2) confirmation of prestige for the elite by whom the artisans are invited. One could add to that the admiration, recognition and power that may be bestowed upon these artisans following their return from 'distant' and exotic places, which have been imbued with meaning. They thus come back with unique knowledge and stories, which they can report to many others.

Referring to the second point of this section, the context where painted plaster has been found seems to point consistently towards elite buildings such as palaces, villas, large well-built mansions or large elite tombs. This form of decoration, therefore, must have fulfilled specific roles within such contexts. I will focus on the precise circumstances in which the artisans of such decorative craft/art found themselves within this elite environment. As stated before, all painted plaster under study for this book shares some fundamentally important issues: its iconography (allowing for certain differences most probably due to personal/local taste) and its apparent technology (see Chapter Six for more detail). I have already considered travelling artisans when I discussed the artisans themselves. In this section, I will focus specifically on the employers of such artisans, the elite or rulers, and on what their relationship entails.

Helms (1993), who examined the dynamics of skilled artisans that produced only expensive and prestigious items within an elite context based on gift exchange and a prestige economy, identified specific reasons why skilled artisans were so valued by the elite. First, they could produce specific desired items, made of expensive and rare materials, circulating in the right social circles (Zaccagnini 1987: 58), and involving a great deal of knowledge and skill, that could be used by the elite for various status-enhancing reasons. By this I mean that the elite could use such objects and thus indirectly such specialized artisans, to (re)confirm their own social status within society, and they could hence legitimize social order and inequality, and ideology, dictated by them and in support of themselves (Knapp 1998: 195-96). So artisans played a role in creating non-verbal forms of communication to unite as well as to differentiate socio-economic classes (Kenoyer 2000: 88; see also Costin 1998b: 129). The actual legitimization of inequality in and between social classes may have been possible due to a strong interrelationship between the secular and religious roles these elites played out in their society. These roles may have been enhanced through the overt public and private display of vast numbers of, or elaborately made, objects with a supernatural connection (Neitzel 1995: 395-96). Artisans and specialists were themselves very sought after for what they could produce, their knowledge and their 'exotic' nature (e.g. Zaccagnini 1987: 59-60). As such, Peregrine (1991: 1) called such artisans 'political personnel'.

A good example, demonstrating *exactly* how important the artisans were, comes from the Near East where artisans were sometimes chained up to prevent them from running away because they were too few in number and constantly in demand. For the same reasons, they always had to return from missions to distant lands, or they were taken as booty from war raids (Zaccagnini 1983: 247-48; Barber 1991: 164, 284). We also know that elite offspring was occasionally taken away as booty, subsequently trained in specific technologies such as warfare, diplomacy etc. within their new elite context. Once they were old enough, they would be sent back to their country of origin in order to take over the local throne, but they had been 'brainwashed' into the socio-political sphere of their dominators to whom they were indebted (Thyrza-Sparks 2004; Moran 1992: 23-29 on Amarna letter EA 296). In other contexts, artisans were relocated from rural areas to urban centres where they could be closely controlled by the elite (Wattenmaker 1998: 49).

Another reason why they were in demand by the elite was that the artisans, having come from distant lands, were themselves an 'exotic item', which was an attraction in its own right. Again this shows the power of the elite in their ability to have artisans work for them or to send them away (Helms 1993: 13; Zaccagnini 1983: 253, Wattenmaker 1998: 51). The conceptual, not just the real distance (Knapp 1998: 193), in fact formed the added value of whatever arrived: the gift or the artisans as gift or social currency. The knowledge of such 'far away' (unknown, dangerous?) places was also considered a tool of power and was often employed by elites to legitimize their politico-ideological status (Knapp 1998: 195), and could be employed by artisans too in their own social circles.

Since all social relationships are based to some extent upon power differences and conflicts of interest among individuals (Saitta 1994: 203), we can see the social dynamics between elite and artisans in such a light. So clearly, not just the item or the craftspeople were important, but also the tight control over them. This control was especially important in order to restrict or deny access to such expensive and prestigious items because the more such items spread, the more they lost their prestige value. The elite may have strengthened their control by excluding an entire craft because it threatened their hierarchical role (Pydyn 1998: 102). Specific items may also have changed value from economic to prestigious. This is nicely illustrated by the presence of amber in the Baltic area versus its presence in Greece during the Late Bronze Age (Pydyn 1998: 99). In fact, Peregrine (1991: 3) argues that by controlling the artisans and their production processes, the elite inherently controlled the access to wealth items at the same time (see also Kim 2001: 447-48). This makes perfect sense and illustrates how important human actors are in the entire production process. Via a cross-cultural study, taking into account the origin of materials, the complexity of the production process and the actual craft technology as parameters, Kenoyer (2000: 91-92) has shown that the more easily something is controlled, the more likely it is to become chosen for production as a wealth item. The lack of knowledge about (1) materials, (2) how to make things, or (3) people who can do it, is easily controlled by the elite and, therefore, the value of unknown items (e.g. from far away) is increased (Pydyn 1998: 99).

Being in power and showing this power was clearly important for the elite, but even more important must have been finding the means to maintain these power positions. Kim (2001: 466) mentions that, next to control over crafts, taxation, military force, law and systems of bureaucracy are effective ways of keeping the population under control. To that list, I would also add the control over exchange systems and trade being in place in order for the elite to get their necessary supplies of exotic goods. This in fact meant a level of control over those involved in such exchange and trade, namely other elites (see Kohl 1987: 16, for the situation in the Near East, seen in a context of core-periphery elite groups). Furthermore, if the elite could also remain economically powerful (e.g. by enforcing taxes upon people in the form of subsistence produce), they were consequently wealthy enough to ensure continued access to wealth items.

This line of thought can equally be reversed. Elites may also possess wealth items first, by which means they ensure their high hierarchical position within society and can hence give themselves access to or claim resources of subsistence. It was important to demonstrate one's high status, via the access to prestigious items, to be able constantly to obtain 'the latest fashion', because accumulating existing items is only one aspect of wealth and some may sooner or later become 'public domain'. In contrast to that, it is exactly the potential of accumulation of subsistence produce that demonstrates power because this surplus can then be used as currency to obtain other things, to remunerate employees (Saitta 1994: 217), or to have a surplus when there is scarcity (on primary and secondary goods see also Knapp and Blake 2005: 12-13). In such a context, painted plaster fits very well in the 'desired' Late Bronze Age assemblage of prestige items in order to show one's status, be that political, economic or social. Moreover, the elite's control over technologies, artisans, their knowledge and access to raw and exotic materials (even though not always, see Peregrine 1991: 7) all form essential ways to stay in power.

Most of the literature related to this specific topic demonstrates an unequal power balance in favour of the elite (Saitta 1994: 204) greatly exploiting the artisans (Zaccagnini 1983: 249): 'elite-sponsored production is *attached* production' (Costin 1996: 211, author's emphasis), while there are also different power balances in between elites themselves (Zaccagnini 1987: 60, 63; Liverani 1987: 66-67). In several societies, the remuneration the artisans received in return for their work actually shows how much their elite employers valued them (Costin 1991: 3; Wattenmaker 1998: 49-50). Highly specialized artisans in some societies may have achieved high status and a cult of personality may have flourished (Latrap 1983, cited in Roe 1995: 54). But the picture painted by power relations is far more complex than the one-to-one relationship between elites and artisans, especially when we consider consumption patterns of (luxury) goods (see Costin's work on the Inca Empire [1996: 214] mentioning the military, the civil administration, the state religion institution and the elites themselves). In contrast to the upper strata of society, the pure fact that the elites (and possibly others) depended a great deal on these artisans gives a different picture of power. Artisans as designers may have been able to make themselves very attractive to the elite, and even if they worked under the elite's control, they would still render their end product and its inherent message 'through the filter of her or his own worldview and life experience' (see Costin 2005: 1055, citing Williams 1985 and Durland 1991). In almost all cases, the artisans knew their work far better than the elite did. Costin (1998b: 133-34, 137), however, showed examples where the elites themselves were involved in crafting, both in the prehispanic Andes and in Mycenaean Greece. Another form of power for the artisans was their extra knowledge of distant places compared to the people in their homelands (see above). Moreover, complex interactions within a hierarchically structured network of artisans may have been in place too, even within individual crafts. Hodder (1986: 8) pointed out, for instance, that the specific use of style in artifact categories can be manipulated by artisans in order to negotiate a higher position or status in that society. In such cases, the object plays an overtly active role in the social interplay between groups.

Highly specialized craftwork in action could be perceived as a performance. The artisans acted out their knowledge of materials and work procedures, possibly combined with ritual behaviour involving all the senses, realizing smell (of incense), hearing (music such as rhythmic drums) and taste (possibly intoxicating beverages or drugs), and so amazing the onlooker who was viewing something totally new. In that *sense*, the actual power of the artisans also lies in the fact that they are creating devices that enhance, stimulate and thus 'manipulate' the senses of the elites. What we sense and feel are emotions and these can be very powerful. Emotions are felt individually but created socially (Gosden 2004: 34), and this can be well illustrated by people's reactions towards a work of art, for instance. All of us may have different reactions to it: admiration, disgust, sadness, anger; but these emotional reactions are created as a result of us socially interacting with the work of art. Artists and artisans know this and can use this knowledge to manipulate people's emotions.

Because knowledge (esoteric or practical) is power, by performing the artisan also acts out, to a certain degree, his power over the consumer. This may especially have been the case for specific performing crafts such as playing music, singing, dancing and painting as all of these may enhance each other in the intensity of the experience. Roe (1995: 56) sees a specific style within the presentational context or the 'stage' for which the objects are made. Even though the elites may have invited artisans to execute a specific piece of art, there could still be a tremendously powerful, captivating and magical experience which emanated from the combination of seeing or experiencing the performance of the object being made *and* obtaining the end product, especially if that continues to change over time (see Chapter Seven on the 'becoming' of paintings done *al fresco*). Helms (1993: 22) mentioned the magical power that was interwoven with the craft of painting and even with paint itself. The meaning of the objects produced, therefore, lies in both the activity and performance of production on the one hand, and in the product itself on the other (Lechtman 1999: 224). The importance and the power of the performance lie in both what they perform and how they perform it. Part of the performance is, in the case of the paintings under study here, the themes and motifs and the technology underlying them, and the emotions that these may create (see also Gosden 2004: 39).

The last point is to try to understand the role of material culture in general, whether objects or features, as active elements in relation to human actors in a complex set of relationships and expressions of power. It would be very limiting to see only human actors as being active in the relationships played out in prehistoric societies. Material culture may both be an active messenger of social, political, economic and cultural identities and a passive indicator of those. Conkey, in her study about style (1993: 13), sees material culture as an active constitutive element of social practice, where style is more about context than about groups and where style studies inform us about the contexts in which groups of people are brought into existence. She goes on to state that in understanding style one can comprehend how it becomes active (from one context into another), how it is generated and when, and how it is used. In my view, one can apply this line of thought to the entire concept of material culture. In a similar vein, Gosden (2004) reminds us that at the heart of our social lives lie our emotions and that we can act out these emotions, in relation to people *and* objects. As such, objects shape our relationships and us. So while showing their active character in the formation or destruction of relationships, objects, and material culture in the broader sense, constitute power. Because material culture is temporal, emergent, active and contested, as style is (Conkey 1993: 13), material culture is not static but changing and dynamic. This may constitute power again, especially if we consider material culture in relation to elites who are in power and aim to stay there by constantly adapting to and adopting new material expressions of wealth (see above). One can, therefore, say that objects become active, and thus are powerful, not just when they are produced or woven into performances, or through their local or distant distribution patterns, but they were also powerful when they came to life through their consumption by elites and others.

Power lies in objects, materials and technologies, since all may play a role in social interaction patterns. The way objects are shaped or styled may be dictated by their technology in combination with the materials used. Roe (1995: 44) illustrates this point by mentioning the 'tyranny of warp and weft' in weaving but the same can be said of many products. There is also the 'tyranny of damp lime plaster' in *al fresco* painting. In that sense, materials and techniques shape the outcome of the object; they are omnipotent and powerful and contain the restrictions within themselves. Another tyranny we may associate with painted plaster is 'the tyranny of its architecture' in which the paintings are created. Maran (2006a: 11) states that a dynamic interrelationship between architecture and social practice starts with the actual building. On the one hand, the building containing the paintings forms the background for action, while, on the other, a building and its interior furniture (e.g. the paintings) are a product of social practice. He goes on to say that there is, in fact, no architecture free of power since the act of building creates asymmetries of power (Maran 2006a: 12), and we can extend this statement to the act of making the paintings *and* the paintings themselves as objects too.

In terms of what is depicted on objects, we should consider for whom it is depicted and for what reason, because therein lies a whole complex set of meaning construction. Since we are working with figurative paintings within an elite context, one can be confident that the imagery was meant to be viewed and enjoyed by the elites in the first place, so they must have known what that imagery was associated with and what, therefore, its power was. Artisans, however, who painted this imagery must have understood part of it as well, purely by having been in contact with it and, in the eyes of the non-initiated or those who do not have access to this, by having depicted it. Morris (2004: 32-34), in a study of the archaeology of the senses in relation to Minoan religion, discusses how elites may have been in touch with supernatural powers via shamanic practices. It is possible that certain depictions of figures or features in their elite socio-political atmosphere referred to ritual activity, or even the supernatural with which they associated themselves, again to express and legitimate their power positions. In this context, both objects imbued with such associations and the artisans who produced them form powerful entities in relation to the elites who depended on them. Moreover, the meaning of certain objects, especially if they are more permanently materialized (e.g. a painting still visible versus the consumption of an alcoholic drink), may have served as memory devices of a recurring or unique ceremony or ritual, potentially 'recalling' all other sensory stimulations that may have been experienced during that ceremony or ritual. Whether such all-encompassing sensory experience was meant to be collective (in the public arena), or was meant for a private selected group, does not change the impact it may have had on those who experienced it. Morris (2004: 35) mentions the importance of a nuanced interplay between potential multi-sensory experiences and cultural contexts through which embodied encounters find shape and meaning. The techniques employed (drinking, smoking, dancing, fasting) particularly those that invoke such multi-sensory experiences, are culturally and historically situated (Morris 2004: 35). 'Smoking pot' or chewing qat may be a routine and accepted way to experience things in one culture but not in another.

In fact, by looking into sensory experiences and giving material culture and objects their appropriate place in social relationships as a way of trying to understand the meaning of things, one may better grasp how people interact with their material culture, how they may have felt about it and how that, for instance, may subsequently have influenced the relationships between elites and artisans. As such, looking at the role of non-human actors in relation to human actors in a complex set of relationships and expressions of power draws all relationships together in one large interconnecting web.

## Human Action at the Epicentre of Technology: a Labyrinth of Relationships

People and things are interwoven and should not be seen in isolation; a lot of sense or meaning would be lost if this interwovenness could not be contextualized within a web of interrelationships, whether these relationships are between human actors, or human and non-human agents. In the context of craft specialization (and not only there), there is a clear inseparability between technology, social agency and social identity and within the framework of this book the following six important relationships emerge (see Figure 2.1):

1. Artisans in relation to elite
2. Artisans in relation to their society
3. Elite in relation to their society
4. Elite in relation to other elite
5. Artisans in relation to other artisans
6. Human actors in relation to objects/material culture as actors

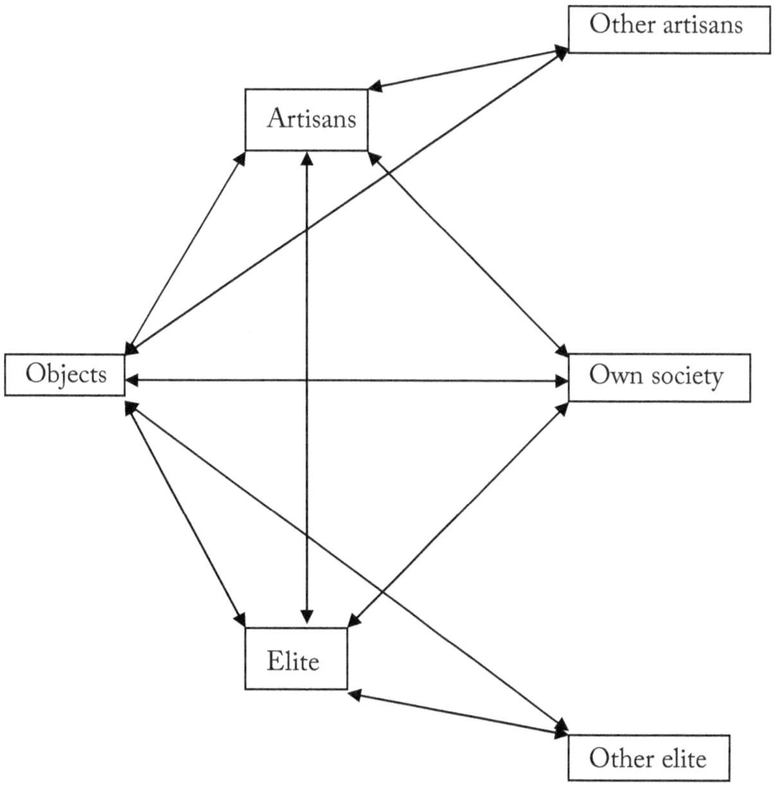


**Figure 2.1** The relationships between each agent.

If the purpose of this chapter was to provide a theoretical framework for the painted plaster case study in order to elucidate the social dynamics of the people involved in this craft, then these six relationships form a practical model to do this. These six relationships are of course very much related to this study, but at the same time I believe that some are also applicable more widely (see below).

My point here, however, is threefold. First, I want to show how studying the relevant relationships may provide answers to social and archaeological questions, via a study of material remains. I believe that in every study one can extract similar relationships. By keeping humans or actors, individuals or groups (of people and objects), as the main focus, each such relationship will reveal insights into the 'hows' and 'whys' of human and other (inter)actions.

Secondly, by considering relationships between the important 'actors' on the scene as a way of approaching social dynamics between them, I feel that there is no limit to such a model. This model does not shape the data, rather the data in relation to the researcher dictate the extent of the model and make the researcher refer back to the data (discursive approach between data and theory).

Thirdly, since the core of archaeological research is to investigate how people lived in the past, independent from the methodology used, I believe that making human and non-human actors central is the key. Moreover, focusing on people's most fundamental behaviour, such as forging and breaking both natural and cultural relationships with each other and the world around them and extracting and studying these relationships, offers useful ways to approach these people.

There is one cautionary note to be made here. Not all the questions can be fitted into just one clear relationship. Just as humans cannot be seen in isolation, neither can relationships. There is bound to be overlap between relationships since humans and other actors have multiple social identities, each associated with another relationship. Such questions are very important, because they show the interconnectedness or the web of the relationships that makes it even more complex and real. These questions are posed within the interface of several relationships in order to be able to provide meaningful answers to them.

Since it is impossible to extract all details from every relationship for this study, the most important questions emerging from each relationship are outlined below. The questions are related to the importance of the relationship and so questions *have* to be related to both (or more) parties involved. Only in this way can the relations be viewed in a balanced way, which has not always been the case in past studies. These questions are discussed as a function of the data collected and presented in Chapter Seven.

1. *Artisans in relation to elite*
   - Were the artisans involved in the painted plaster craft in the service of a specific ruler or elite group, and how is this visible in the archaeological record (attached versus independent, full-time or part-time)?
   - Where did they come from and where did they go?

- Who held the power in this relationship and why? How was that visible?

2. *Artisans in relation to their society*
- Who were these artisans involved in the painted plaster craft, and to whom did they pass on their knowledge? (gender roles, social status)
- Why are they not mentioned in any contemporary document from the Aegean or the Levant/Egypt?
- How did society view the artisans?

3. *Elite in relation to their society*
- How did the elite show their leading position and their controlling power to the population? (political status)
- What was the role of society in the elite's efforts to stay in power?

4. *Elite in relation to other elites*
- How did the elites express their power and diplomatic position(s) vis-à-vis each other?
- Can we see these power relations as a function of larger versus smaller states, and why or why not?

5. *Artisans in relation to other artisans*
- What was the nature of possible contact between artisans involved in the painted plaster craft and other artisans?
- How and where was such contact likely to take place and why?
- What was the broader effect of such contacts within the society, and how did it affect the individual crafts?

6. *Human actors in relation to objects/material culture*
- Where within materials/objects lies the power that limit people's action in making, exchanging and receiving these objects?
- Who influences whom in terms of shaping identities?
- Do materials/objects take an active part in the production of meaning? How?
- What is the power of stability in material culture?
- What is the power of change in material culture?
- How are artisans organized vis-à-vis the work they carry out for their different employers? (involves relationships 1, 2 and maybe 5)
- What instruments, tools and installations did they need for their work? (involves relationships 2, 5 and 6).

To sum up, this chapter has provided a social framework in which the main aim was to place people in the scene of crafting and technology, and to show that technology is in essence a social practice. Such a framework informs us about how to read and interpret the archaeological, analytical and experimental data collected for this study, particularly if we want to understand more about the human forces that underlie technology and, more specifically, technological transfer of the painting tradition under study.

Artisans have multiple social identities, and these emerge as a result of their interaction with each other, with other people and with their material world through technological actions. Each interaction forms a relationship, and through these we can investigate the social dynamics between these people as forces behind the technological transfer. Having looked into the interconnection of people, their identities and their social relationships, it is to be expected that the painted plaster interconnects in an equal fashion. By investigating a range of approaches to the visual aspects of material culture in Chapter Three, we can allow interconnections between them, and we can understand what power(s) hold(s) them together and what powers are formed.

# 3 Technological Style, and the Power of Technology and Knowledge

## Introduction

As stated in Chapter Two, scholars seem to be divided, mainly into two groups, over the current debate about the direction of transfer of the painting tradition under discussion. There are, however, reasons why it is still worthwhile holding a discussion of transfer, and this was also outlined in the previous chapter. The next step is to have a closer look at the concepts of iconography, technology and style and, perhaps most importantly, the nature of their interrelationships and where their power lies.

This chapter, therefore, deals with issues of iconography, technology and style in visual art expressions. It briefly touches upon what each of these is, how they were studied separately in the past, and why they can and should be treated together when investigating visual art forms. Technology is seen as an integral part of the aesthetic forms of any material culture, as much as iconography and style are, and their interrelationship is indicated with the concept of 'technological style'.

## Iconography, Technology and Style

Even though this study does not employ iconography as a tool to answer specific questions, iconography cannot be seen in isolation from technology (Ingold 2001; Dobres 2000 on this division). Moreover, it does show that iconography plays a crucial part in understanding the role of technology in their mutual relationship. After all, it was the iconographic similarities and parallels within the paintings that led to the thorough study of the technology of these paintings.

What then is 'iconography' and how does it differ from technology and, perhaps more challenging, from 'style'? In the *Hutchinson Encyclopedic Dictionary* (Croft *et al.* 1994: 406), 'iconography' is defined as follows:

- illustration of a subject by drawing
- the study of portraits especially of one person
- in art history, significance attached to symbols that can help to identify subject matter and place a work of art in its historical context

In archaeological practice, people who work with iconography tend to combine both points 2 and 3 together to arrive at meaningful explanations about the images that they study. For example, the study of the Prince of the Lilies (Knossos) (Niemeier 1987 with bibliography), or the Ladies in Blue or Red (Knossos) (Cameron 1971) falls under point 2. The meaning, however, given to the imagery of houses, boats, people looking at the boats, etc. (miniature painting, Akrotiri) (Morgan 1988), or human figures in physical exercise with a bull (Knossos, Tell el-Dab'a) (e.g. Marinatos 1994), fits in with point 3. Nevertheless, it is by combining both ways of looking at images and specific elements of scenes that most information can be gained from this endeavour.

Technology and its meaning have been explained in Chapter Two and so will not be repeated here. Suffice it to state that a clear relation exists between iconography and technology in any aspect of material culture. This point is explored below.

A third concept, that of 'style', needs to be considered. Style is used by archaeologists, art historians and many others (e.g. Burke 1999; Carr and Neitzel 1995) as a 'tool' to help us understand the materials we study and to understand aspects of the past. People in the past may have decorated objects or constructed their houses in specific styles that they liked, that demonstrated they belonged to a certain group, or that they were permitted or able to employ in their specific socio-political context. *Hutchinson Encyclopedic Dictionary* (Croft *et al.* 1994: 815) defines 'style' as follows:

- a kind or sort, especially in regard to appearance or form
- the manner of ...doing something, the distinctive manner of...a period
- elegance, distinction
- shape, pattern, fashion (in dress etc.)

A partial understanding of style in an art context forms a combination of the first two points and possibly also of the third and forth. Style may not only encompass *what* (i.e. flowers, animals, humans) is depicted but may also involve *why and how* this was done, as just one way of looking at style (see Conkey and Hastorf 1993 for several papers on the variety of understanding and uses of style in archaeology). This method, therefore, may be considered elegant (according to personal taste) or may demonstrate specific patterns or shapes (i.e. flying gallop, repetitive elements such as rosettes, etc.). Specific combinations of iconographic elements may represent 'exoticness' (Lilyquist 1997). How things are depicted or done is influenced and limited by a wide range of factors. These include: the iconographic elements of what is depicted, interacting cultural groups, belief systems, technological possibilities and limitations, how artisans felt about the material they were working with, the skills they posessed to carry out what they did, what had already been done before and what was new, the socio-political contexts in which these actions took place, and many more. This is where iconography and technology meet, in 'style'. Hodder (1993: 45) states that style is a relational property and the specific relation that matters to us here is the one between iconography, technology and style.

Another important aspect of style is its inherent characteric of being communicative. Style and communication, often non-verbal, are strongly interlinked and there is no style without communication between individuals nor can people communicate without using one or another style to do so (Renfrew and Cherry 1986: vii).

Interestingly, when exploring the meaning of iconography, no explicit reference has been made to aesthetic meaning or values, while 'elegance' has certainly been part of the explanation of 'style'. Clearly, values, and more specifically 'aesthetic value', lie not so much in what is depicted, but in how that was carried out, style-wise, no matter how subjective a category 'aesthetics' may be. In some ways, one could therefore argue that aesthetics are more connected to technology because it specifically seems to come into play once iconography is combined with technology. Moreover, the aesthetic aspect of an object may provide the first attraction, even though the object will be experienced through a variety of senses. Who likes to eat a plain tasteless cake, listen to a poor musical performance, or look at faint and discoloured photographs? The link between aesthetic value and technology has been beautifully illustrated by Winter (2003) recently. Through the investigation of textual sources from the Near East (mid-third to early second millennium BC) she aimed to understand how specific terminology used within these texts revealed a link between the production of objects and the subsequent appreciation of them. It was clear that aesthetic value lay in both the skills and knowledge of artisans, and the embellishment and decoration of the end product (Winter 2003: 408, 412, 416).

The personal experience of aesthetically pleasing or stylish objects or features is of vital importance to the value of that object. In modern-day experiences and global market economies, one cannot afford not to give priority to aesthetically attractive products. Aesthetics, particular the visual aspects, may be the first thing that a consumer will experience; only subsequently may quality and price be considered. 'Without style we have little or nothing to say' (Conkey and Hastorf 1993: 2), and aesthetics is the way people communicate through the senses. It is 'fundamentally the art of using line, form, tone, colour and texture to arouse an emotional reaction in the beholder' (Postrel 2003: 2, 6), at least for visual experiences. In this latter quote, aspects of both technology and iconography come together. Thus aesthetic value may lie in both technology and iconography and may, therefore, communicate specific values intrinsic to the artifact in its relation with the outside world. Alternatively, as Gosden (2006) sees it, values are given to objects by people in relation to their sensory experiences about these objects. In other words, he believes that humans do not directly ascribe value to objects, but, following an experience with an object, a human may then attach (or not) a value to that object. This fits entirely with the concept of appreciation that is discerned in ancient texts (Winter 2003: 415).

Furthermore, aesthetic embellishment may also transform ordinary materials into non-practical goods, again adding value to the basic materials. Depending on the type of embellishment, the effort undertaken and who the artists and customer were, the beauty of specific materials may form the distinctive common factor between elite prestige objects and the objects in daily use possessed by commoners. Beauty itself exercises power because it has the ability to attract (Helms 1993: 67). Beauty, as part of the aesthetic attractiveness of objects, can be *over*powering: we become very quiet, get goose-pimples, smile or even cry while experiencing something really beautiful, regardless of whether this is a visual, auditory or any other sensory experience. Objects may have power over us as individuals because of the values we attribute to them, the way we experience them, or simply because we like their style. Style is linked to power and to differential power relations (see e.g. Morris 1995; Maran 2006a: 12 in relation to architecture). Styles and iconography are said to have had an organizing, and so an active role, in social relations and were part of the political realm of legitimizing social inequality during the rise of early civilizations (Earle 1993: 73; Burke 1999: 35). Those effects and roles, however, are said to operate only within a personal and cultural context. What about experiencing, for instance, a beautiful foreign music performance that gives us a heart-moving feeling even when we are not familiar with the cultural context in which it belongs? It is possible that the experience we have is very different to that felt by its own players or people, but that is hardly an issue or even surprising since there is a clear differentiation to be made between etic and emic experiences, especially when the present-day researcher studies past materials. S/he can experience one thing and be explicit about this, but cannot have access to the context of prehistoric people in order to be explicit about their sensory experiences. In some cases, however, we see that feelings about beauty and its related value may transcend time altogether, and that the words used in the past to describe masterpieces skilfully crafted in luxurious materials are words we would use ourselves to describe these very objects (e.g. the lyre of Ur: Winter 2003: 408).

## Technological Style

Within the research field of Bronze Age painted plaster, no study has tried to integrate iconographic, technological and stylistic approaches beyond the mere descriptive level. I want to demonstrate that one cannot be divorced from the other in continuity nor in changes. 'Making things beautiful or interesting is as valuable, and necessary, as making things work' (Postrel 2003: 179), or, as the anthropologist Boas (1955: 10) stated: 'the judgement of perfection of technical form is essentially an aesthetic judgement'. This point is very much supported by research done in other fields. Through her thorough study of both Andean metal (1996: 509-10; 1999: 224-28) and textiles (1993: 272; 1999: 227), Lechtman has been able to demonstrate how the decorative aspect of these materials, dictated by cultural phenomena and reflecting cultural preoccupations, was part of the technical make-up of the actual production process. As a result, the completed decoration could not be separated from its substrate, since this substrate already bore the essential traits for the decoration. In fact, decorative design and structural design of the cloth were literally interwoven (Lechtman 1993: 272). These examples are what she calls a 'technological style'. A technological style exists when this technology is culture-specific, which means that it has enough traits within it to make it recognizable as belonging to this specific culture (Lechtman and Steinberg 1979: 154). Style, therefore, is embedded in, and produced by, human cultural practice (Conkey 1993: 14). In modern terms, 'style' is a 'good design' and there are many different styles coexisting in our lives. An industrial designer sees it this way: 'Good design is not about the perfect thing anymore, but about helping a lot of different people build their own personal identities' (Postrel 2003: 9).

Style, in general, is the harmonic principle that unites works of art into collective groups and corresponds to the anthropological theme of 'culture' or, as Conkey and Hastorf (1993: 2) describe it: 'the very process of cultural transmission [is] through the style of things', but also through social actions, and through the recognition and use of materials. Culture is style (Gell 1998: 153), and style is culturally determined as long as we do not see culture as a static object to be described, but as 'being in production' or 'always in process of formation' (Drummond 1981, cited by Conkey 1993: 12).

A good example of technological style has been described by Reents-Budet (1998: 73) concerning the Late Classic period in Mayan culture. Personalized ceramic vessels, decorated in a very complex fashion, which required sophisticated technical skills, served to boost the elite's prestige during politically difficult times. The style of these vessels then functioned as a visual index of social cohesiveness. This fits very well with Wiessner's (cited in Burke 1999: 25) argument that 'style becomes a medium for identification via comparison, and a means by which people negotiate and communicate personal and social identity *vis-à-vis* each other'. Style, thus, is about being part of a larger entity, about belonging to a group, a culture. So style is as wide and as plentiful in range as there are multiple identities and may thus be socially determined (see for instance the concept of artistic *koine*). People's expression of identity through style, therefore, is as multifaceted as the expression of identity through group membership (Burke 1999: 27) and varies according to the context in which these social practices are situated. Initially, style studies were concerned with patterns of similarities and differences which were subsequently linked to group relations. This forms Conkey's concept of pattern recognition (Burke 1999: 27). Secondly, we need to follow this up by looking at how and why those groups generated these patterns, and what these specific expressions of social identity may reveal about people's construction of the world and their relative positioning within it (Conkey's pattern generation: Conkey 1993: 15; Burke 1999: 27). This is fundamentally why style is important and why technological style, as part of the general concept, matters in this context (see also Chapter Seven).

Iconography can also be viewed as a technology in its own right: a social technology or the technology of enchantment (Gell 1998: 74). Gell sees decoration as essential to the psychological functionality of artifacts. This aspect should not be dissociated from other functionalities that artifacts possess, such as their practical or social functionality, or even the power they exert over us. As Sackett (1993: 34) argues: 'style and function are not distinct, self-contained, mutually exclusive realms of form in themselves, but instead complementary dimensions ...of variation that co-exist within the same form'.

I would like to suggest that such technological styles and changes were also present within the make-up of the painted plaster, especially once the figurative themes were introduced at the start of MM III (±1740–±1700 BC) on Crete. Seeing technological styles and changes are two different things. It is, therefore, crucial first to understand the material well enough to distinguish the differences and hence how they may have had an influence upon cultural development (Lechtman and Steinberg 1979: 143). Secondly, we must consider the technical and social factors that influenced those changes. This latter issue may be far more complicated to determine, especially if we do not have all evidence from each step of the production process. Again, ethnographic work and experiments may help because they give us, as previously stated, partial insights into people's role at every stage of the performance. In this context, Wright (1986: 4-5) illustrates and discusses how technology, as a cultural variable, may have an impact upon stylistic changes.

At this stage, we also have to allow for the active inventiveness of the artisan who, no doubt, experiments with his or her materials and causes changes this way. Those artisans that master their craft completely can allow themselves to 'play' both with form and content, to find out what will happen (see also Wright 1986: 18). Both changes and continuities in technology, or indeed in any cultural subsystem for that matter, involve choices.

A set of constraints dictating possible technological choices has been summarized by Bleed (2001: 154) as follows:

- Material properties and production tasks
- Wisdom, lore and knowledge of the toolmaker
- Physical constraints: laws of physics
- Somatological constraints: the human body

In my view, the last two fit with the first two (body and mind are inseparable, see Ingold 2001: 18 discussing exactly this separation). Moreover, Bleed (2001: 157) does not take into account any other constraints, such as cultural, social and symbolic ones, which the artisan surely also had to live with (on boundaries of technological and stylistic changes: Wright 1986). The main reason is that he sees technology as an extension of the human body and as an application of effort through it. For example, a student painter, no matter how talented, may not have been in a position to paint the most complicated of scenes. Their young age and inexperience may have been reason enough for 'master' painters to let their students carry out only the simpler parts of the painting until the student is familiar enough with those to move on to something more complicated. Moreover, it may also be a prestige matter for the 'master' to paint the most difficult part himself. In such a way, he can 'show off' his status of 'master' painter towards the viewer, which may affect his relationship with his employer.

As long as there are technological choices, there will be constraints. The main thing is to understand the constraints, see how and why people deal with them (discursive communication of artisans with their material culture, or the materiality dictating what artisans may be able to do with what they have in front of them, depending on how one looks at it), and how such responses are reflected in the material remains we investigate.

This chapter has shown the necessity of approaching technological studies integrated with their related iconography and style if we want to understand the craft in its entirety. We noted a similar web of interlinking relationships on a social level in Chapter Two. It is, therefore, timely to have a closer look how this can be done in practice through the investigation of the material, and, in order to unravel these relationships and their interconnectedness, Chapter Four explores and outlines the interdisciplinary approach that has been employed in this study.

# 4 Archaeometric Approaches to Technologies and Materials

## Introduction

By adopting an approach in which people, their identities and their social organization are central to considerations of technology and technological activities, we can see the main features within a framework of complex relationships whereby social identities change and shift according to the relationship in view. An interdisciplinary approach appears to have revealed multifaceted identities and relationships in the specific case under study, and this chapter demonstrates how experimental work and scientific approaches can contribute to the discussion of social issues.

How, though, are such multifaceted relationships revealed? In Chapter Two we touched upon a range of potential and real-life relationships that may have existed in people's lives and environment, and looked at how the material world played a crucial role in making or breaking these relationships.

The concept being explored here is that an interdisciplinary approach can reveal multifaceted identities and relationships because it looks at a subject from many different angles, allowing a variety of information to emerge. For example, one may investigate the phenomenon of football hooliganism from many viewpoints including that of the football players, the hooligans themselves, the other fans, the financial implications of damage done or accidents caused, the police forces who need to maintain order, the social relationships involved between these interest groups, and the socio-political debates that result from violent events between football teams, etc. Each angle will provide a different set of data and most data sets will reveal a certain amount of overlap, so confirming each other. I believe that multiple confirmations of specific data indicate the strength of these data as reliable, or at least as more than purely hypothetical. The more such data are confirmed, the more they can be used as evidence to back up specific statements. Wylie (2000: 232) refers to this method as 'triangulation' as a way to strengthen interpretations resulting from converging arguments that were each built up in a different way and based upon their own results and conclusions. This wide data range with overlap between data sets, and their subsequent interpretation in context, is what I have tried to achieve in this study.

This chapter outlines and explains how an interdisciplinary study of the technology of painted plaster by means of a scientific multivariate analysis, combined with experimental replication, may reveal multiple strands in social relationships influencing and criss-crossing each other. The core of these relationships reflects human action in all its variety and colour, and it is exactly this human 'way of life' that we want to understand better in archaeological research. Ultimately, one of the reasons for studying human action is to attempt to answer questions about *why* people acted the way they did. The point is to explain the social phenomena that are expressed in the material culture under study (see Chapter Two), to understand the processes of interaction between society, culture, artifacts and landscape and, by so doing, to also refine our ways of understanding.

This chapter seeks to account for certain phenomena. Consequently, the researcher is forced to investigate the underlying assumptions that s/he has vis-à-vis the data and what they mean, by constantly going back and forth between a range of approaches in order to enrich our insights into the matter under study. People's motivation for action needs to be placed within its own context. Human action is not purely individual because no individual acts completely independently from any of his/her social, political, religious or economic environments. Therefore, relationships of human actors within (and outside) their society and context need to be seen as a constant and enriching 'tacking back and forth'.

I stated earlier that isolated technological studies on painted plaster have been carried out before. This study, however, forms the first work that incorporates and combines, comparatively, a range of different methods emphasizing its interdisciplinary nature:

- Non- or micro-destructive instrumental analysis (this chapter and its results discussed in Chapter Six)
- Non-destructive experimental replication (this chapter and its results discussed in Chapter Six)
- Documentation by means of slides, drawings and digital photography
- A social approach, incorporating insights from anthropology and social science related to the role of technology in people's daily lives (touched upon in Chapters Two and Three)

The rest of this chapter is divided into two main sections. The first section deals with aspects of the scientific analyses employed, though without describing each technique in detail (for detail see Brysbaert 2004b). The aim of this section is to demonstrate the multivariate nature of the scientific analysis and how this has yielded the current data. The second section explains some of the theoretical issues relating to experimental archaeology as an approach and the nature of the actual experimental replication carried out.

## Instrumental Analysis and Sampling Strategies

This study has adopted three strategies relating to scientific analysis:

1. Conservation and preservation considerations and condition survey of the material
2. Sampling strategy
3. Adopted instrumental and analytical techniques

The first strategy involved both examining and considering the condition of the material under study, including any current or previous conservation measures. This demonstrated how such measures had a varying influence upon the material and hence on the subsequent strategy of sampling and analysis. This in turn had a varying influence upon the results obtained, some of which was anticipated. Because of varying results, and because this study is comparative in nature, it was necessary to explore these influences in more detail as part of the attempt to adopt a rigorous methodology.

Second, it was evident that a sampling strategy based on a range of different permits allowed different outcomes. A specific methodology—from macro to micro—was adopted in order to achieve a responsible sampling strategy, taking fully into account the three guiding principles of conservation: reversibility, minimal intervention and longevity (Caple 2000: 63-69; but see Munoz-Vinas 2004 who challenges these classical principles). Furthermore, the sampling permits also influenced people's roles in the sampling and consequently how work had to be conducted. This, indirectly, influenced the subsequent analyses and results.

Third, the study employed a wide range of instruments, whether for analytical or for investigative purposes. Most of these were micro- or non-destructive or, if a sample was required, its size was kept to an absolute minimum and it was prepared and analyzed in such a way that these analyses are repeatable.

The points above indicate the strongly interdisciplinary approach that has been adopted, which deeply influenced the data and sampling strategy adopted.

Many studies have mentioned the advantages of an interdisciplinary approach (Bradley 1997; Jones 2002) or the disadvantages of not applying it (see for instance Manning *et al.* 1994: 222-26 on the impact of neglecting social theory). Practically, however, an often uncritical division of materials according to different (unexplained) criteria (termed 'materials' fragmentation' by Jones 2002: 40-46), and unexplained labelling of materials in the literature (Televantou 1992: 148) can be misleading for researchers who want to study specific technical aspects of a craft (for instance 'frescoes'). Dabney (1997: 467-71) has noted single-approach work, whereby technological studies on crafts mainly focus on production and distribution while the archaeological remains mainly show consumption. In order to make the link between all three phases, the need for an interdisciplinary approach is clear.

A final consideration is that the available analytical instruments affect the data and sampling strategies. Factors include what a given instrument can measure (materials identification, technical features), the sample size, the accuracy of results, the level of detail of the analysis (which influences the level of detail in the answers, see Jones 2002: 66) and what we finally decide upon in terms of which techniques to use in combination.

## Conservation and Preservation Considerations and Condition Survey

Certain problems in the study of incomplete and fragile primary material of craft production cannot easily be resolved. These include the following:

- The unknown or unquantifiable difference between materials produced and archaeological remains recovered from sites (see Caple 2000: 587-88; Raiswell 2001: 595-603)
- The amount of re-use and recycling of materials
- The impact of recovery methodology in past excavation practices (complete, selective or random)
- The impact of recording the recovered materials.

All this influences the quantity and quality of materials, how they have been processed and hence our knowledge about it. This is why all other controlled factors were defined; these are outlined in Brysbaert (2004b: Chapter Three).

For this study no samples have been taken from *in situ* extant structures. Most fragments were found as deposits or dumps in secondary contexts within or outside structures in the excavated areas of a site (e.g. Brysbaert 2003b for Gla and Knossos; Bennet 2004: 99-100 for Pylos).

*Condition of the Primary Material*

A condition survey of the painted plaster was carried out, both macroscopically and, where possible, microscopically. The different condition of the finds and their state of preservation depend on a range of factors, especially the extent to which these were originally evaluated, since this influences the treatment such fragments receive. Equally important, their state of preservation is also influenced by the length of time of their deposition (Taylor 2000), by the excavation methods employed, by the degree of care taken during subsequent processing and by storage conditions.

Although all these factors are interwoven and influence each other, the easiest way of describing deterioration processes is to divide the causes into three sections (Table 4.1).

| *Physical* | *Chemical* | *Biological* |
|---|---|---|
| Abrasion of paint | Pigment alteration (pre/post-excav.) | Organisms leaving pigmentation |
| Spalling of layers | Organic filler rotting | Droppings covering the decoration |
| Salt effervescence pushing layers | Leaching due to water percolation | Trampling on site, in storage |
| Soil seeping in cracks | Past recycling of older material | Post excavation handling + storage |
| Crumbling of plaster | Excavation | Conservation (mis)treatment |
| Past recycling of older material | Post excavation handling + storage | Root growing on site |
| Excavation | Conservation (mis)treatment | Discoloration of plaster and paint |
| Post excavation handling + storage | Discoloration of plaster and paint | |
| Conservation (mis)treatment | Environmental (air pollution) | |
| Discolouration of plaster and paint | | |

**Table 4.1** Physical, chemical and biological state of the painted plaster in this study.

*Plaster.* The main cause of damage to plaster is inherent in its composition and the nature of the material. The factors are twofold: external and internal. Constant interaction between materials, techniques and environment occurs as follows:

*External* (surrounding environment)

- The effect of microflora in porous materials (Caneva *et al.* 1991: 87)
- Presence of insoluble salts at the surface or in the interior of the material (Palaikastro: Brysbaert 2000: 55)
- Damp soil causing softening of the matrix and rotting out of hair and straw (Cronyn 1992: 119)
- Deleterious effect on structural stability of the growth of certain floral species on site prior to excavation (e.g. Caneva *et al.* 1991: 127-34).

*Internal* (strength of material relating to its technical quality)

- Leaching out of certain components due to percolating water (Palaikastro, Monastiraki, Tell el-Dab'a, Miletus).
- Waterlogging: plaster becomes rock-hard when its wet state is combined with the presence of insoluble salts and burning (Tell el-Dab'a)
- Pores filled up with soluble salts that react upon changing relative humidity so weakening the matrix (Alalakh, Tiryns, Palaikastro) particularly when in combination with a moisture barrier (e.g. wax as a consolidant on Alalakh plaster).

*Pigments.*

*External*

- Chemical alteration of both organic and inorganic pigments due to high light levels (organic), or changing pH (organic and inorganic) (Cronyn 1992: 118), contact with water and oxygen causing corrosion of metallic compounds (El Goresy 2000: 63)
- Staining or pigmentation caused by low-level organisms, often confused with paint, for instance at Palaikastro, Qatna and Tell el-Dab'a (certain calcicolous micro-organisms, fungi, bacteria, algae and lichens, see Caneva *et al.* 1991: 88-98)

*Internal*

- Loss of coherence of the pigment layer, influencing the loss of the paint. This may be related to the painting techniques.

Furthermore, organic binders may dissolve and pigments may become loose or dislodged (Cronyn 1992: 119). No positive signs of organic binders have yet been published on Bronze Age painted plaster in the Aegean and eastern Mediterranean though this may change soon, as far as the on-going study at Pylos is concerned. Reference has been made to organic binders for the painted plaster at Tell el-Dab'a (Seeber 2000: 115; on organic binders experiments see Brysbaert 2004a) and in relation to tin-covered Aegean Late Bronze Age pottery (Gillis 1994: 57-61, n.11). On the one hand, many specialists believe that such organic binders cannot survive burial for several millennia under constantly changing soil conditions (Perdikatsis *et al.* 2000). Ronca (1994: 115) reported that a specific minimum level of amino acid should be present in order to analyze intentionally added proteinaceous binders via extraction methods. This stands in contrast to detecting its presence as part of contamination (see also Gillard *et al.* 1994 on organic dyes and pigments). Photos-Jones *et al.* (2003), furthermore, reported on the application of Fourier Transform Infra-Red Spectrometry (FT-IR) to analyze organic binders on painted plaster from Chania (negative) and later Byzantine material (positive). The organic traces that have been used as evidence (at Tell el-Dab'a) may equally have been introduced or deposited on the material by organic-rich soils. Unfortunately, we do not have details on the amount of amino acids that were detected there, on the extraction methods used, or on soil conditions/compositions analyzed. Potential soil contamination (introduction of nitrates, amino acids, lipids, etc.) may be especially the case in soils that were used agriculturally before being excavated (e.g. fertilizers, rodent action) or if there was refuse in the vicinity of the finds leaving nitrate deposits (Massari and Massari 1993: 12-13). On the other hand, from conservation research on the interaction between specific pigments and their binding media, we can see that certain mineral pigments, such as hematite, inhibit oxidative polymerization as a result of the air oxidation of oily binders (linseed oil) (Meilunas *et al.* 1990: 43). We may therefore be more hopeful of finding traces of organic binders in the future, provided the analytical equipment is up to the task and such research is sufficiently contextualized within its pedological and geomorphological investigations.

### *Effects of the Condition upon the Material and its Expected Results*

The way painted plaster material has been recovered, stored, catalogued, labelled and made accessible for study (partially or completely) varies from site to site, sometimes even within one site, if several directors were at work. Furthermore, the level of collaboration between the various people and specialists who process and study the material is vital, but again this varies a lot. All these factors had a serious influence upon my sampling strategy, which had often already been dictated and made explicit by the permit specifications.

As I am also trained as a conservator, when submitting my permit requests, I requested the minimum sample size required for the analysis I intended to carry out without causing damage and loss. Occasionally, some samples were so small (1 x 2 mm) that the X-Ray Diffraction (XRD) analysis planned could not be carried out. This left a gap in the data when it came to comparative work using XRD data. Other problems encountered were past overcleaning of painted fragments, or uncleaned fragments obscuring surfaces and thus making their study difficult (personal observations). Some material from older excavations had also lost its labelled context information.

## Sampling Strategy (Table 4.2)

Sampling and its careful preparation is of crucial importance to the outcome of any analysis, and it involves taking into account a range of conditions and making careful

Selection of sites in the Aegean (10), Turkey (3), Egypt (1), Syria (2) and Israel (1)
↓
Selective preservation of the material per site after its burial
↓
Selective excavation of the material
↓
Selective recovery of the material upon excavation
↓
Selective storage of material + its conditions
↓
Accessibility level to all material
↓
Number + type + size of samples that can be taken
↓
Sub-sampling from each sample for each analysis
↓
Final analytical results

Table 4.2 Conditions and choices taken in the process of sampling for this project.

choices at several levels (Table 4.2). The outcome of these analyses, however, may often not be reversible at the sample-taking level. Possible contamination, therefore, present on archaeological materials due to its long burial conditions (in agricultural and fertilized soils, for instance, or by past conservation products) should be carefully considered because these materials will have an influence upon the results, especially in attempting to analyze materials of an organic origin. They should, however, never have an influence upon our interpretation because we should be able to filter them out. While the actual results of the observational techniques are discussed in Chapter Six, it is useful to consider their role within the sampling strategy adopted because a clear interface exists between observational analysis and the actual sampling strategy *before* any sampling was carried out.

*On-site Macroscopy*

By studying each batch of material both macroscopically and microscopically (see Ciliberto 2000: 5), the necessary sampling could take place responsibly avoiding well-decorated fragments or reconstructible parts of the painted plaster and reducing the number and size of the samples. Moreover, this combined macro- and microscopic study brought together two sorts of information:

- All technical phenomena visible via a surface study (Brysbaert 2002b: 95-107 and Chapter Six)
- An aid in deciding what to sample as a form of control for data optically collected

The result was a limited set of randomly chosen samples that was representative of the research questions posed (Figure 4.1). In the process of studying more material, an adaptive sampling strategy was adopted in which most parameters were kept very similar in order to achieve a sample range as homogeneous as possible, facilitating comparative work afterwards.

If material answered a specific technological query by means of macroscopic investigation, this was noted down and no sample was taken, though, in a few cases, I decided to sample anyway. This was done, first, to be able to confirm my observations by analysis and, second, to demonstrate what analytical results it revealed so this could then be helpful for other less clear samples (see issues of fresco painting and pigment penetration, Chapter Six).

*On-site Microscopy*

When fresh breaks existed, a macroscopic study was sufficient to see multiple layers of both paint and plaster. More often than not, however, only a microscope could give clear answers since it provides more detail. The surface and section were carefully cleaned and inspected under the microscope before any conclusions were drawn and samples were chosen. Furthermore, it was also very helpful to work with the microscope on site because it helped in reconstructions. In the past, joins were made macroscopically between fragments that did not join at all. Once they were inspected under the microscope, it became clear that one of the two fragments actually had a damaged top paint layer and was matched to the other fragment based on the wrong coloured paint layer. This is not only misleading for reconstruction, but also for the choice of samples since I did not want to sample anything that could be reconstructed iconographically.

When no microscope is present one can be misled by fragments painted with mixed colours. These can be confused with contaminated samples as opposed to the colour that was originally intended by the artist at application (observed at Palaikastro, Tell el-Dab'a, Thebes and Monastiraki where fragments rubbed against each other in their packing and subsequently bore traces of different colours: see also Lee and Quirke 2000: 113).

*Actual Sample Taking*

Finally, the parameters that were adopted in sampling had to be in line with the aims of this study (see Chapter One) since the results of the planned examination and analysis should give answers to the archaeological questions posed. Based on the aims of the research, these parameters can be summarized as follows (see also Brysbaert and Perdikatsis 2008):

*External*

- Samples are taken from a *geographically* representative spread across the Aegean: Crete, Cyclades, Boeotia and the Argolid.

- Samples taken from eastern Mediterranean sites where painted plaster was found bearing *iconographical* themes and motifs similar to those in the Aegean.
- *Chronologically*, the sites cover the period from the MM I-II on Crete until the end of the LH IIIB on the Greek mainland as far as the Aegean proper is concerned (18th century BC–13th century BC). The eastern Mediterranean sites fall nicely within this timeframe.

*Internal*

It was neither possible, nor was it my aim, to sample everything. The aim was to sample *painted plaster*, so the emphasis, therefore, was placed on sampling the present *colours, multiple layered plaster and pigment structures* and all possible *fillers*. On a few occasions samples from *pigment lumps* (Figure 4.2) were taken for analysis. This was done for two reasons: a) to study the composition and general nature of materials produced, and b) to compare these with what was observed on this site's painted plaster. Lastly, the *content of certain ceramic vessels* filled with either plaster or pigment was sampled in order to compare the composition of such materials with the painted plaster fragments.

Conserved and consolidated fragments and those stained with pigmentation of micro-organisms were avoided where possible because their contamination would affect their subsequent study.

The sample taking was usually a very rapid process. The selected fragments of the correct size were packed in soft tissue and stored in a labelled film canister. No samples were damaged during transport from the museum or site facilities to the laboratory where the sampling preparation and sub-sampling took place.

## Adopted Instrumental and Analytical Techniques

Preusser (2000) has stated that analyzing for its own sake is meaningless unless the methodology reflects the question posed. This is of particular importance when studies based on destructive analyses are undertaken. In employing instrumental analysis, it is equally important to allow communication between the observer or analyzer, the material and its context.

A strategy, therefore, can be put forward consisting of consecutive steps, preferably all carried out by the same person or a collaborative team, in order to keep a consistent data collection pattern based on accumulated knowledge of the material. In the context of the specific case study presented here, this strategy started with an intensive macroscopic study, followed by the use of microscopes and instrumental analysis. Working from macro- to micro-scale analysis allows for discussions of the results both on a regional and/or micro-scale contextual framework such as specific intra-site phenomena. It also allows us to see results in terms of past social practices at a much larger scale upon which local practices may have had a direct impact (Jones 2002). In any given period the scale of people's interactions are multidimensional and so is the material expression of it.

The choice of analytical techniques carried out took into account a range of specific factors (Table 4.3):

- The type of information provided
- Their availability and cost
- Their past well-known, standardized use and results which are useful for comparative work (published results)
- Their minimal destructiveness
- Speed.

Furthermore, some techniques were not used routinely throughout the study, but served as pilot projects in order to be able to compare the data generated with the data obtained via more traditional techniques. The purpose of such pilot studies employing both Laser Induced Breakdown Spectroscopy (hereafter LIBS) and micro-Raman Spectroscopy (hereafter MRS) was to investigate how far these latter techniques could replace more destructive techniques, or whether, by using them, the number and the size of the samples could be reduced further in future projects (for full publication of these pilot studies see Brysbaert and Vandenabeele 2004; 2006; Brysbaert *et al.* 2006). In the cases of Polarized and Reflected Light Microscopy (hereafter PLM, RLM), Energy Dispersive X-Ray Fluorescence (hereafter ED-XRF) and Electron Probe X-ray Micro Analysis (hereafter EPMA), these techniques were employed to clarify some of the blue pigments analyzed by means of X-Ray Diffraction (hereafter XRD) and Scanning Electron Microscopy (hereafter SEM). All these data are fully outlined and tabulated in Brysbaert 2004b.

| *Technique* | *Sample type* | *Result* | *Cost* | *Availability* | *Reusability* |
|---|---|---|---|---|---|
| *PLM* | Pin head | QL | + | Easy | Yes |
| *XRD* | 10 μg powder | QL, sQN | ++ | Easy | Yes |
| | Small frgm (NP) | QL, sQN | ++ | Easy/medium | Yes |
| *SEM (EDAX)* | Cross-section | QL, sQN | +++ | Medium | Yes |
| *LIBS* | Object | QL, sQN | ++++ | Hard | Yes |
| *M-Raman* | Object | QL | ? | Hard | Yes |
| *EPMA* | Cross-section | sQN-QN | +++ | Medium | Yes |
| *ED-XRF* | Small frgm (NP) | QL | +++ | Medium | Yes |

QL = qualitative sQN = semi-quantitative NP = no preparation needed

**Table 4.3** Analytical techniques and the reasons why they were employed in this study (for abbreviations see earlier in this chapter).

After the macroscopic investigation, I looked at the material by means of a range of microscopes and different light sources. Some types of microscopic investigation require sub-sampling (SEM for instance) while non-destructive stereomicroscopic work combined with raking light on fresh breaks may reveal specific technological features (paint and plaster sequences), and thus help to reduce further sampling.

The reflected light microscope together with a digital camera set-up to produce photomicrographs of polished cross-sections of the painted plaster fragments proved to be a good combination. Through combining these images (at different magnifications: x 100–x 400) with image analysis techniques and software, various features could be quantified.

In employing an image analysis technique, the greatest size of any feature was measured as a matter of consistency. In some cases this only made sense, for instance, when measuring the top paint layer of a sample. The thickest area was measured since erosion and time influenced the preservation of that surface. Under the microscope, hardly any paint surface was perceived to be intact. This inevitably also influenced the colour appearance as well as our present perception of it compared with how it was meant to look. Image analysis allowed us to measure:

- Thickness of paint layers
- Pigment grain size
- Primary calcite grain size
- Grain size of areas enriched with secondary calcite.

Once all preliminary microscopic work and sub-sampling were conducted, XRD provided qualitative analysis involving mineral phase identification of crystalline materials (mineral-based pigments, certain synthetically made pigments, rocks), and provided for each a unique XRD profile or X-ray diffractogram. If materials have different phases, their relative proportions to each other can be determined from the intensities integrated in the diffraction pattern. The shape of the peaks also provides information on the different types of impurities, an aspect relevant to this study. Various degrees of samples can be subjected to the XRD, ranging from a single pigment grain to a tiny painted plaster fragment that can subsequently be used for other investigations. Non-destructive surface analysis has recently been carried out successfully and this technique was adopted for this project (see Perdikatsis *et al.* 2000: 105). To the extent that all these sub-samples can be reused and reanalyzed, the technique can be considered non-destructive. For this reason, all the sub-samples were kept and archived under proper storage conditions.

In this study the SEM was used where specific minerals could not be readily identified. SEM formed a very useful tool to perform elemental analysis and often provided the information necessary to be able to go back to the XRD and apply this to the unidentified peaks. Moreover, as a microscope, the SEM provided the possibility of studying a wide range of materials under high magnification and with a large depth of field. Direct examination of surface morphology and topography of the specimen was possible and a semi-quantitative chemical composition of different elements was obtained when the instrument was fitted with an Energy Dispersive X-ray Analyzer (EDAX).

Both EPMA and ED-XRF techniques were employed for one specific reason: to analyze the blue pigments from several sites, and compare the results. This way I found out: (1) the identification, to the trace level, of their metallic compound and, (2) where tin was detected in the Egyptian blue, whether there were certain trends of transfer or spread detectable both geographically and/or chronologically.

A study of organic binders was not undertaken because my sample strategy took into account the non-destructiveness to iconographic scenes, and further took into account the official permissions. This strategy thus did not allow for such investigations, both in numbers and in location of the actual samples to be taken. Moreover, on the sites investigated for this book there were many indications that the *al fresco* technique was used to paint even small details of many paintings (see Chapter Six). Finally, certain techniques that were tested on similar material did not show positive results (Photos-Jones *et al.* 2003). I investigated only small samples that were not part of any iconographically reconstructible unit. I, therefore, do not exclude the possibility of *al secco* additions to specific iconographically rich areas of some of these paintings.

The relation between multiple analyses and their multivariate outcomes in terms of technological data sets can now be integrated into wider socio-economic debates. To illustrate this, for instance, by finding copper alloy with tin phases in the artificially produced Egyptian blue and discovering that glass production places and metallurgical workshops may be located side by side (at Amarna in Egypt for instance, see Chapter Six), one may derive, with some confidence, forms of social interaction between two workshops, or two artisan groups, on at least one level: sharing or exchanging raw materials for specific purposes. Many more such examples may be provided, and all serve to demonstrate that social issues may emerge from multivariate approaches to material culture studies (a more synthetic analysis and interpretation is provided in Chapter Seven; see also Brysbaert 2007a).

## The Methodology of Experimentation and Replication

In introducing the methodology of experimental archaeology, I want to clarify the recurring concepts in experimental archaeology with 'experimental replication' and 'ethnoarchaeology'. As outlined before, the correct use and interpretation of specific terminology employed in this book is important because its proper understanding avoids confusion and misunderstanding. Moreover, the role of ethnoarchaeology and experimental archaeology needs to be clarified, especially in relation to how they may be relevant for archaeological understanding and interpretation (Skibo 1992: x) and, more specifically, how they play a role in facilitating our understanding of social issues and the wide networks of social relationships between people and their material world.

### Experimental Archaeology

Experimental archaeology can be defined as the 'creation of an artificial system to explore archaeological material or processes' (Skibo 2000: 199; Shimada 2005). As a sub-discipline of archaeology, experimental archaeology aims to study the relationship between material culture and human actions by recreating (replicating with different levels of control), in the present day, the human-made material worlds of past societies. In contrast to ethnoarchaeology (see below), the actor (experimenter, scientist) is the one responsible for the creation of an artificial system. Consequently, experimental results can answer and assess pending research questions and create new hypotheses. Mathieu (2002: 1-2) states that experimental replication can then provide analogies for archaeological interpretation because he sees all replication work as experiments. Such replicas can be illustrative or comparative, whereby the illustrative type allows for differences between the replica and the archaeological record, while the comparative replica aims to replicate the original as closely as possible.

Past phenomena have left their material traces in the archaeological record, with their state depending partially on their preservation. Experimental archaeology does not try to recreate the material remains, but instead it attempts to recreate the material world and past contemporary phenomena as they once existed in people's lives, from the moment of their creation until their final deposition at the end of their lives and use. More specifically, experimental archaeologists recreate objects, structures, constructions and phenomena in order to study all technological aspects of (1) past creation processes that resulted in this material world and (2) uses, modifications and functions of these material creations. According to Skibo (2000: 199), experimentalists may also want to understand (3) how these materials, once discarded, behave in their depositional environment, how they deteriorate and how quickly. This interest is the main difference between the approaches adopted by Binford (1983) versus Schiffer (1987), the latter considering formation processes of vital importance in archaeological research (Gould 2000: 183).

It is of vital importance that this recreation is only based upon well-informed and documented knowledge of the material world (Table 4.4). Therefore, it is crucial to know not only about the material aspects that one wants to recreate, but also how materials were used, what their (multiple) functions and meanings were within the society for which they were created, used, discarded or deposited, and finally, which processes the material underwent before it was excavated. This raises the problem that such knowledge is unavoidably biased by what has been found during excavations, and it is important to know whether there are contemporary documents that may clarify certain issues. Clarification of issues such as (1) how the archaeological record was recorded, preserved, and documented upon excavation, (2) how it was put in its relational context, and (3) how it was subsequently understood, interpreted and published by the archaeologists and finds specialists are crucial.

Experimental archaeology, therefore, requires the following:

- An idea or concept of what needs to be recreated
- The experimental design, including parameters and variables and a plan about how to perform the experiment
- Raw materials and a set of tools, instruments, and energy sources

| **knowledge is built up based on existing literature** |
|---|
| ↓ |
| experiments follow from this knowledge |
| ↓ |
| more knowledge obtained from experiments which are tested against analytical results/arch. evidence |
| ↓ |
| more experiments follow from this knowledge |
| ↓ |
| **knowledge is built up based on existing literature** |
| ... |

Table 4.4 Cycle of research-knowledge-experiments.

- An agent(s) to recreate (and who decides the level of control over the variables)
- An observer(s) of the recreation process who documents every step.

In order to make the experiment useful for testing existing research questions and creating new ones, a specific level of control over certain variables is necessary in order to judge the effect of this control in the outcome, because the answer to the research question depends on it. The variables and their level of control should all be defined in advance as part of the experimental design, but the latter should allow for a certain level of flexibility as well. Furthermore, the more tightly controlled the variables, the more artificial the system becomes in the hands of the experimenter. This sort of laboratory-controlled experiment, however, can create very insightful situations, especially when they are material-based and hence aim to extrapolate information on production modes and techniques. Both strictly controlled laboratory experiments (Kiriatzi 2003) and 'real life' replication (Jones 1998; 2000; 2003; 2005) have their plus and minus points.

## Experimental Replication

In the context of experimental archaeology, replication is the end result of an experimental project that recreated a part or parts of the material world of a given society. The more familiar examples include chipped stone tools, clothing, boats, an entire village inclusive of defensive walls, and rubbish pits, etc., while the typology presented by Mathieu (2002: 2-6) gives many more categories. By constructing or producing these items and by using them, the experimental archaeologist hopes to understand better the technology behind the object, together with the labour and organization that were required to obtain a finished product with the means people then had. S/he may also come to a better understanding about whether the skills and knowledge achieved by the artisans were achieved either empirically or accidentally or both. This in turn may also make the experimenter understand the inherent importance of these objects and constructions to the maker, user or destroyer. Experimental archaeology thus fulfills an educational role (see e.g. Stone and Planel 1999).

The experimenter may be able to replicate certain aspects of material culture exactly as found upon excavation by using the same materials, tools or techniques. S/he, however, will always live with the unfinished picture of the surroundings of the agent(s) or actor(s) involved. By studying material remains, archaeologists lack what anthropologists often have available, namely the chance to observe and interact with the people who are responsible for the choices they make within technological production modes, whilst set within their cultural and social contexts. Archaeologists may not know how many people were involved in a specific craft, who they were (men, women, children), how dextrous and knowledgeable they were, when the stages of the work were carried out over the course of the year, whether people shared their technological skills and knowledge across crafts, which factors dictated the choices they made, how technical knowledge and skills were passed on in time to the next generation, and, finally, what social status these groups and/or the individuals had who knew specific technological processes (Lemonnier 1993: 19). If we study technology from an agency-centred point of view where both people and objects are social agents in relation to others, however, technological studies can become more effective and even more revealing.

I would, therefore, like to stress the role of experimental work, based on and informed by traditional crafts still visible today (see below), as a way of showing people's active, intentional and social involvement in any technology (Gooding 1990). I believe that studying pre-industrial technologies or, more specifically, the technology of painted plaster *necessarily* includes experimental and replication work in order to complete the chain of interdisciplinary work. Each experiment shows how many people may get actively involved in the possible technological procedures of the technique or object under study. In certain cases, the actor (agent, experimenter) and observer of the experiment may be the same person. In my opinion, this should encourage experimental archaeologists (whether actor, or observer or both), when writing up their work for publication, to use active sentences. The actor (experimenter) can also be observed by one or more other observers, in order to gain insights both into social aspects such as time use and management, labour input and decision-making (making choices), while interacting with the materials, and into mechanical aspects such as technical achievements and/or failures. Gosden (1999: 11, 57) discussed the act of participant observation as a standard method in ethnographic and anthropological fieldwork. In discussing the social aspects of a technology we should see, first, the specific groups involved: (1) those that procure raw materials, (2) those that produce and supply finished objects (while we stay aware of possible cross-craft interaction and task sharing) and (3) the people who consume the end result of the production. Second, when experiments and experimenters are observed in order to extrapolate information on time, labour energy, skill and decision-making, there is an undeniable hiatus between what experiments and experimenters can do and how skilled craftspeople carried out tasks. Accordingly, time measurements in this context may depend on the experience gained by the experimenter.

## Relation between Experimental Archaeology and Ethnoarchaeology

Gould (2000: 18) explains ethnoarchaeology as 'the study of contemporary, ethnographic human behaviour as a basis for explaining patterning and variability in archaeological associations'. This is only one way of defining ethnoarchaeology and, as stated by David and Kramer (2001: 12), such a definition is very much dependent on who formulates it. These scholars describe the sub-discipline of archaeology or anthropology as a research strategy that employs a range of approaches to understand the relationship between material culture and cultures as entities, whether archaeological or living traditions, and to use these insights to improve archaeological interpretation (David and Kramer 2001: 2). The interrelationship of the materials and people within their original context, comprising a range of elements such as belief systems, socio-political and economic context, and their religious and social lives, is of great importance in ethnoarchaeology.

According to Skibo (2000: 199), both ethnoarchaeology and experimental archaeology have similar theoretical underpinnings, in that they both attempt to understand past phenomena by exploring present-day material culture and human behaviour. In my view, the main difference in approach is the role and general background of the agent(s) (whether social, political, educational, religious, etc.), and this needs to be defined specifically in order to avoid confusion. Mathieu (2002: 6) discusses ethnoarchaeology and ethnography as 'system replication', but feels that these disciplines form their own approach in generating analogies for archaeological interpretation.

### *Ethnoarchaeology*

In ethnoarchaeology, the scientist (i.e. ethnoarchaeologist, ethnographer) 'goes out into the field' with the intention of studying, observing and recording, via informant interviews (London 2000: 2), a range of different practices and behaviour employed by contemporary, most often non-industrial, societies that may produce material worlds similar to those seen in the archaeological record. By observing and interviewing people in these societies, one may extrapolate information as to the function and use of archaeological objects and structures (see e.g. Verhoeven 2005; Watson 1999). As case studies have shown (e.g. London 2000; case studies in *Near Eastern Archaeology* 63.1, e.g. Kardulias 2000), some observations made in such contexts may have very wide applications beyond the specific context in which they were made; others, however, need to be very carefully employed and may not be applicable at all to any other situation or context, simply because they are very much culture-specific (David and Kramer 2001: 4). In addition, the strength of ethnoarchaeology lies in the fact that its data are often analogous to archaeological material and may thus be used to study and explain technological change (see Longacre *et al.* 1991; Schiffer and Skibo 1987). As in all research fields, however, the actual strength in using analogies surfaces only when the right research questions are asked and a solid methodology in the field is employed (London 2000: 6-7). Asking the right question is indirectly hinted at by Sutton (1994: 313-35) in discussing how archaeological surveying and ethnography coinciding alongside the surveys can serve each other in the study of settlement patterns and perceptions concerning Greek villages.

In ethnoarchaeology the observer is the scientist-ethnoarchaeologist, who has virtually no control over the situation (*contra* the experimenter who creates the artificial system), and the actor is the member of a living society that performs specific actions of interest to the observer. It is not possible for the observer and actor to be the same person (but in some cases one can ask who observes whom) since they have very different backgrounds, skills and cultural ties. Therefore, the actions performed by a member of a particular society are considered to be genuine, set as they are in their specific, but changing, cultural context, which gives the full meaning to the action and its consequences. Adaptive behaviour towards an 'intrusive' or 'eager' ethnographer has, however, been remarked.

The outcome of each type of action, whether in ethnoarchaeological or experimental archaeological context, is dependent on the level of control that surrounds the action, the actor and his/her background. In the case of ethnoarchaeology, the only control I can see is the pre-knowledge and skill of the performer although s/he may decide whether to perform well or not. David and Kramer classify ethnoarchaeology under what they call 'actualistic studies', and it thus plays a major role in recognizing the relationship between human behaviour and material culture. Experimental archaeology, however, does not seem to fit into this group simply because the latter is carried out typically with strict control over the relevant variables (including the same actor for series tests or repetition tests), and it is lacking context even if boundaries are crossed (David and Kramer 2001: 13).

## Experimental Work in Aegean Bronze Age Painted Plaster Studies

While an overview of experimental work in Aegean Bronze Age studies would be useful in demonstrating their strengths and weaknesses (but see Brysbaert 2004b), this book focuses only on those relating to the painted plaster. The ones outlined below are the first and only experiments carried out in this field as far as the Aegean and eastern Mediterranean Bronze Age is concerned. They are, therefore, important because they indicate the start of a more interdisciplinary approach to this subject and some are of direct relevance to the work presented here. Moreover, I want to use these studies as a starting point from which to demonstrate my own contributions to and limitations in experiments. This section, therefore, is divided into three main sub-sections:

- M. Cameron's and E. Chryssikopoulou's experiments are summarized indicating the achievements and limitations of the previous researchers
- My own experiments are contextualized.

In the final section I, first, briefly identify which gaps exist in the field of painted plaster experimentation. I then demonstrate where my experiments fill some gaps (and create new ones), but I also discuss their limitations. Subsequently, I outline the methodology applied in my experiments. Since the research questions for each experiment, the experimental design, the work procedure and the results of each experiment have been published already, none of that will be repeated here (see Brysbaert 2003a; 2003b; 2004a; 2007b).

In Chapter Seven these experimental results are, however, contextualized together with the results of the analytical work in order to show their contributions and limitations within the broader framework and questions posed in this book.

*Mark Cameron's Experiments*

The work carried out by Cameron, which was the first of its kind, had as its main goal the replication of Minoan 'frescoes' (my quotation marks). The planning, its methodology and the results of the six-month project have been conveniently summarized by Jones (2004) and will not, therefore, be repeated here (see also Cameron *et al.* 1977: 166, n. 94; Jones 1999: 142, 151). The outcome of these experiments was successful and revealing for anyone with an interest in technological issues of Bronze Age painted plaster, and they are, in my view, most important because they set a trend in experimental archaeology in this field and in Bronze Age Aegean Archaeology. Some limiting factors, however, are summarized here:

- As Jones stated correctly, Cameron's work had been biased towards the *al fresco* technique from the planning stage, and this excluded any other approach (such as *al secco* or any other possible painting technique).
- A great disadvantage of this interesting early work (conducted in 1976) is the fact that, due to the untimely death of the main researcher, it has never been published in its entirety and it only touched upon certain areas. The Cameron archive (kept in the British School of Archaeology in Athens) is vast and, from this, Jones has extracted and summarized useful information, mentioned in two papers (see above).
- All work was carried out in Canada at Cameron's home university in western Ontario. The climatic circumstances are different from those in Greece where the painted plaster was found. This inevitably has consequences for the results obtained (Jones 2004: 221-22; Chryssikopoulou *et al.* 2000: 119).
- Not all the materials Cameron used were close to the materials originally used. Some of the pigments he applied were synthetically produced and some procedures in executing the experiments were not realistic (e.g. the covering of the applied plaster layers).

*Chryssikopoulou and Team*

A more recent small experimental study focused specifically on the complete wall paintings from the site of Akrotiri on Santorini and was published in a detailed manner (Chryssikopoulou *et al.* 2000: 119-26). Its main aim was replicating the wall paintings as they were found on the site, using the ancient techniques that had been employed. The replication was prepared based on the conclusions of the analytical studies conducted on the original material. The replication was subsequently compared to the originals. The outcome of this project was successful, in particular because it was carried out in Athens, much closer in climatic circumstances to the originals. In contrast to Cameron's work, however, Chryssikopoulou and her team did not manage to polish the painted surface without obscuring or interfering with the applied pigments.

The limitations of this project can be summarized as follows:

- This published work has been a 'one-off' piece of work. After personal contact with the main author, it was clear that no continuation was planned.
- This work focused mainly on the plaster from Akrotiri. Although some of the results may be applicable to other Bronze Age painted plaster remains as well, this has not been tested.
- Not all materials used were close to the original materials and not all were explained in detail (for instance the cement of the experimental plaques, Chryssikopoulou *et al.* 2000: 122, possibly affecting porosity, hence drying rates etc.), so these experiments cannot be reproduced by anyone else.

*Experimental Programme of this Research*

*Introduction.* The present experimental programme, combined with interviews with traditional *fresco* artists and plasterers, is a continuation of previous research and aims to answer open questions about painted plaster technology. This programme, thus, manifests itself at two levels (see below): (1) methodological and (2) practical.

1. Through a survey of the existing literature on experiments within Aegean archaeology and more specifically relating to painted plaster, it became clear that there were three main problems:

    - There is no standardized procedure (following a proper design that can lead into multiple testing and series-experiments) for any given experiment and so its value is not easily assessed,
    - The work, no matter how small, does not get properly published,
    - Not enough attention is given to experiments in this specific field.

Consequently, the methodological use of experimental archaeology needs to be tightened up in order to be useful to archaeological interpretation of the original material.

2. Coles (1979: 249) states that, although a lot of experimental work has already been done, there is still space for more. Emanating from this positive comment and based on past work, I summarize the immediate practical gaps in painted plaster technology studies, and demonstrate where my work fits in:

    - *Chaînes opératoires* involved in obtaining and using lime plaster: quarrying, (transport to oven for) calcining (building and operating a kiln for calcining), slaking and maturing (building pits for slaking and maturing lime putty), mixing with filler, and on a large (real) scale, i.e. out of the laboratory and its controls
    - Carrying out and monitoring the different features related to plastering

itself: applying the plaster (vertically for walls, horizontally for floors because they have different wear and tear patterns) onto a range of different backing supports, comprising all possible variables (plaster composition, seasonal preference) found in the analytical and technical literature

- *Chaînes opératoires* involved in obtaining usable pigments (finding usable sources, quarrying, transport, crushing-grinding-sieving-floating-firing-chemically treating, applying onto the backing support) (see also Jones 2004)
- Plastering and painting, related to seasons, onto the different types of backing support
- Sequence of applied colours in a composite, multi-coloured design
- Replication of life-size and miniature-size paintings in vertical position onto a suitable backing support
- Means of transport and labour intensity (time and people input) related to providing fuel for the calcining, slaking, plastering, pigment production (see also Jones 2004)
- Replication of Mycenaean wall paintings (Jones 2004)
- Replication of specific features related to painted plaster other than from Knossos and Akrotiri, which, until recently, were the only paintings focused upon (but see Brysbaert 2003a; 2003b)
- Artificial ageing experiments (ideally done by conservators or conservation scientists) to test deterioration behaviour of organic pigments and binding media in their archaeological contexts

*Contributions and Limitations.* The aim of my own contributions was to fill in gaps, both at a methodological and practical level, with potential for both larger-scale (*chaînes opératoires*-type of work) and smaller-scale experiments (by answering very specific questions).

From a practical point of view, this involves considering the steps and materials taken in collecting and manipulating raw materials into usable products, and evaluating all implications and applications. To date, this has resulted in four *chaînes opératoires*-type experiments and a material-specific experiment:

- The steps and methods involved in replicating Bronze Age painted plaster from the Aegean and eastern Mediterranean, resulting, on the one hand, in replicas of the Minoan Dolphin (Queen's Megaron, Knossos) (first project), and the Tripartite Shrine (Palace, Knossos), and, on the other, the *sgraffito* design (Palaikastro, Building 6) (second project), (Brysbaert 2003a).
- The procurement of coloured earths from the island of Melos, Greece, which are subsequently manipulated into pigments and applied in a range of media to be applied as paints (Brysbaert 2004a: 9-15) (third project).
- Calcining, slaking and applying shell lime plaster (Brysbaert 2003b; 2007b) (fourth project)

- The testing of other plaster materials and paints applied in order to understand their effect upon the painting technique: the gypsum plaster tests (fifth project).

From a methodological point of view, in order to maximize information from relatively small experiments, I have attempted to identify the specific features of each task, and I designed a standard outline based on this task identification. Having followed this simple system I have managed to reach a clearer understanding of what each experiment entailed, what it had to offer and what its limitations were. I also understood where I made mistakes and what the next step for each of them was. Finally, working within a certain standardized system also stimulates very systematic work, and the subsequent information proved more relevant to the original research question. The system employed is outlined and discussed below.

Apart from experimental replication work, the ethnoarchaeological literature was consulted where relevant, in order to understand better certain technological aspects which, otherwise, would be far more complex to explain. For example, I am aware of the use of different binding materials for *secco* painting and I have seen it in use. Traces of organic binders in the Bronze Age wall paintings in the Aegean have not yet been published (see Militello 1998: 231-32; 2001: 180-81; Photos-Jones *et al.* 2003), while such work is under way. This absence does not, however, exclude its past use and existence. From the interviews and conversations I held with traditional *fresco* artists about binding materials, it is this specific ethnographic knowledge that forms the basis for my experiment and my choice of some of the likely organic binders. The interview I held with traditional plasterers about mixing and applying plaster informed me of the practical and social aspects of the plastering work. In addition, analytical literature, mentioning contemporary organic residues and uses, was consulted. Moreover, newly applied analytical techniques (micro-Raman Spectroscopy) were better able to detect certain organic materials than was previously the case (Wehling *et al.* 1999; Vandenabeele *et al.* 2000; Vandenabeele and Moens 2003; Brysbaert and Vandenabeele 2004; 2006). Those tested in experiment 3 (Melos and paints) were the possible and most probable products used in that period and that part of the world, i.e. products that were known to be in use by those cultural groups for other applications. This specific experiment, therefore, tried to discover the possible workings of such binders. The experiment can consequently be used to test for deterioration rates by inducing artificial ageing tests (see above). As a consequence of employing ethnographic methods, it becomes clear that many societies, whether past or present, can mutually throw light upon each other through human interchange and cultural reciprocity (David and Kramer 2001: 419), as long as we keep the contextual and other differences in mind when drawing analogies.

No such experimental work can be conducted without certain anticipated limitations with regard to the design, the conduct or the outcome of the experiment. These can be grouped into four entities (1) the experimenter, (2) practical implications—time, finances, levels of control, environmental conditions—(3) the starting knowledge that (in)forms the experiment and (4) the experiment itself.

(1) As is well known, every study is limited or enhanced by the capabilities of the researcher, including this one. This involves every possible aspect of the researcher, including being well-organized and having the appropriate skills to recognize what is developing within the experiment if s/he conducts it him/herself. I realized that, with every new experiment I set up, its outcome was more significant because the design was more refined, better organized and followed up.

(2) First, funding and time factors play a serious role in what can be achieved from the very start of the project. Coles (1979: 41-44) suggested some very interesting ways of reducing costs involving a set of adaptations of the scale of experiments. This may certainly form a solution as long as we realize that the scale is not similar to the original one and conclusions are drawn accordingly. Most of my experiments can be seen in this reduced form, at least within the framework of this research. Secondly, although it may be the aim of every experimenter to replicate, as closely as possible, all variables involved in a laboratory-controlled experiment, the raw materials used, for instance (geological provenance, manipulation), may not be exactly the same (loss of control over certain variables). This can involve various reasons including the materials no longer being available, the knowledge about their source being lost, it being too expensive to work with original source material, etc. Thirdly, climatic and environmental circumstances in which the experiment is undertaken are important. The closer it can be carried out to the assumed original, the better it may be for certain research questions, but this may be impossible for practical reasons. Moreover, no indication for the seasonality of this activity may exist. Within this context, embedded activities may overrule our idea of 'the optimal seasonal conditions' anyway.

(3) Another important limitation is that the experiments conducted were based upon what was found in the literature (very basic as shown above) and on very fragile and fragmentary material. The positive point is, however, that it creates, as it were, a '*chaîne opératoire*' within the methodology of the experimentation (Figure 4.3).

The triangular repetitions between two or three fields depend on the range of variables to test, on the level of our questions and the level of answers they get, on other related research going on at the same time (by the same person or team or independently), on the financial and practical resources, and on the researchers' level of interest. The referring back and forth between the different fields serves to test the results obtained and to refine working methods as often as required, until one can return to the research questions about the archaeological remains. Such is the design of a series-experiment, not yet carried out within the present research.

(4) A final restraint is inherent in the experiment itself because the results remain, at most, approximations of the reality. The experiment-versus-reality factor of the experiment is an important limitation that should not be overlooked.

*Applied Methodology*. Every well-planned and organized experiment has to consist of at least five main components outlined below, several of which are subdivided. This outline has been followed for each experiment published:

1. Definition of the research questions
2. Design of the experiment
    - 2.1. Identification of parameters and variables
    - 2.2. Decision upon the level of control over variables
    - 2.3. Planning for phase 3 (choose, provide and pay for materials) and 4 (how to proceed, which step, level of documentation)
3. Description of the used materials
4. Outline of the work procedure
5. Analysis and evaluation of results

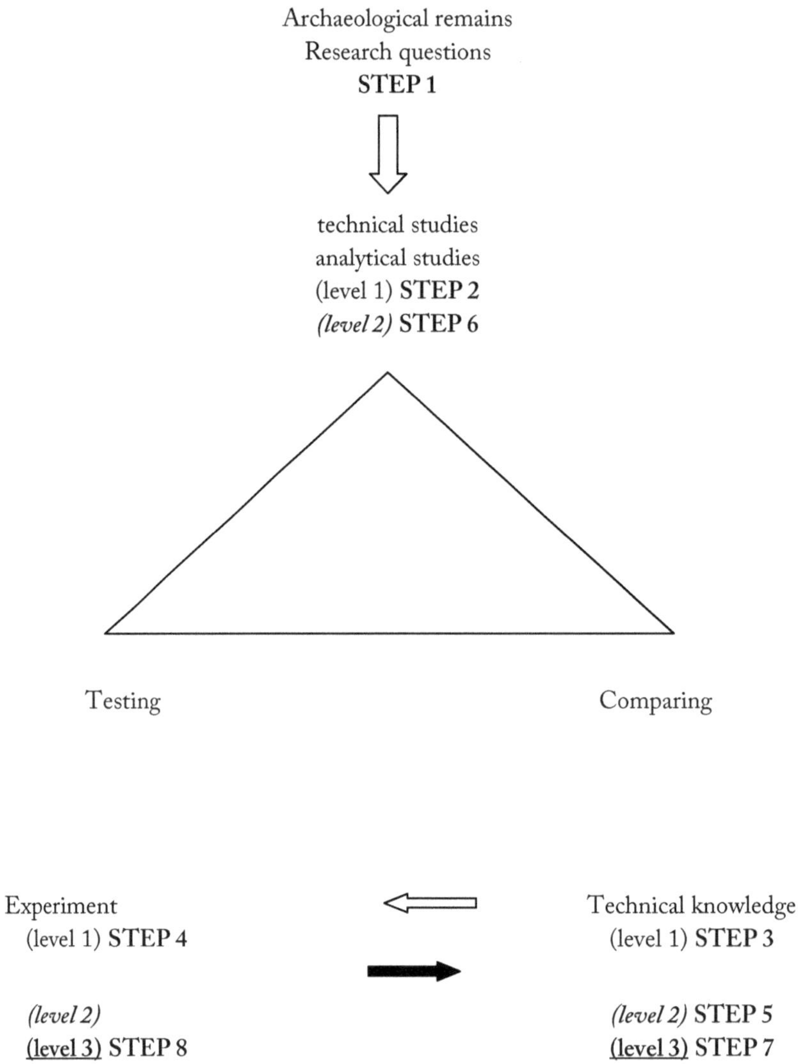


**Figure 4.3** The methodology of experimentation.

*Design:* The experiment's design is as important methodologically as the work procedure and its results. The better the design, the more the problems are foreseen, the better the outcome, even if negative results are provided (because they are indicators as well) (see also Knutsson 1988).

Many classifications for the design phase or any other phase of the experimental procedure have been described in the literature (e.g. Coles 1979: 61; Reedy and Reedy 1992). Table 4.5 below demonstrates where my own experiments fit in with Mathieu's typology, which I found the most suitable (Mathieu 2002: 2-7).

| *Main category* | *Sub-category* | *Examples* |
|---|---|---|
| Object replication | Visual replicas | Dolphin, tri-partite *fresco* |
| | Functional replicas | |
| | Real artifacts/'full replicas' | |
| Behavioural replication | Functional replication | |
| | Comparative experiments | *Sgraffito* experiments |
| | Phenomenological studies | Choosing materials for pigments: seeing |
| Process replication | Formation process replication | |
| | Technological process replication | Pigment making+painting |
| | | Shell into lime plaster |
| | Simulation studies | |
| System replication | Ethnoarchaeology | |
| | Ethnography | |

Table 4.5 Hierarchical table of experimental archaeological research (based on Mathieu's typology, Mathieu 2002: 2-6), fitting in my experiments.

This hierarchical typology is based on the increasing scale of types of past phenomena that can be simulated or replicated, including objects, behaviours, processes and systems (Mathieu 2002: 2). I personally do not feel comfortable with his category of 'behavioural' replication. The category itself makes sense but calling it 'behavioural' refers, in my view, to much more than just carrying out physical actions resulting in material changes as he describes them. Behaviour has also roots in the psychological and emotional aspects of people, which cannot be simulated or replicated based on scientific research, although attempts in this field by phenomenological studies have been promoted. This work can in fact be extremely promising as long as we see the clear limitations, i.e. drawing analogies with intangible phenomena such as what went on in people's minds in the past when living with their material culture and each other. Can we really go as far as conducting a sort of experimental psychology and relate or incorporate that into archaeological research? This is where I make the distinction between 're-enacting agency' versus 're-enacting the actions of the agent' (Brysbaert 2003a).

*Actual Experiment:* All the steps designed should be run as planned, allowing, however, for necessary adaptation, change and creativity. Everything needs to be recorded properly by keeping a notebook and by documenting everything visually. As part of good planning and design, data input and documentation during the experiment are of the utmost importance for subsequent data processing. All findings should be put onto a database, and linked to digital images and video recordings where required.

*Assessment of the outcome:* The results, whether positive or negative, need to be properly assessed to give value to the outcome. Important aspects to keep in mind are:

- That the results are repeatable
- That there is no one single response to the question posed, rather the results may eliminate some of the options.

## Conclusions

The second part of this chapter offered clarifying definitions and explanations concerning the difference and the relationships between experimental archaeology, on the one hand, and ethnoarchaeology and the use of ethnographic methods, on the other. By providing an overview of experimental archaeology projects relating to painted plaster technology in the Aegean and eastern Mediterranean, it was possible to bring this information together allowing the gaps in this field to be highlighted, and it was based upon these gaps that some of my own experiments were set up and conducted.

Moreover, this section also offered some further ideas about how to proceed towards a sound and already existing theoretical and methodological framework in which such experimental replication should take place if it is to be meaningful within an interdisciplinary approach to specific archaeological questions. This discussion, therefore, contributed to the general approach of experimental archaeology, but focused specifically on technological and social issues in the craft of painted plaster. The concept of the *chaîne opératoire* within such experiments (Figure 4.1) showed how the link between technical data and social issues and relationships can be made (see now also Brysbaert 2007a). Finally, this discussion provided a clear idea about the contributions and limitations that such experimental work may have within the larger framework of interdisciplinary research in the field of archaeology.

# 5 Painted Plaster in the Aegean and Eastern Mediterranean

## Introduction

Within a wider debate about aspects of material culture, this chapter looks first at some trends that seem inherent in the growing cross-cultural contacts during the Middle and Late Bronze Ages in the eastern Mediterranean. I focus here solely upon craft traditions within their geographical and chronological context. From that summary ensues the more detailed discussion about the east Mediterranean painted plaster as a phenomenon in its wider context, the meaning it may have had and the role it may have played in its societies: architecturally, socially and practically. It then focuses on the material of this specific study in two further sections. The first places the material in its archaeological context both spatially and chronologically and presents a summary overview of each site from which material was analyzed or studied. It describes thus the find contexts and how the dating of the plaster was achieved. Only a brief mention of its main iconographic themes and motifs is made since this aspect has been extensively covered by Immerwahr (1990) and many others and is the topic of a large ongoing project set up by S. Hiller and F. Blaholmer in Austria (the Fresco Atlas). The geographical and chronological relations between the sites are represented and the importance of the painted plaster on those sites is discussed within the context of its overall architectural and socio-political functions. The purpose of this book is not to present a new Bronze Age Aegean overview, and therefore this study focuses solely upon the sites of relevance to the discussion here. Since the chronological framework here (Middle and Late Bronze Ages) is under intense debate, a brief overview of the relevant discussion is presented. This study, however, does not attempt to solve any chronological questions through its technological focus but, somehow, has to take a standpoint in the debate. Therefore, the last section outlines the chronological framework chosen for this work and briefly explains its arguments.

## Cross-Cultural Contacts and Crafts in East Mediterranean Societies

Considering the extent of literature and scholarship regarding contacts between the Aegean and the eastern Mediterranean during the Middle and Late Bronze Age, it is impossible to give a complete overview of the socio-political, economic and cultural situation in this region. A selective literature review with a focus on craft traditions and the role they played in past societies, however, suggests that certain trends seem to have emerged from contacts in a period of intense socio-political, religious, economic and cultural change in the eastern Mediterranean. None of the trends was steady or homogeneous, however, and each region still needs to be viewed within its own context and for its unique character.

Most recently, Rahmstorf (2006: 50, 61, 80, 84) argued that in discussing the frequency and intensity of contacts between the Aegean and western Asia, we cannot limit ourselves simply to examining imports that came from western Asia into the Aegean. By analyzing and comparing a *range* (my emphasis) of finds from the Aegean and eastern Mediterranean, he concluded that there were clear contacts and interactions between the regions involved already in the Early Bronze Age II (2500–2300 BC). His conclusion is based upon (1) the imported finds themselves, (2) their location *and* how they were used and integrated in local behavioural patterns in each region, and (3) how they consequently testify to the use of common knowledge and knowledge transfer between these regions. The above-mentioned points also provide a major clue: we can no longer justify looking for patterns of production and distribution alone, but instead need to investigate patterns of consumption if we want to understand more fully the range of contacts and the motifs behind them. Bearing this argument in mind, which is also reflected in cross-craft interaction studies (Brysbaert 2007a), it can be demonstrated that the trends summarized and exemplified below are very much interlinked.

During the formative periods of the Middle Bronze Age in the eastern Mediterranean there seems to be a level of transfer of iconographic motifs and stylistic features across media, on both mobile and immobile phenomena of material culture (e.g. Crowley 1989). Simultaneously, technological changes and transfer, both in production and consumption modes of objects, are also visible across a range of crafts; at the same time they also *created* more contacts as the result of shifts in desire for specific objects. These changes and transfers (iconographically, stylistically and technologically) became, later on, interwoven with each other in certain craft traditions and are, furthermore, seen in connection with changing distribution patterns. Transfers in iconography, style and technology, whether simultaneous or separate, brought about—in a certain group of luxury objects and features—a fascinating level of hybridity in appearance *combined* with the same level of complexity in their meaning-making (see further in Chapters Seven and Eight).

These levels of transfer continued well into the Late Bronze Age and beyond. The trade in textiles connecting the eastern Mediterranean and the Aegean forms a good example whereby Aegean textiles may have travelled to the Near East and Egypt (where the actual objects perished but whose patterns were copied onto Theban tomb ceilings), even though such items may have been produced locally. Their intrinsic value may have lain in the fact that they were 'foreign' or of better quality than the local produce (Barber 1991: 311, 357). Other examples are well illustrated by Feldman (2006).

Cross-craft interaction between artisans (Brysbaert 2003b: 176; 2007a) forms an interesting phenomenon visible across objects of different media and includes both transfer of iconographic and thematic aspects but also of technological features. Such iconographic and thematic interactions have been highlighted by Barber (1991) and others with respect to the connection between textile, pottery and paintings, and to the connection between glass and gold beads in jewellery, and between mould-manufacturing and gem-cutting (Laffineur 1995: 189, 191). As far as cross-craft interaction is concerned, technologies were transferred next to stylistic and iconographic elements, and this was visible, for example, in the origin of glass-making in Egypt (Nicholson 1997: 378, 380), in ivory-carving (Steel 2004: 165) or in any form of skeuomorphism between media, and I return to the links between technology, iconography and style in the context of transfer in Chapters Seven and Eight. The hybrid nature of the goods found at the Uluburun shipwreck (see e.g. Pulak 2005), Marsa Matruh in Libya (White 2003), Kommos, Tiryns and many other locations (see Cline 1994: 258-77), illustrate such international contacts and trade. Warren (1995: 2-3) has also noted other transfers such as knowledge, symbols and decoration. In some cases, the transfer of objects and ideas was linked and goes hand in hand (Maran 2004b). There was also evidence of knowledge transfer between Egypt and other regions, as attested in the Amarna letters (e.g. EA 10 Moran 1992: 19-20), and Canaanite technological terms found their way into the Egyptian language (Redford 2000: 7), while Keftiu spells were transcribed into hieroglyphs (Strange 1980: 99, 101). Egyptian-style objects and architectural elements entered Canaan as imports or local imitations (Killebrew 2005: 64-65). In Cyprus, centrally located in the eastern Mediterranean, early contacts with surrounding areas resulted in a range of changes in technologies relating to pottery production, metallurgical and agricultural activities, weaving and burial practices (Steel 2004: 126-27, 143). Close contacts between the Aegean, Cyprus and the Levant, especially during later phases of the Late Bronze Age, are mirrored in almost all material remains and crafts (Portugali and Knapp 1985; Knapp 1986; Cline 1994; Steel 2004: 161-71; Feldman 2006: 25-58, 169-75).

Geographical positions of specific locales and people's level of mobility are crucial in making, keeping and breaking contacts. The Mediterranean, more fragmented than its neighbouring regions, consists of a conglomerate of small sub-regions (e.g. Cyprus, the southern Aegean) and groups of such sub-regions (e.g. Cycladic islands). Even though its physical geography is split up into a myriad of islands, with the sea in between, one of its distinctive characteristics is connectivity (Horden and Purcell 2000: 172, 343-44), which is partially dependent on physical aspects (e.g. inter-visibility of land) and partially a matter of the patterns of human mobility as a response to the physical environment. Furthermore, both ethnic diversity of the population and a specific Mediterranean cultural homogeneity are the most obvious correlates of mobility (Horden and Purcell 2000: 395, 400). During MM IIIA-LM IB, for instance, Minoan overseas contacts seem to have been established via at least three major trade routes within the Aegean, along which crafted goods would have travelled (for earlier Minoan contacts see Davis 1982: 33-41; Davis 2001: 26). Equally, even though no painted plaster of this period was found on Cyprus, the island did play a pivotal role in the interconnections between regions in the eastern Mediterranean Middle and Late Bronze Age (see discussion about *Alashiya* in the Mari tablets [Muhly 1986: 39; Knapp 1996: 3, 5; Goren *et al.* 2003: 248-52]). Some of the Aegean and other foreign imports that came into Cyprus appear to be the types of luxury goods that may indicate the interest of local elites in demonstrating their social status and which emphasize growing social differentiation (see e.g. Steel 2004: 143). The long-distance trade links that Cyprus had with Egypt, the Levant and the Aegean were especially visible in the wide copper distribution and export of Cypriot pottery to those regions (Gitlen 1981: 49-50), in a large importation of Mycenaean pottery to the island (Steel 1998: 293-96), and the presence of Egyptian objects in Cyprus (catalogued in Jacobsen 1994). Ideally placed geographical locations aided polities such as Alalakh (Turkey), Yamhad (Aleppo), Mari, Qatna, Ugarit (all in Syria—see e.g. Akkermans and Schwartz 2005), Byblos (Lebanon), Tel Kabri, Hazor, Megiddo (Israel/Palestine—see e.g. Killebrew 2005), and Tell el-Dab'a (Egypt) in the establishment of their economic and political power and made a range of contacts possible in several directions. The Uluburun shipwreck should be mentioned again as an exemplary testimony to the extent of mobility of objects and people between many locations.

Socio-political and economic shifts with an emphasis on a combined presence of hierarchies and heterarchies rather than just a growing hierarchical system with a dominant political control and presence have been noted in the Aegean (see Schoep and Knappett 2004) and seem to have been stimulating environments for the growth of craft production and consumption. This local development of peer polities (Cherry 1986: 38-39) and some of the more locally developed hierarchical polities on Crete may have been emulators of Near Eastern status symbols in order to secure their higher inter-polity status, visible in the long-distance exchange of luxury commodities in the Middle Bronze Age between Crete and the Near East. Whatever the direction of the first contacts, they stimulated, among other things, more commercial exchange and especially the growth of craft specialization from the Middle Bronze Age onwards and into the Late Bronze Age on Crete, and later also on the Greek mainland. Examples worth citing are: purple dye production (Burke 1999), wheelmade pottery production (Knappett 1999), Kamares pottery imitated in Egypt (Kemp and Merrillees 1980). A similar situation of heterarchical organization leading to competing social groups (elites) on Cyprus is argued by Keswani (1996: 219) and may be linked to a wide variety of well-developed craft traditions on the island towards the end of the Late Bronze Age, when Cyprus took a pivotal place as a diplomatic power within intense Near Eastern trade networks (Steel 2004: 187).

During MM IIIA-LM IB, Minoan cultural influence increased somewhat in comparison with earlier phases (Old Palace Period) and can be seen on the Aegean mainland and islands (Melos, Samothrace, Kea, Thera, Rhodes, Kos, etc.) and in Asia Minor (Iasos, Miletus, etc.). Little MM III material evidence, however, was found in Egypt. Some Minoan products reached as far south as Palestine (Manning 1999: 109), but no LM IA pottery was found in Egypt, apart from that found at Tell el-Dab'a (but in a secondary context, see Manning 1999: 107-108). Following on from Manning and

Hulin (2005: 279, 283), who looked at the actual quantities of foreign objects in places, not a great deal of Aegean objects found their way 'abroad'. Later on in the Late Bronze Age III phase, more Mycenaean pottery was found throughout Egypt and the Near East, but again, not in the massive amounts that allows them to be interpreted as a sign of commercial trading (Manning and Hulin 2005: 284-86). We may assume Cretan contact with Egypt via diplomatic contacts such as the reception of Minoan officials at the court of Hatshepsut (Barber 1991: 331), and also through earlier Minoan evidence in Middle and Upper Egypt (Manning 1999: 109). At Alalakh level VII, for instance, a Minoan stone vase fragment and painted plaster may indicate a common tendency for elite emulation through hybrid artistic expressions spanning the eastern Mediterranean during the Middle Bronze-Late Bronze Age phases.

Seen from the opposite side, imported goods throughout the Middle Bronze Age and Late Bronze Age from Egypt and the Near East, such as the Kassite *lapis lazuli* seals from Thebes, may have reached the Aegean via Cyprus (Porada 1981–82: 68-70; for other imports, see also Manning 1999: 113-19; Cline 1994), but in much lesser quantities than Aegean goods in the east, if we think about omnipresent amounts of Mycenaean pottery. During the later part of the Middle Bronze Age phase in the Near East, several major and new Palatial centres appeared in the Levant (Oren 1997a: 257), and the artistic influences on the Aegean from the Near East and Egypt are clear and have been much studied (e.g. Davies and Schofield 1995; Laffineur 1998; Phillips 1995; 2001; 2004).

Not all contacts, however, may be accurately described as pure commercial and trading contacts. While several types of exchange mechanisms have been described, to which I will refer in Chapter Eight (see Knapp and Cherry 1994), some of these systems may have existed together in the same venture (Knapp and Cherry 1994: 125-51). The extent to which commodities moved as 'gifts', on the one hand, and certain forms of diplomacy between rulers and elite, on the other (e.g. the request for artisans via official letters between elite rulers), led to the movement of people (Cline 1994: 85; Moran 1992: 112).

During LMII/LH IIB-IIIB, the cultural and politico-economic domination in the Aegean shifted to mainland Greece. Fortified palaces, often located on an acropolis (e.g. at Mycenae, Tiryns, Thebes, Orchomenos, Gla, Pylos, Midea, etc.), controlled the economy as redistribution centres for agricultural produce and may have functioned equally as control posts on trade routes (see Sherratt 2001). These palaces also seem to have fulfilled a religious role, where there may have been a joint political and religious rulership (Maran 2001a). This close relationship between religious and secular spheres also demonstrates the importance, for these non-industrial societies, of esoteric contacts with non-human/non-physical entities via the ruling elites. Through such contacts, power relationships were established between and among the different groups of each society, and patterns of inclusion and exclusion were established, negotiated, confirmed and disputed. That this aspect of life was important to craft traditions may be visible in the presence of workshops in palatial contexts at Thebes, Tiryns and other Mycenaean palatial settings where religious functions and rituals took place (see also Chapter Seven).

Craft centres and their activities were described to some extent in the Linear B tablets, and there was much pottery production of pictorial vessels for export, notably in the Argolid (Mommsen and Maran 2001). A hierarchy among the neighbouring centres, such as Mycenae and Tiryns, must have existed, reflecting the existing social ranking, which was also visible in tomb types and grave goods throughout. Large-scale overseas export of pottery containing oil and wine took place and small Mycenaean stirrup jars were known all over the eastern (Leonard 1994; Hankey and Leonard 1998; Hankey 1995: 172) and western Mediterranean (Buxeda I Garrigós *et al.* 2003).

The presence of natural and/or human resources seems to play a crucial role in the production of goods, in the mobility of craft traditions in specific regions and in the final consumption of these goods. We need to consider, however, that certain materials are better preserved in the archaeological record than others, based on their material existence (e.g. inorganic copper and tin versus organic wine, oils and perfumes), production processes (wooden furniture, unfired pottery versus stone mortars and cast metal), distribution patterns (mobile pottery versus immobile paintings in architecture), and their consumption patterns (edible versus inedible, usage through handling versus adoration through vision, hearing and scent). Natural resources seem somewhat easier to detect than human resources, especially if no textual evidence illuminates the latter. Copper as the main natural resource on Cyprus, on the one hand, is self-explanatory in terms of its importance during the Late Bronze Age (on copper oxhide ingots: Gale 1991). Human resources, on the other hand, are more complex to trace and were not always a voluntary resource as is illustrated below. Slavery and captivity of human booty during war campaigns are phenomena known from documentary evidence in the eastern Mediterranean, Egypt and the Near East. Examples from Egypt, showing how the former looked upon and used Asiatics during the Middle Kingdom period for all sorts of reasons, can be read in contemporary and later papyri and inscriptions (Redford 2000: 7). Amarna letters testify to Asiatic young elites who were taken to Egypt to be educated while local princesses were married off into the Egyptian court, all tying local courts into obligations and loyalty to Egypt (Thyrza-Sparks 2004; Moran 1992: 23-29 EA 296). Linear B tablets referred to foreign female slavery in Mycenaean textile workshops (Killen 2001: 172), and the Mari tablets mention artisans that were tied up to keep them where they were wanted (Zaccagnini 1983).

The mobility (and equally the production and consumption) of wealth and prestige items should not and cannot be separated from mundane goods distribution (Horden and Purcell 2000: 347), since one type seem to have brought about the other. Examples are known from Egypt, where Cretans (and others) may have shipped olive oil to (Serpico and White 2000b: 399), and Mediterranean-wide (see e.g. Cyprus in connection with Syria, Cilicia, Egypt and the Levant—Eriksson 1995: 205). On Cyprus, a dual craft (specialization) system seemed to have taken shape side by side: (1) the production of everyday household objects and (2) elite luxury items, both systems implying different uses of labour in terms of time and organization (Steel 2004: 161).

It seems that the drive behind contacts between groups of people cannot just be explained by trade, thus insinuating specific needs only. The mobility of raw materials and finished objects seems to be driven by a wide range of contacts and reasons for these, thus reflecting these shifts in changing socio-political build-ups of regional smaller and larger polities and people's changing desires and access to things. This was visibly so in Cyprus at the end of the Late Bronze Age, for instance, where the strength of its continued trade lay in the added value of finished objects in contrast to the earlier raw materials (Steel 2004: 181). The painted plaster can be seen very much in the same vein, a finished product, made by artisans from elsewhere, important for its consumption value.

Thus it seems that the actual drive behind contacts between groups of people can only be understood by investigating all aspects of the *chaîne opératoire*, both technologically and socially and including the consumption of goods, of as many crafts and technologies as possible, whether mundane or elite-controlled. Furthermore, during the Late Bronze Age eastern Mediterranean politics were very much intertwined, and a few additional but important factors seem to tie all the above together.

First, by the start of the Late Bronze Age, many documents that inform us about the political, social and cultural situations in both the Near East and Egypt were written in a *lingua franca*, Akkadian, used extensively between Egypt, many Levantine city-states and Cyprus 14th century BC (Van de Mieroop 1999: 37). This was the case with many Hittite texts in the state archive at Hattusha where tablets were translated into Akkadian. The state archive at Alalakh was in Akkadian and the archive in Ugarit contained texts in both Ugaritic and Akkadian. In addition, the Amarna letters in Egypt between the Egyptian court and, for instance, the king of *Alashiya* in Cyprus were all written in Akkadian cuneiform (Van de Mieroop 1999: 37). Second, especially during the Late Bronze Age, the Mediterranean world may be characterized in part by its coastal towns, maritime ports, multi-ethnic populations, merchant and artisan quarters, and storage facilities for both ships and goods (Knapp 1998: 196). Third, another of the strong characteristics of this period is an elite artistic *koine*, which seems to have been appreciated by many regions involved. Hybrid decorative elements applied in labour-intensive techniques onto luxury materials were used in elite contexts of only the most powerful rulers of the Late Bronze Age. Feldman's (2006) in-depth study about the phenomenon of such luxury objects, their artistic hybrid appearance and their implied meaning-making suggests that these objects appeared in the specific contexts of diplomatic relationship building, negotiating and confirming. I return later to the socio-political meaning of such objects, in Chapters Seven and Eight. These three factors mentioned above can be seen as man-made facilitators, active aids, to stimulate ongoing and new contacts in the region, each within their own social context.

The painted plaster found at Alalakh, Hattusha, Qatna, Tel Kabri and Tell el-Dab'a fits very well in the context described above, in that it displays hybrid features, iconographically, stylistically and technologically. They are located in geographically important locations that simultaneously required the mobility of human and natural resources, and were very much part of a socio-political and cultural context where craft traditions had a place to develop, interact and explore. Painted plaster with possible

Aegean-like motifs were also found at Malkata (Nicolakaki-Kentrou 2000; Kemp 2000), and at Amarna (Kemp and Weatherhead 2000; Weatherhead 2007). The specific role of painted plaster from 17 sites (see Figure 1.1) investigated will now be explored in more detail.

## The Role of Painted Plaster in Eastern Mediterranean Societies

Painted plaster in the Aegean and eastern Mediterranean cannot be seen separately from the existing traditions in contemporary Egypt and the Near East. The technologically and iconographically different nature of the huge number of Egyptian paintings and the very few known Near Eastern paintings is made clear in Chapters Three and Six. Even though many studies point out the relationships between Aegean, on the one hand, and Egyptian and Near Eastern iconography, on the other (e.g. Kantor 1947; Morris 1993; Stevenson Smith 1998; Feldmann 2003; Collon 1994; 2000; Davis 2000b; Shaw 2000b; Hiller 2003; Warren 2005), this study suggests that the assemblage of both iconography and technology presented within these paintings, called a 'technological style' (Lechtman and Steinberg 1979), is Aegean in origin.

This has serious implications for the role that painted plaster played both in the Aegean and the eastern Mediterranean. Painted plaster, its producers and its patrons formed a dynamic and socially interacting package within the Late Bronze Age, establishing meanings and values during, and as a result of, these social interactions (see Chapter Seven). Even though the spread of painted plaster within the Aegean is far wider than it is outside, it always appears either in palatial or in elite architecture, and these paintings consistently bore a hybrid iconography which described visually their belonging to a social group spread all over the area rather than belonging to one specific cultural group (following the reasoning of Feldman 2006: 5). The presence of such paintings, together with a whole range of other objects like metals and textiles, sheds light upon some of the elite contacts that took place in the eastern Mediterranean Late Bronze Age.

The painted plaster craft consists of many different components, each of which contributes a specific part of the technology. Many of its components do not just belong to this specific craft; their knowledge is also applicable to several other crafts. This suggests a possible level of cross-craft interaction (Brysbaert 2007a) which may have played a role at different intra- and inter-regional levels, as is suggested in this study.

The appearance of figurative painted plaster as an art form came rather late in comparison with most other Aegean crafts, from which it may have been adopted and adapted. Being part of an international artistic *koine* it also became immediately an elite artistic expression (Manning 1999: 137-38 mentions other *koine* elite expressions; see also Feldman 2006 on portable objects hybrid in iconography and style). The fact that these paintings belonged in the elite atmosphere and depicted imagery at home in the hybrid international style may indicate the elite's constant need during this period for an iconography that expressed a specific vocabulary of power to the onlooker. The clearly visible and large-scale iconography on paintings (in contrast to miniature-scale carving on seals for example) seems to have been both secular and religious, and fulfilled several functions (Morris 1989; 1993; 2000; Hiller 1990; 2000; Shaw 2004). This does not mean, however, that where these iconographic motifs and themes were used in distinctly different areas (e.g. Egypt versus Qatna), their meaning and value were understood in the same way locally (Pydyn 1998).

These paintings were part of the international elite 'must have' assemblage (Cline 1994: xvi; Knapp 1998: 198 on Helleno-Semitic prestige goods) of the time and, as such, they played four other roles. First, they may have fulfilled the role of gift or exchange value between elite groups that may have sent artisans from one place to another to let them execute the paintings. This gift or exchange product may have ensured specific diplomatic contacts between different regions, necessary to allow other requests or exchanges to take place (Feldman 2006: 17; Liverani 1990 on gift exchange). Second, they also brought the possessors of such paintings closer together through a sense of belonging and distinguished their social rank from that of other people. Third, these paintings were part of the architecture in which they fulfilled a decorative role. The location of these paintings within the actual building, the iconography and themes they portrayed, the efforts it took to produce them, their coexistence with other architectural features, and several other aspects, imbue these paintings with a specific function and role of power upon the onlooker and even the owner (see Maran 2006b; Markus 2006). This last aspect is discussed fully in Chapter Seven.

The fact that these paintings are rather widely but thinly spread outside the Aegean makes them a good starting point from which to study concepts of maritime space and the role of conceptual distance and the exotic within the Aegean and eastern Mediterranean. This approach has been foreshadowed by Knapp (1998), and also by Broodbank (1993), who applied the ideas of M. Helms (1993) within the Aegean and eastern Mediterranean. As such, these paintings make scholars look at the Late Bronze Age social dynamics in a different way. Due to their reasonable visibility and preservation as archaeological remains, these paintings are the material traces of a much wider and interconnected Mediterranean network of people. As such, my research takes the form of a case study to approach the wider topic of social dynamics within the Mediterranean during the Middle and Late Bronze Ages.

The final and often overlooked role of these paintings is a technological and practical one: they are surface covers. Although their aesthetic function was prioritized in the Late Bronze Age, they still had an architectural function. Most certainly, the earliest examples found in EM Knossos, Vassiliki and Myrtos Fournou-Koryphi were applied very much as a combination of protection and decoration and were painted just red or yellow (see Kingery *et al.* 1988: 222). Being permeable and flexible (Gibbons 1995: 1), plaster seals the stone or mud brick wall (see Palyvou 2004: 189) while regulating the ambient humidity levels within a specific space. Moisture within a building may evaporate through lime plaster over stone quite naturally. Moreover, because a building can breathe when plastered, plaster protects other architectural materials such as wooden beams from rotting and further destruction (Gibbons 1995: 3). Such functional use is very important in situations of rising ground water, capillary action and wind-driven rain penetration.

It is now no longer possible to maintain that the present Mediterranean weather has not changed since the Bronze Age (see most recently Moody 2005: 443). Moody's study discusses new techniques of data collecting (through looking at solar cycles, orbital parameters, relating different palaeoclimatic systems of the Aegean and east Mediterranean to each other and placing them in their wider and global palaeoclimatic contexts) and a subsequent better understanding of the palaeoclimate of the Mediterranean region. She (2005: 443, 460-70) demonstrated that there were fluctuations even between the different periods within the Late Bronze Age and these must have had a profound impact upon crops, crop cycles, animal husbandry and architecture. Since plaster surfaces have a moisture-regulating function within architectural units, we should, therefore, consider seriously the climatic circumstances and their effects on daily life and their production. All places where these paintings were found had a mild Mediterranean seasonal climate with changeable rains in autumn, winter, spring and even summer (see specifically Moody 2005); hence humidity regulation through plastered surfaces would have been most welcome.

## The Archaeological Context and the Importance of the Material per Site

In the following discussion organized per site, the emphasis falls on the material sampled for this study while related material from the same site is briefly mentioned for reasons of completeness. This section is divided into three broad regions: Crete and the Cyclades, the Greek mainland and the eastern Mediterranean. Each individual site is treated separately under the regional heading. Finally, painted plaster from other relevant sites but not part of this study is mentioned for completeness.

### Crete and the Cyclades (Figure 5.1)

*Knossos*

Knossos is best known as the largest multi-period Minoan and Mycenaean palace on Crete. Pottery evidence (local production increase, increased quantity of imports from elsewhere on Crete and beyond especially in EM IIA) demonstrates that a strong central authority was in place at Knossos by EM III, resulting in well-organized and ambitious building plans, possibly reflecting changes in social organization (Wilson 1994: 42). Furthermore, the pottery imports from the western Cyclades in EM IIA reflect Knossian contacts possibly in connection with the metals trade (copper from Kythnos: Davis 2001: 24-26; Kassianidou and Knapp 2005: 220). Even though Knossos does seem rather isolated during most of the EM I-MM IA period, its constant local

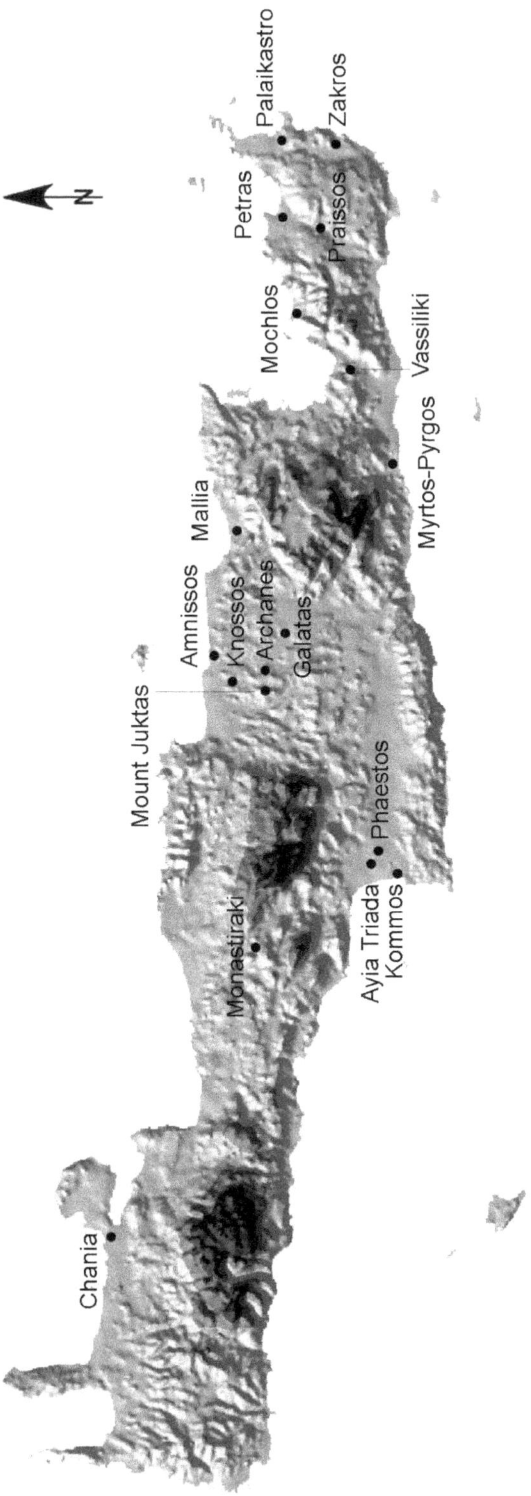


Figure 5.1 Map of Crete indicating the Cretan sites studied.

stability must have influenced the early emergence of a palatial society there. The periods, dated by EM IIA and EM III pottery sequences, formed periods of change and growth and paved the way for the rise of social complexity at Knossos in this period. During these periods, Knossians started to produce painted plaster in dwellings (Momigliano and Wilson 1996) (see also early painted plaster [EM II-III] elsewhere at Knossos [Cameron *et al.* 1977]). Samples from the EM III foundation trench of the north wall of the south front house and from trench V (EM III in date) were collected for this study (see Momigliano and Wilson 1996: 44, 49-50). This early material bore no decorative patterns, but was painted yellow or red and was very well polished (see also material from Myrtos-Fournou Korifi [Cameron 1972: 305-14], Vassiliki [Cameron 1972: 304-305; Shaw 1973] on Crete, the House of the Tiles at Lerna [unpainted] [Peloponnese] of EH II date [Cameron 1972: 310]). Even though the EM phase is of great importance to Knossos' settlement history before MM IB-II (First Palace Period), it is a difficult period to study due to its poor preservation, largely as the result of later intrusions and building activities and also due to poor documentation by its early excavators (Wilson 1994: 23; Momigliano and Wilson 1996: 1). The 1993 excavations at the South Front of the Palace at Knossos, from which the samples were collected, clarified many of the earlier existing problems (Momigliano and Wilson 1996: 2). Apart from the EM III painted plaster, I also studied macroscopically material from Evans' palace excavations stored at the stratigraphical museum at Knossos, but did not carry out further sampling and analytical work (well covered by Cameron *et al.* 1977). Many scholars (A. Evans [1921–35], N. Heaton [1910; 1912], M. Cameron [e.g. 1975], S. Immerwahr [1990], more recently M. Shaw [2000a; 2004a; 2004b], A. Chapin [e.g. 1997], S. Hood [2004], and L. Morgan [2004]) have conducted iconographic and other studies of the huge repertoire of painted plaster from Knossos. Overviews of Knossian themes and motifs, such as for instance the extensive miniature, life-size and three-dimensional scenes (some of which were found both in palatial and non-palatial contexts), can thus be found in those publications so no repetition is required here. Well-known for its long painted plaster tradition, Knossos has been considered the birthplace of this craft, and not without reason, as will be shown further on.

*Palaikastro*

Palaikastro, situated on the east Cretan coastline, is the location of one of the larger Bronze Age town settlements with a peak sanctuary (Petsophas) on the island. Early excavations (undertaken during 1902–06) uncovered the central part of the MM III-LM I town and its reoccupation (from LM II onwards, Driessen and Macdonald 1997: 228). Figurative scenes (3D arm of female), floral motifs (crocus blossoms) and decorative striped bands painted on plaster, and plaster-working tools were found and published (Bosanquet and Dawkins 1923: 148).

By the LM IA period, Palaikastro became a robust and prospering town in Aegean terms, demonstrated by its large and well-built houses in a local style (see for instance LM I Building 1) intersected by wide paved streets. There were strong indications of cultic activity (see for instance in Building 1, 5). In total, close to 40 Neopalatial houses were investigated during the old and new excavations. Some of these show evidence of potential multiple floors (e.g. staircases, a series of finds, fragments of a painted relief plaster and other painted wall plaster in Block B, some wall plaster in house Chi1-17), possibly used as domestic shrines, especially during the advanced Neopalatial period (Driessen 2005: 85-88).

The focus of my study and sampling at Palaikastro fell on the painted plaster material from Area 6 which has been uncovered since 1990, including red plaster floors (MM II), fine floral plaster (LM I) and painted imitation of veined stone (MM III) (MacGillivray *et al.* 1992: 128; 1998: 239-40; Brysbaert 2000). This painted plaster has been dated through its stratigraphic association with datable pottery typologies. Area 6 contained both the remains of a major mansion (southeast zone) and of industrial production (west) (Driessen and Macdonald 1997: 233). The 1994 and 1996 seasons clarified the architectural outline of Building 6 and its stratigraphy, and during both seasons more plaster was unearthed. The excavators regard Building 6 as palatial in architecture and function for several reasons (its *propylon*, a double colonnade, a system of pier-and-door partitions, or *polythyron*, and the possibility of reconstructing a Minoan hall system with three axially situated areas). The unparalleled structure of Building 6 at Palaikastro may relate to Knossian connections for its architecture and plaster decoration; indeed central Cretan imported pottery has been identified on the site on various occasions (MacGillivray *et al.* 1998: 240).

*Myrtos-Pyrgos*

Myrtos-Pyrgos, a small provincial settlement, was continually occupied from the EM II until the LM I period and again sporadically in later times (Cadogan 1992: 202). Its location was easily defensible in a very good agricultural setting with a river nearby. Apart from the settlement founded in EM IIA (at the same time as Fournou-Korifi, see Warren 1972), a large two-storeyed house-tomb (founded in EM III/MM IA or Pyrgos II) with adjacent ossuaries was in use and contained burials from the MM IB-II phase (Pyrgos III) or LM IA (Pyrgos IV). Other traces of Pyrgos III on the hilltop were a tower and a large cistern, but were most probably part of a larger building, later incorporated into the LM IA as the 'Country House' (Cadogan 1992: 202).

The architecture of the Country House, making use of ashlar blocks, purple limestone and gypsum, including store rooms, is of palatial quality, and its painted plaster with elaborate and colourful floral motifs and decorative bands amply demonstrates this technologically (my personal study of this plaster). Furthermore, Linear A tablets, clay sealing, fine faience and ritual objects were found as part of a shrine context. Only this large house, in contrast to the smaller ones around, was burnt in LM IB and subsequently abandoned (Cadogan 1992: 203). The date of the painted plaster thus falls within a narrow period between the LM IA construction of the Country House and its subsequent destruction in LM IB. From the storage areas, short-lived species provided $^{14}$C dates for the different phases of its occupation (Housley *et al.* 1999). So far, the painted plaster material from Myrtos-Pyrgos has undergone a detailed macroscopic study in 2001, but no samples have been taken from this site and, hence, no analytical data are available.

*Monastiraki*
Monastiraki, which is situated on the Charakas hill in the Amari valley, Central Crete, was occupied mainly during the First Palace Period. Founded in MM I, this collecting centre was destroyed by fire and already abandoned in MM II (Kanta 1992: 194-97). Monastiraki's monumental architectural remains, found during the new excavations (since 1981) which concentrated in three distinct areas, comprise two, possibly three, building phases (Watrous 2001: 204; Kanta 1992; 1994–96). The latter phase of the Greek excavations on the east side of Monastiraki includes architecture comprising extensive storage rooms, large habitation areas and impressive archives of clay sealings. This same large building also contained deposits that may have been ritual in function (Watrous 2001: 206). The massive archive and storage facilities, in particular, point to palatial architecture, whilst its location on a natural route between north (coastal plain of Rethymnon) and south Crete (Messara) may reinforce such an interpretation. The pottery found, including Kamares Ware, is typically MM or Old Palace style and shows close similarities to the assemblages from Phaestos, as does its administrative system for storage. The presence of a Linear A document indicates some degree of literacy at the site.

The latter phase of the large building on the southeast corner of the summit (Greek excavations) fits into a well-defined period (MM I-II) and dates the painted plaster I studied (Watrous 2001: 205-206). The samples were taken from this group of material while other material was studied by an Italian team (Kanta 2002: pers.comm.). The decorative motifs of this plaster were limited to coloured bands of a limited colour palette.

*Phylakopi (Melos)*
Phylakopi, on the island of Melos in the Cyclades, was excavated for the first time around the turn of last century (Atkinson *et al.* 1904). Occupation of Phylakopi was known throughout the entire Bronze Age and was defined in three 'Cities' (Renfrew 1978: 404-405; 2007; Whitelaw 2005). Whitelaw (2005) recently discussed the evidence for the different stratigraphical phases relating to the 'cities', based both on the most recent excavations (1974–77 by Renfrew) and the old notebooks by McKenzie who dug around the turn of 19th/20th centuries. Phylakopi demonstrated physical 'urban' evidence as a result of nucleation (already started in the Middle Bronze Age) in the Late Bronze Age. Whitelaw (2005: 38) saw this as an independent development from Cretan influence and possible Minoanization which took place during the Early Bronze Age, as Melos's population continued to be involved in larger off-island economic and political systems throughout the Bronze Age. He thus saw two different developments taking place simultaneously.

The most important find for this study was the discovery of an administrative mansion dating to LM I which predates a Mycenaean megaron. Pottery finds indicate that this mansion may have gone out of use in LM IB. Renfrew's team also unearthed a well-constructed Mycenaean sanctuary, with an east (LH IIIB) and a west (LH IIIA) shrine which is roughly contemporary with the Mycenaean megaron (constructed in 1400 and destroyed around 1120 BC).

The LM I mansion contained a pillar room lavishly decorated with painted plaster. The bulk of the painted plaster fragments came from levels with mixed pottery mainly dating to LC I (or LM IA) with some LC II and were found in rooms immediately to the east of this pillar room (Davis and Cherry 1984: 154). Miniature scenes and scenes of flying fish, women (rooms 6 and 7, building G 3) dressed in elaborate skirts and associated with a young girl acting as a votary were some of the highlights of the decorative compositions of these paintings, next to swallows (building J 3), rosette bands (room 14, building G 3), the white lilies on a red background motif (room 11, building G 3) (Immerwahr 1990: 48, 62, 161, 189), and decorative coloured bands (Morgan 2007: 372-95). The plaster samples I took at Phylakopi come from the excavations conducted by Renfrew, east of the pillar room in the areas G 3 and J 3. These fragments join some of the paintings found in and near the older excavations of the pillar room and helped in dating the earlier found material correctly to LC I or LM IA.

Whitelaw argues that the level of Minoanization, as has been ascribed to Phylakopi in general, was far less 'overall', but instead focused specifically on the Pillar Room complex and its architectural construction and decoration. Its elite inhabitants may have used Minoan fashions in a range of expressions in order to manifest specific power and ideological strategies towards the rest of the community. From this evidence, it is clear that this phase of Minoanization was rather short-lived and it disappeared at the end of City II (transition Middle to Late Bronze Age, associated with MM III-LM IA pottery on Crete) in what was potentially a site-wide destruction.

## Greek Mainland (Figure 5.2)

### *Mycenae*

The period of interest for this study at Mycenae starts at the time of intense contacts of the Minoans with the mainland where MH III-LH IIA (or Shaft Grave Period) coincides with MM IIIA-LM IB (or Second Palace Period). Apart from influences from the east (French 2005), the Minoan contacts and influences, especially in the Argolid, are clear from the range of finds such as pottery, faience objects, ivory, stone vessels, and the precious metal objects found in the Shaft Graves (Dickinson 1994: 137, 140-42, 303). No painted plaster, apart from the earliest plaster from Mycenae (see below), seems to date from this period.

The subsequent period of importance on the mainland (LH IIIA-LH IIIB, roughly the 14th century BC until 1200 BC), coincides with the rise of the Mycenaean palaces, as economic, political and religious centres. Most were decorated with painted plaster demonstrating individual taste (but see Sherratt 2001: 215, 230). In this period, Mycenae

Figure 5.2 Map of the Greek mainland indicating the sites studied.

and its sphere of influence had supplanted Minoan influence in the Aegean, and its own power and contacts became visible in the spread of materials over the entire Aegean and eastern Mediterranean (French 2002: 51), although it was not the only centre of importance in the Mycenaean world.

The palace at Mycenae (see Wace and Lamb 1921–23; Wace and Williams 1963) had five phases in total ( brought together in French 2002: 95) with a post-Palatial phase 6, and the top of the acropolis seemed to have been inhabited from the EH period onwards (French and Shelton 2005: 175). Painted plaster was found in many areas at Mycenae, in palatial and non-palatial contexts, and iconographic scenes and programmes are as abundant as at Knossos, and therefore also very well studied (for an overview see Immerwahr's catalogue 1990: 190-94), and still under constant study (e.g. Iakovidis' painted plaster from the Plakes House). The plaster which I sampled for this study came from Palaces II (from East Lobby, LH IIA, date of the 'early palace' destruction), III (use of palace: LH IIB-IIIA1), and IV (palace built in LH IIIA2, heavily damaged in LH IIIB) (French 2000: pers.comm.). French (2002: 44-47) argues that there was a central building at Mycenae serving as the 'early palace', predating the full-scale Mycenaean palaces (LM II/LH IIB-IIIB). Refuse containing pottery and painted plaster fragments was found on the upper Acropolis slopes and must have come from this early building. Moreover, French and Shelton (2005) recently discussed three deposits dating from the early palatial period at Mycenae (MH III/LH I to LH II); the area from below the East Lobby dated to the later pottery phase. Apart from the actual palaces, Mycenae comprised, during LM II/LH IIB-IIIB, several clusters of houses, some purely domestic, while others had evidence of workshops and storage facilities and were probably part of the palatial administration (e.g. House of the Oil Merchant: Tournavitou 1995). Several of these houses were decorated with painted plaster, and the House of the Oil Merchant also had painted plaster fragments in the fill beneath its floor which was sampled as well. This material dated to the LH IIIA2 pottery phase (French 1965).

Apart from domestic dwellings, irrespective of their scale, Mycenae's religious centre, epitomized in its 'cult centre' (LH IIIB), was equally beautifully decorated with painted plaster (e.g. 'Room of the Frescoes', see Immerwahr 1990: 115, plates 59-61: 118-21), depicting goddesses, processions, the famous 'Mykenaia', etc. Moreover, several tholos tombs at Mycenae were decorated with painted plaster (French and Shelton 2005: 181, n.13-16, 182). Other tombs with painted features existed at Argos (Deiras V, VI, XII), Tiryns and Kokla (Voutsaki 2006: pers.comm.). To my knowledge, none of these tomb paintings have been investigated for their technology so far.

### *Tiryns*

The multiple-period site of Tiryns, only 15 km from Mycenae, formed the second Mycenaean centre in the Argolid (final reports on old excavations: Schliemann 1886; preliminary reports on recent excavations: Kilian 1978; 1979; 1981; 1988c). Some argue that it may have been subordinate to the control of Mycenae (Dickinson 1994: 78), but it was, certainly at the end of LH IIIB and in LH IIIC, the leading settlement of the Argive plain and surroundings (Kilian 1980; 1988a; Maran 2001a: 120). Its massive early fortification, built up in three main phases (Müller 1930), like the one at Mycenae, is considered to be an expression of power rather than of a defensive nature. This function may have changed in its later phases around the Middle and Lower Citadel (Maran 2004a: 261-66).

The palatial structure, consisting of two main phases (LH IIIA1/2 with repairs, rebuilding and additions in LH IIIB1, and LH IIIB2, plaster samples from both phases were taken), and a later reoccupation (Building T) (Maran 2001a: 113, 119; 2001b), was founded on the Upper Citadel (Kilian 1987; Shelmerdine 2001: 335). These later building phases erased most Middle Helladic and Early Mycenaean building remains (Maran 2001b). The actual palatial building consisted of a great *propylon* or entrance to the inner palace. The latter had a small and large megaron, and megaron-like apartments, all decorated with painted plaster on walls and floors (Rodenwaldt 1912), the latter being a typical Mycenaean feature. Tiryns' contacts with the broader Aegean and eastern Mediterranean are evident, for instance, in its pictorial vessel distribution, although this was less intense than the corresponding production in the area of Mycenae and Berbati (Jones and Mee 1986; R. Jones 2004: pers. comm.; Mommsen and Maran 2000–2001). It has been suggested that the decline of Tiryns and the collapse of its palatial system were caused by a series of earthquakes during the 12th century BC (Kilian *et al.* 1981: 192; 1996).

The painted plaster from Tiryns forms a wealth of material, both from older and more recent excavations (see latest 1999–2001 Greek excavations of A. Papadimitriou of the western staircase area), and plaster was uncovered from both palatial and non-palatial contexts (most recently Maran 2007 from the Lower Town). Famous scenes from Tiryns include animal hunting, a lifesize bull's head, friezes of women in different sizes, figure- of-eight shields, architectural elements and dado imitations, the toreador scene, etc. (Reusch 1956). Most of this material is housed in the National Museum in Athens. The material that I sampled for this study was dated by stratigraphy and pottery typology, and came predominantly from the Small Megaron (LH IIIA2 or earlier), the last phase of the Great Megaron (earlier than LH IIIB2), and from the megaron-like Apartments, all located on the Upper Citadel (earlier than LH IIIB2). Four samples came from the Greek excavations on the west slope, where lots of plaster fragments were found in the epichosis, and have been dated by their excavator to LH IIIB. Four others came from the Lower Citadel settlement excavations and date to LH IIIA until LH IIIC or earlier, depending on their context specifics. The total settlement area in the LH IIIC period is estimated at 25 hectares (Kilian 1978: 468, fig. 26).

*Thebes*

Excavation activities in Thebes are problematic because the hill, on which the remains of the seven-gated Bronze Age city were situated, is entirely covered by the modern town. What has been recovered comes from the early excavations run by Keramopoulos since 1906 and rescue excavations since 1960 and is scattered over the whole area. Nevertheless, two richly furnished buildings have been uncovered and were called 'The House of Kadmos' (1906) and 'The Treasury Room' (in 1963 and 1997). They date to two different consecutive periods (LH IIIA and LH IIIB respectively), and both buildings were slightly differently orientated (Dakouri-Hild 2001: 83). Chronologically, Thebes was inhabited throughout the Late Bronze Age but fewer remains seemed to fall into the LH I-IIIA1 and the LH IIIC/post palatial periods (Dakouri-Hild 2005: 208).

Both buildings (House of Kadmos and the Treasury Room) and several others all over Thebes have produced painted plaster. The fragments from the HK seem to come from two or three different phases of its use. One group came from the fill of a renovated floor (in Room N), and was thus not affected by the fire that destroyed the building from which the third group came (and, as such, comparable to the fragments also found in the fill of Room Π). The latter two groups were blackened by fire and were found in the destruction fill covering the floors of Rooms N and Π. A second group was embedded within the floor overlying the first group (Room N) and may belong to that first group (Dakouri-Hild 2001: 92-93, 99). The painted plaster from the TR most likely came from the upper floor (Dakouri-Hild 2001: 104). The pottery of the House of Kadmos dates its destruction to the late LH IIIA2 or early IIIB phase (Mountjoy 1993: 17). Some painted plaster samples were taken from the House of Kadmos and were dated, through a reassessment of the stratigraphy by Dakouri-Hild (2001: 99), to LH IIIA. Other samples came from the Treasury Room and were dated to LH IIIB1, and E. Kountouri dates the material from the OS excavations (1997–98 seasons) tentatively to LH IIIB; this material was sampled too. Some of the famous scenes that decorated Theban architecture are: female friezes, griffin fragments (early HK phase), figure-of-eight shield fragments, a human eye (Room Π), and a warrior's head (Immerwahr 1990: 200-201; Dakouri-Hild 2001: 93). The House of Kadmos material is now prepared for final publication by N. Dakouri-Hild.

*Gla*

The site of Gla is situated on a rocky eminence that is surrounded by a plain as part of the Kopais basin in Boeotia. The function of the site may have been to house the officials overseeing the important Kopais basin drainage works and, as is suggested by storage jars with grain remains, for storage of the resulting crops. In contrast to the other Mycenaean palaces, Gla is, therefore, considered a specialized administrative centre. Inside an enclosure the architectural remains are divided into two parts.

The northern enclosure consists of two similar wings in terms of size and construction. Each wing consists of a series of 'apartments', corridors and a megaron-shaped room at its end. This part, therefore, is sometimes referred to as a 'palace' or *melathron*, but there is otherwise no evidence for columns or a central hearth, although an interesting *polythyron* system was found between Room 2 and 3 (Iakovidis 1983: 99; Shelmerdine 2001: 340). From this area came painted plaster fragments and De Ridder reported finds of relief stucco in the shape of fluted half-columns, some *in situ*, on the walls

(Iakovidis 2001: 38-39 reporting De Ridder's findings). Such a feature is unique to Gla together with the find of Horns of Consecration (Iakovidis 1989: 289; 1998: 274).

The southern enclosure or 'Agora' comprised two long and narrow buildings with two free-standing buildings, H and A, at each end. H contained beautifully decorated plaster fragments (Iakovidis 1998: 265-66), and also within and close to room N (possibly residential in nature) painted plaster fragments were found. These free-standing buildings probably served as barns and storerooms for the produce of the area and may have contained workshops for different crafts, and living quarters. Themes represented at Gla are: marine scenes with dolphins, rosettes, veined stone imitation, intricate cloth patterns as part of human figures (see purple-painted scenes: Brysbaert 2006: 253), etc.

All plaster remains from Gla are dated to the same period (first half of the 13th century BC) because all buildings were erected at the same time (Iakovidis 2001). The site was occupied until some point in LH IIIB2, when the buildings were destroyed by fire and when the drainage system was probably deliberately obliterated. The material I sampled from Gla is from three areas: building H, the *melathron* area, from within and close to room N (Brysbaert 2006: 255).

*Orchomenos*

Orchomenos, situated in the Kopais region in Boeotia, was excavated in a few seasons at the beginning of the last century by Bulle (1907) for the German Archaeological Institute, and again in the 1970s and 1980s by T. Spyropoulos. Orchomenos was mentioned by Homer in relation to the Trojan War, which may be an indication of its importance and power. None of the excavations were well published, leading to difficulties in extracting information about this site, although some of its pottery has been studied and published. Mountjoy (1983: 11) stated that the stratigraphy of the site was not discussed by Bulle in a manner that could help the pottery study. The majority of the sherds kept were LH IIIB, although others were datable to the whole period from LH I to IIIC. The dates of the palatial building are not at all clear, but Mountjoy (1993: 22) maintains that Orchomenos was eventually abandoned in early LH IIIC.

The early excavations revealed parts of what was probably a large palatial structure (Area T) and a huge tholos tomb called 'the treasury of Minyas' which was probably the funerary monument for this palace ruler. This tomb, comparable to the Treasury of Atreus at Mycenae in splendour, was probably of similar date (soon after 1400 BC). The palatial building was probably built in the vicinity of the church of Skripou where Bulle found plaster fragments and where the megaron-like building with three wings was found in the 1970s. The painted plaster from the potential palace building has been dated, on stylistic grounds, to the LH IIIB period, that is, before the building was burnt down (see burnt plaster fragments in Immerwahr's catalogue, 1990: 195, 125-27). The splendid scenes are rather fragmentary but the following can still be discerned: scenes of human figures in architectural settings possibly related to 'swimmers', horse and chariots, boar hunt, human combat scenes, spirals and half-rosettes, etc. The material I sampled from Orchomenos came from the palatial structure excavations conducted by Spyropoulos.

## Eastern Mediterranean (Figure 5.3)

### *Alalakh*

Alalakh (Tell Atchana) is a site of great importance to this study because it has been considered by certain scholars (Mora *et al.* 1984) to be the place where the oldest 'frescoes' in the world were made. Furthermore, it may have played an important role in the first contacts between the east and Minoan Crete, independent of the direction of these contacts.

The site, situated in the Amuq plain near the river Orontes and situated on one of the principal trade routes from Mesopotamia to the Mediterranean, was first excavated by L. Woolley (1955) in two campaigns during the 1930s and 1940s. It has recently

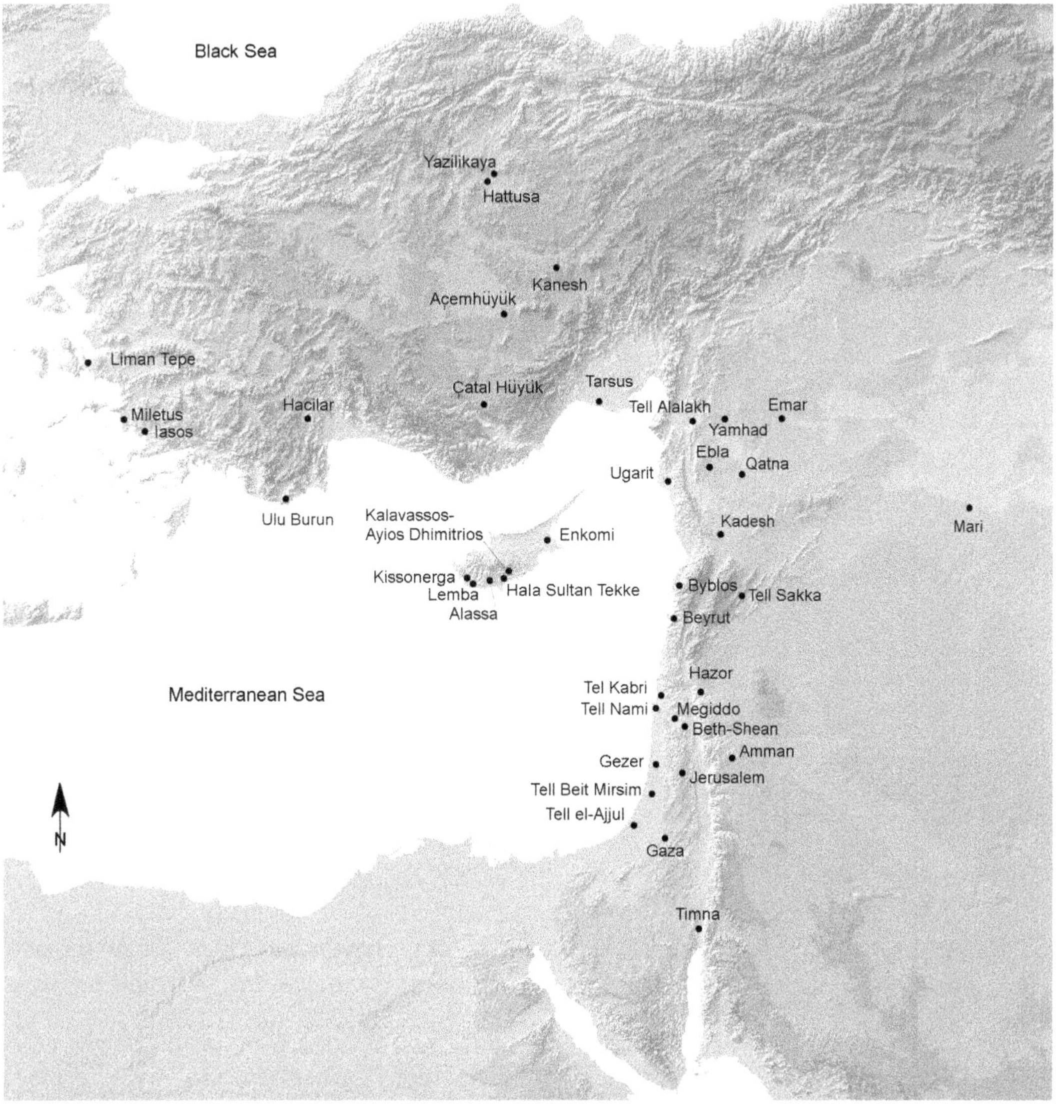


Figure 5.3 Map of the eastern Mediterranean indicating the sites studied.

been reinvestigated as part of a large-scale survey in the region (University of Chicago, Oriental Institute, under the direction of A. Yener). Moreover, the imported Cypriot pottery, key to the dating of Middle Bronze Age/early Late Bronze Age Alalakh and found on the site by Woolley, has been subjected to a recent thorough study (Bergoffen 2003). The huge multiple-period tell (17 strata, 22 hectare) revealed an unusual temple of the 24th century BC and two palaces, one in level VII (MB II), and one in level IV (Late Bronze Age).

The level VII palace, spanning about 3000 square feet and constructed along three terraces (Bergoffen 2005: 16) was dated by Woolley to 1780–1730 BC; this date has proved to be too early and has been reconsidered (Niemeier 1991: 190-91; Bergoffen 2003: 396, 2005). The start of level VII is now dated to the start of Yarim-Lim's reign at Alalakh at *ca.* 1710/1650 BC (Bergoffen 2005: 56). This palace of level VII which was destroyed by the Hittite king Hattushili I, probably around 1650–1575 BC or around 1561 BC (Bergoffen 2005: 56-57), was erected two to three generations earlier around 1700 BC (Niemeier 1991: 191).

Studies concentrating on the architectural similarities between Crete and Alalakh, as suggested by Woolley, are not convincing due to many differences in architectural outline of the palaces in both areas (see now Feldman 2007 who compares Alalakh's palace with Tel Kabri and Qatna). Other possible contacts were shown in the presence of a Minoan stone vase fragment found in level VII.

Important for this study are the painted plaster fragments from level VII now housed in the Ashmolean Museum. Veined stone imitation fragments were found in Room 5 or 'Chamber of Audience' (Woolley 1955: 92), while griffin wing fragments, plant motifs in motion and a possible bull's horn in association with coloured bands came down from a large room above magazines 11-13, or 'the Great Salon' (Woolley 1955: 94). Level IV at Alalakh also produced painted plaster in the shape of architectural designs of imitated orthostats, and wood imitation resembling the wood imitation scenes from many Mycenaean sites such as Thebes, Orchomenos and Thebes (see Woolley 1955: plate XXXIX). Alalakh forms, together with Qatna, Tel Kabri, Tell el-Dab'a, Miletus and Hattusha, a string of sites that contribute to the discussion about iconographic and technological transfer (Niemeier 1991: 196), and may help to resolve the issue of the direction of technological transfer from the east in relation to Crete. The material I sampled was on display at the Ashmolean Museum, and minute samples were taken from the material originating from Room 5 and the Great Salon.

### *Miletus*

Miletus, on the coast of Asia Minor, is an important place in understanding Aegean influences or 'colonization' in western Anatolia. Mycenaean contacts with Miletus have been understood since 1907, and Minoan contacts through pottery finds (mainly of LM IA-IB date) have been clear since 1938 (Niemeier and Niemeier 1999: 543-44). At that time, three prehistoric levels were uncovered in the area of the Athena temple, the first was Minoan (Miletus IV), the other two Mycenaean (Miletus V and VI).

A Middle Bronze Age stratum (Miletus III) contained MM IA pottery together with a Kamares MM IIB vessel and two MM (Old Palatial) seals, a clay sealing (MM IA late-IB, IB-II), a sherd hearth with Messara sherds, and Minoan domestic wares locally produced (Raymond 2005: 186-88). This material was found together with a higher ratio of local pottery (Niemeier and Niemeier 1999: 546, 553). This Middle Bronze Age phase was destroyed during the MM IIB period, the same period as the destruction of the first palaces on Crete.

The architectural house remains of the 'first building period' (LM IA-IB or Miletus IV) are similar to those found on Crete (e.g. Block B at Palaikastro). Minoan fine wares, including LM IB palatial pottery, were found, but the vast amounts of Minoan conical cups and cooking ware confirmed the actual Minoan presence to the excavators (see Kaiser 2005). This was further confirmed by evidence for Minoan-type subsistence and diet and the combined presence of Minoan-type loom weights, Minoan-type ritual furniture grouped around a mudbrick altar, found together with conical rhyta, bronze figurines, triton shells, clay bull figurines, Linear A on roundels and inscribed on pottery vessels, Minoan weights, and finally, the painted plaster finds from Building A (Niemeier and Niemeier 1999: 547-50; Niemeier 2005: 200-202).

The above-mentioned finds show that Miletus IV may have been under strong Minoan influence, or, according to its excavators, a certain number of Minoans may have lived at Miletus. Hittite texts mention Millawanda (understood as Miletus by the excavators) as a vassal of Ahhiyawa, considered by many scholars a place on the Greek mainland (Niemeier 2005: 203; 2006; but see Dickinson 1994: 253).

The samples I took at Miletus came from Building A, dated to LM IA. Scenes represented at Miletus are: notched plume griffin, blue plants in motion, white lilies on red background, coloured bands, etc.

*Qatna*

Qatna (Tell Mishrife), situated about 20 km from the town of Homs in Syria, in the 'Homs corridor', had easy access through a pass in the mountain range (Anti-Lebanon) to the relatively nearby Mediterranean Sea (90 km). It was situated at the crossroads of two major trade routes, (1) east-west: Mesopotamia-Syrian Desert-Mediterranean Sea, and (2) north-south: Anatolia/north Syria (Alalakh-Aleppo)-Palestine-Egypt (Novák 2005). The immense city of 100 ha was surrounded by an enclosure wall, up to 20 m high and still largely preserved in the landscape.

Minoan fineware pottery was found from Middle Bronze Age occupation levels at Qatna (Cherry 1986: 40). In the 16th century BC, the city became a vassal of the Mitanni state. Its palace was destroyed most probably in the 14th century BC by the Hittites and saw only a small and less significant reoccupation during the Iron Age (Novák 2005).

Early excavations in the 1920s (Du Mesnil du Buisson 1935) and modern reoccupation of this area have merely removed all palatial upright structures, but the intact foundation walls, excavated since 1999, allow for the complete reconstruction of the building's layout (Novák 2005). One of the most spectacular discoveries was the unrobbed tomb found in 2002 underneath the Palace floor. The art and skill with which the precious metal objects from the tomb are combined with semi-precious stones is exceptional and demonstrates a hybrid mixture of Syrian, Egyptian, Mesopotamian and Aegean influences.

In the west part of the Palace a large cistern with internal staircase was uncovered, into which fell, from room N, about 3000 painted plaster fragments (Novák and Pfälzner 2001: 179, and 2002: 224-31). More fragments came from Room U, and isolated painted plaster fragments were also found in Rooms P and AG. These are associated with Middle Bronze Age and Late Bronze Age pottery, which now seems to date them to the Late Bronze Age (15th–14th centuries BC, U. Sievertsen 2003: pers. comm.) (*contra* previously Middle Bronze Age or 18th–17th centuries BC).

The material I sampled from Qatna came predominantly from Room N. Scenes depicted were: a wide range of spiral motifs, stone imitation dadoes, a miniature scene with date palm, fine floral motifs, garlands, turtles and crabs, etc.

*Tell Sakka*

The site of Tell Sakka is situated in an arable region, on the east side of the Ghouta oasis of Damascus, served by the waters of two rivers, and on the edge of the Syrian desert. The excavations, started in 1989, have revealed four levels reflecting four different periods. Level 4, MB II (Taraqji 1999: 35), dated by Tell el-Yahudiyeh ware (late Middle Bronze Age, before 1600 BC, Manning 1999: 62, n. 269, 132) among others, is the most important one for this study.

An important large structure of palatial nature, built in mud brick joined with a white mortar, shows an unroofed hall with four central pillar bases (and two-by-two smaller ones at the southern end) and a covered gallery around it. The walls were covered with a mud plaster and then whitened with a lime wash. The south part of this hall consisted of a room for large storage *pithoi.*

In this hall close to the west wall much fragmentary painted plaster with a figurative iconography, hybridizing Egyptian, Syrian and Canaanite elements, was found. The material I sampled for this study came from this MB IIB hall. Its hybrid iconographic style is reminiscent of the typical style of the seal impressions found by the hundreds at Qatna from the Middle Bronze Age II and Late Bronze Age periods (Elsen-Novák 2002: 265; 2003: pers. comm.), and some of the themes on these paintings are not unlike some found in Mari (goat standing up against a tree, see Moortgat 1969: 83, for a chronological discussion of the different Mari scenes), and now from MB II Tell Burak (south Lebanon) (goats standing up against tree of life, among others: Kamlach 2007: pers.comm.). In a few adjacent rooms north of the hall, other paintings were found still *in situ* showing simple red dot motifs on a white background. Their complete disorder when found is probably due to the intentional destruction that this building underwent at the end of its existence (end 18th–early 17th century BC).

The excavator asks which painting tradition is represented here in view of both its iconography and its technology: local, Mesopotamian, Egyptian or Caananite (Taraqji

1999: 37). This question is addressed in this book in Chapters Six and Seven. The paintings are extremely important here because they form the only comparisons for the other eastern Mediterranean paintings that are studied in this research programme. It would be very interesting to see how these paintings compare in terms of technology and iconography with the more Egyptianizing paintings on mud plaster found in Tell el-Dab'a in Palace G, to those found in Mari and Tell Burak, and how they may relate to each other in chronological terms.

*Tel Kabri*

Tel Kabri, a large multiple-period tell in northern Palestine, was situated on the Via Maris, one of the most important trade routes in the Near East (Niemeier 1991: 196). During the (Palestinian) Middle Bronze Age, Tel Kabri was a large fortified city and probably to be identified with the Canaanite city state of Rehov (Niemeier 1991: 196). The destruction and abandonment of the palace and the city can be placed in the pre-Bichrome ware phase of MB IIB or around 1600 BC (Niemeier and Niemeier 2000: 769; 2002: 262), which is contemporary with the destruction of level VII at Alalakh (Niemeier and Niemeier 2002: 270).

Area D contained the remains of the Middle Bronze Age palace of about 2000 sq m, including painted plaster fragments which I have sampled and studied for this research. In this area, a plaster floor, painted *al fresco* with a chequer-board pattern filled in with iris-like flower buds, and a threshold (all floors in the palace are plastered, see Kempinski and Niemeier 1993a: 182) were uncovered in ceremonial Room 611 (Kempinski and Niemeier 1991: 192). The pottery dates this material to the MB IIB phase, and I sampled some of this as well. More painted plaster finds of the same period were uncovered in 1991, showing miniature town scenes with boats, notched plume griffin fragments and plants in motion comparable with Aegean scenes (Kempinski and Niemeier 1992: 203; 1993b: 258). The renewed excavations started in 2005 under the direction of E. Cline and A. Yassur-Landau, with the intention of establishing the extent of the Middle Bronze Age palace, finding other areas with potential painted plaster finds, collecting data to date the painted plaster fragments accurately and so establishing the correlations between the Aegean and the Levant in the transitional phase from the Middle to Late Bronze Age.

*Hattusha*

Hattusha or Boğazköy, situated in northern Central Anatolia, about 150 km east of Ankara, near the northern border of ancient Cappadocia, was the ancient capital of the land of Hatti for most of its existence and became a significant settlement during the 3rd millennium from when it was continuously occupied. Its significance became clear when a Karum, a trading post of Assyrian merchants, was established there in the 19th and 18th centuries BC (Seeher 2002: 159), transporting and trading metals and semi-precious stones, textiles and local Anatolian produce while they also introduced Akkadian writing to Anatolia.

The site was first destroyed in a huge conflagration around 1700 BC. After this destruction, the city was soon re-erected by Hattushili I who chose it as his royal residence and capital of the country. The extended city walls were decorated with some massively constructed gates of which the sphinx and the lion gate are the most impressive (Neve 1996: 81; Seeher 2002: 51-58, 37-45). The lion gate, together with Chamber 2 and the tunnels under the sphinx tower (a vaulted cyclopean construction), and various other architectural features have been compared with constructions from Mycenaean towns such as Tiryns (tunnels-corridors of the Sphinx Tower, 13th century BC) and Mycenae (Lion Gate) (Thaler 2007). Apart from the rock sanctuary at Yazılıkaya, many temples were constructed in the different phases of its occupation at Hattusha itself.

Painted plaster remains were recovered at different locations on site, more specifically in Temple 9, Rooms 11, 9 and 27 (Neve 1999: 50, pl. 31c), in secondary deposition in a wall fill of Temple 5 (Neve 2001: 29, taf. 48c), and in the palatial complex of Büyükkale (Neve 1993: 639, fig. 16). All were found in very small amounts and in tiny fragments which had to be taken into account in relation to my sampling strategy. Their state of preservation has severely limited their study; nevertheless these fragments are important due to their original find spot and their iconographic representations (half-rosettes, hair or feathers from animal scenes, dado imitation, spirals, coloured bands, etc.). Temple 9, dated to the first quarter of the 13th century BC (Neve 1996: 26, 30; Seeher 2002: 68, 13; Müller-Karpe 2003: 391-92) and was, by the 14th–13th centuries BC, part of an established cult complex comprising around 30 temples. Several temples seem to have been involved in economic activities since they stored foodstuffs and raw materials for artisans' activities. It is clear that this is not the only place where artisans' activities were associated with ritual and cultic aspects of life (see Thebes's tablets on textile working, Dakouri-Hild 2005: 210, 214). In 1996 Neve (1996: 30, fig. 75) published a colour photograph of the fragments, and recently Müller-Karpe (2003: 392) linked the iconography of the fragments from Hattusha with paintings from Bronze Age Mycenae (see now also Özyar 2006: 131-37, on Hittite art, including the paintings from Hattusha), and with the Amarna paintings. The limited publicity given to these fragments is the likely reason why these paintings were never considered in relation to Aegean paintings and technological transfer discussions.

*Tell el-Dab'a (Figure 5.4)*

The multiple-period site of Tell el-Dab'a is situated in the eastern Delta, just south of Quantir, and has undergone major excavations covering an area of about 200 hectares on a natural mound. Its occupation levels extend from the 1st Intermediate (2216/2166–1976 BC) to the 2nd Intermediate Period (1794/1793–1550 BC) and beyond. During the latter period it became the capital of the Hyksos, and thus the centre of Syro-Palestinian occupation in the Delta area, forming a focal point for trade between Canaan, Syria, Cyprus and the Aegean (see presence of Tell el-Yahudiyeh ware as far as Thera, Manning 1999: 122; also Kaplan 1980). Because of this strong factor of foreign trade contacts, Avaris became one of the largest cities in the eastern Mediterranean, and

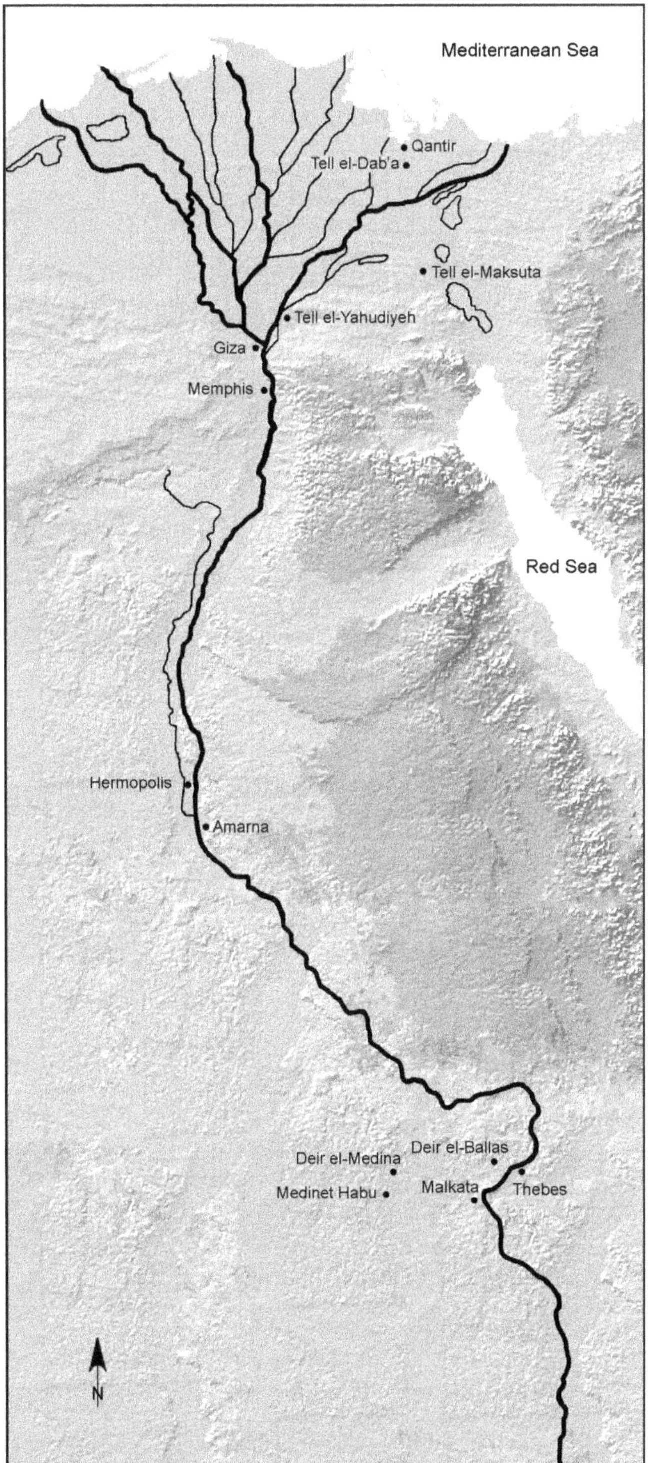


Figure 5.4 Location of Tell el-Dab'a in Egypt together with important sites from the Middle and Late Bronze Ages.

its occupiers built a huge citadel surrounded by a fortification wall, which also enclosed some gardens. This citadel was destroyed after Ahmose conquered Avaris and founded the 18th Dynasty. Some of the citadel's building materials were reused for 18th Dynasty constructions (Bietak and Marinatos 2000: 40-41).

Contacts between Tell el-Dab'a and Crete seem to have materialized in two distinct phases: first in the early 13th Dynasty with finds of Classical-style Kamares pottery and a gold pendant from a tomb (1st half of the 18th century BC), and again during the end of the Hyksos-18th Dynasty period (2nd half of the 16th century BC) (Bietak's ultra low absolute dates: 1995: 19 but see below, and Table 5.1 below).

| *Absolute dates* | *Crete* | *Mainland* | *Period* | *Egypt* |
|---|---|---|---|---|
| 3100–2700/2650 BC | EM I | | Prepalatial period | |
| 2700/2650–2450 BC | EM IIA | | | |
| 2450–2200 BC | EM IIB | | | |
| 2200– 2050 BC | EM III | | | |
| 2050– ±1900 BC | MM IA | | | |
| ±1900–1820/1810 BC | MM IB | | 1st Palace Period/ | |
| 1820/1810– ±1780 BC | MM IIA | | Old Palace Period/ | |
| ±1780– ±1740 BC | MM IIB | | Protopalatial Period | |
| ±1740– ±1720 BC | MM IIIA | | | |
| ±1720–1700 BC | MM IIIB | | 2nd Palace Period/ | |
| 1700/1675–1600 BC | LM IA | LH I | New Palace Period/ | 13th Dynasty (1759–1606 BC) |
| **1665–1595 BC: Thera** | LM IA * | | Neopalatial Period | 15th Dynasty (1637–1529 BC) |
| 1600/1590–1525/1479 BC | LM IB | LH IIA | | 17th Dynasty (1606–1539 BC) |
| 1479–1440/1425 BC | LM II | LH IIB | 3rd Palace Period | 18th Dynasty (1539–1296 BC) |
| 1440/1425–1390/1370 BC | LM IIIA1 | LH IIIA1 | | |
| 1390/1370–1325/1300 BC | LM IIIA2 | LH IIIA2 | | |
| 1325/1300–1200/1180 BC | LM IIIB | LH IIIB1 | | 19th Dynasty (1295–1186 BC) |
| | | LH IIIB2 | | 20th Dynasty (1186–1069 BC) |
| 1200/1180–1070 BC | LM IIIC | LH IIIC | Postpalatial Period | |
| | Subminoan | Submycenaean | | |

* LM 1A advanced

Table 5.1 General chronological overview of Minoan and Mycenaean pottery phases of relevance to this study, in relation to the palatial periods (based on the high chronology, see Manning 1995; Cullen (ed.) 2001b: 106, 222, 332; Ramsey *et al.* 2004), and Egyptian Dynasties (Rehak and Younger 2001: 391, table 1). Note the overlap in Dynasties.

Since the early 1990s the main focus of the ongoing excavations fell on a palatial (H/I-IV) complex consisting of several monumental buildings, where thousands of painted plaster fragments were found. These were not found *in situ* but in dumps on top of gardens of the Hyksos period. These dumps are considered to be from the palatial district of the first part of the 18th Dynasty-early Tuthmoside period (1500–1450 BC).

The actual palatial complex was constructed on top of the Hyksos palatial remains and on installations dating to the early 18th Dynasty (str. e/I—Ph.D/I). Within this large complex, two palaces were constructed (F and G), with a large courtyard separating them, while another smaller palace (J) was built to the south of the former two. Palace F contained most of the painted plaster fragments which were found in two main concentrations (Bietak 2004a: 83). The palatial district indicated two building phases of which the latter, phase C, dates to Tutmosis III-Amenophis II (c. 1479–1400 BC). The exact date of the first construction phase from which most painted plaster comes is less clear, but it is best situated in the period of Tuthmosis I–early Tuthmosis III (1500–1450 BC) (Bietak 2004a: 88). Bietak believes that the larger Palace G belonged to the king of Avaris and the smaller Palace F to the queen. According to Bietak this queen may have been of Aegean origin and may have arrived in Avaris as the result of a marriage. The marriage hypothesis is a possible explanation for Bietak to explain the presence of the 'Aegean'-style painted plaster at Tell el-Dab'a (Bietak 1992: 28). In the debris of the substructure of palace G, painted mud plaster fragments were found as well, revealing Egyptian writing and decorative motifs and the influence of Minoan colour coding and iconography (Bietak and Forstner-Müller 2003: 45, 46-47), while some more Aegean-style paintings were also found (Bietak and Forstner-Müller 2003: 44). These were not as well preserved as the ones from palace F, painted *al fresco* on lime plaster. The mud plaster fragments in fact may be either from Palace G or even from the older Hyksos palace, since similar ones were found there too. To complicate the situation, other painted lime plaster fragments were also found near Palace G and the excavator mentions the possibility of two painting programmes executed *al fresco* on lime plaster (Bietak 2004a: 88-89). Depending on the final dating of the mud plaster paintings we may have either two completely different painting technologies taking place at the same time at Avaris: (1) painting *al secco* on mud plaster and (2) painting *al fresco* on lime plaster. Two other scenarios are also possible: (1) Hyksos mud plaster *al secco* paintings followed and replaced by later lime *al fresco* paintings or (2) Hyksos mud plaster *al secco* paintings followed by later lime *al fresco* paintings while the *al secco* tradition continues to exist alongside the *al fresco* tradition.

The most important points of the painted plaster from this site are its context, date, style and technology. Initially, the plaster found in area H/I-IV was dated before the start of the 18th Dynasty (Bietak 1995: 23), while that found in H/II and H/III was dated to the 18th Dynasty. Later, however, all the paintings were redated to the early 18th Dynasty and, most recently, well into the 18th Dynasty by its excavator (see Cline 1998; Manning 1999; Bietak 2000a; M. Bietak 2002: pers. comm.; Niemeier and Niemeier 2002). This dating has a serious impact upon both our understanding of the direction of iconographic and technological transfer and upon the dating of the Theran eruption, all discussed in more detail in Chapter Six (Manning 1999).

The wealth of the scenes at Tell el-Dab'a is phenomenal and its relation with the Aegean paintings is striking: acrobats and bull leapers, half-rosettes, life-size human figures, blue plants in motion, white lilies on red background, animals in flying gallop, so-called 'easter' eggs, etc. are only a handful of the range now published in final form by M. Bietak (2007). The material I sampled from Tell el-Dab'a comes from Palace F (H/I-IV) and were found in two groups. I have not seen or studied the paintings on mud plaster.

## Other Significant Sites with Painted Plaster Remains

In the site overview above I discussed only the painted plaster in the Aegean that I have sampled. Several sites of significance in the discussion of technological transfer, such as Akrotiri (Thera), Trianda (Rhodes), Ayia Irini (Kea), Pylos (Messenia), were not accessible to me because other researchers are working on them, while some material from specific sites is still undergoing further investigation by myself (Mycenae, Knossos, Palaikastro). To my most recent knowledge, the sites of Kato Zakros, Prasa, Galatas and possibly Archanes (Crete) and Trianda (Rhodes) are at present under study by E. Chryssikopoulou. Together with several collaborators, she has recently published analytical results from the painted plaster remains found on Akrotiri (Chryssikopoulou 2004; Chryssikopoulou *et al.* 2000; Chryssikopoulou and Sotiropoulou 2003), and this formed an invaluable contribution to our understanding of Aegean painting in the Bronze Age. Chryssikopoulou also presented the first results of her study of the Zakros material in the most recent Cretological Congress (held at Chania, October 2006, not attended). Most recently, a mathematically modelled study of geometrical patterns has been conducted and published on paintings from Xesti 3 at Akrotiri (Papaodysseus *et al.* 2006). The results of the painted plaster material from Chania have been studied, analyzed and published by Photos-Jones *et al.* (2003). A technological study with limited analytical results from both Phaestos and Ayia Triada, is now available and published by Militello (1998; 2001), and both are crucial in our understanding of the relations between the different palatial sites. Mallia did not produce vast amounts of pictorial paintings and so limited sample material has been available to be integrated into technological studies by Dandrau (1999; 2000; 2001). These Cretan sites would benefit, however, from being integrated into comparative work; the material from Phaestos versus Knossos especially warrants a detailed study. Finally, at the time of writing the results from the new ongoing study on the material of Pylos in Messenia (under the direction of S. Stoker) are keenly awaited and mention has been made of scenes with large ships in procession and a white 'goddess' figure (Davis 2008). These will enrich considerably our iconographical and technical knowledge of the painted plaster from this site (since Lang 1969).

Outside the Aegean, I should mention the painted plaster from Amarna (see most recently Weatherhead 2007) and Malkata (Nicolakaki-Kentrou 2000), which undergo major studies, and the new finds from Tell Burak, south of Sidon (Lebanon), whose material has so far been studied and analyzed in Germany, and will undergo comparative work by the author.

## Discussion of Painted Plaster and its Date within its Context on Site

| *Site* | *Find spot* | *Relative date* |
|---|---|---|
| *Knossos* | SE houses | EM III |
| *Palaikastro* | Area/Building 6 | MM-LM |
| *Monastiraki* | Palatial structure, east area | MM I-II |
| *Myrtos-Pyrgos* | Villa | LM IA |
| *Phylakopi* | Pillar room, mansion | LM IA |
| *Mycenae* | Early palace (E. Lobby, W. slope) | LH IIA |
| | Below HOM | LH IIIA2 |
| *Tiryns* | Old Palace (Oberburg) | LH IIIA1 |
| | *Epichosis* (West slope) | LH IIIB (tentative) |
| *Thebes* | House of Kadmos | LH IIIA |
| | Treasury Room | LH IIIB1 |
| | Spourli Plot | LH IIIB (tentative) |
| *Gla* | Melathron | LH IIIB1 |
| | Room N 1, Room H | |
| *Orchomenos* | Palatial structure: area T | LH IIIB |
| *Miletus* | Building A | MM III-LM IA |
| *Alalakh* | Palace, Room 5, other | Level VII/MB IIB |
| *Qatna* | Palace, Room N | (MB)-LB |
| *Tell Sakka* | Palace, open hall | MB IIB |
| *Tell el-Dab'a* | H/I, II, III, IV | 18th Dynasty |
| *Tel Kabri* | Palace, Room 611, other | MB IIB |
| *Hattusha* | Temple 5, 9, palace | 1st quarter 13th C. BC |

Table 5.2 Overview of the sites with their painted plaster find spots and relative dates.

As can be seen from Table 5.2 the painted plaster at most sites comes from specific, mostly secondary, contexts (see also Chapter Four; Brysbaert 2004b: appendix 1), but could be dated by stratigraphical sequences and pottery typology associations, combined with iconographic and stylistic criteria (Mycenae, Gla, Thebes, Tiryns, Palaikastro, Monastiraki, Knossos, Phylakopi, Miletus, Alalakh, Qatna, Tell Saka, Tel Kabri, Hattusha, Myrtos-Pyrgos), or based mainly on stylistic and iconographic criteria and associations (Orchomenos). The paintings from Tell el-Dab'a form a rather complicated group in this context (Bietak 1996; Cline 1998; Bietak 2000a; Brysbaert 2002b), but will be discussed in more detail later (see Chapter Seven). It will become clear that this level of dating by context or pottery association is detailed enough for the points presented in this study. Moreover, this secondary context information is *as* useful as primary contexts since it shows the *entire* life cycle of the fragments, and hence the different meanings they may have been imbued with from the time of production until their final deposition. These iconographically based dates can obviously provide no more than a *terminus ante quem*. In the case of painted plaster this is logical, since the fragments that were found, for instance, in a dump on site in a safe context associated with LMII pottery must, necessarily, have been made before the end of this pottery phase. The plaster remains of the Mycenaean mainland form a crucial group from the point of view of technological transfer and its local continuation. Thus samples were taken from the earliest possible phases from the sites of Mycenae, Tiryns and Thebes. These should be the closest possible in date to those paintings from the LM IB-II and LM III phases of the palace of Knossos, Ayia Triada, and some of the Cycladic paintings, especially those at Ayia Irini on Kea (Immerwahr 1990: 100, 103, 109-10). Next to these early samples, samples from all five sites on the rest of the Late Bronze Age mainland complete the study.

The picture outside the Aegean is more complex and scant, notably when we turn to the Near East. Only one publication manages to pull together some of the available technical and other information (Nunn 1988), and, since then, not much has been reported. The painted plaster from Alalakh (Turkey, Hatay region), both from level VII and level IV, was found in a palatial context (Woolley 1955; Niemeier 1991), and the same applies to the Qatna plaster (Syria, Orontes valley). Interesting comparative materials are the painted plaster finds from the main hall of the possibly palatial structure at Tell Sakka, Syria, dating to the MB IIB period (Taraqji 1999). The LM IA finds from Miletus were found in a large domestic building showing, in its entirety, clear connections with Minoan Crete, while its pottery (imported conical cups next to locally made and other imported ceramics, see Knappett 2003), the presence of Linear A on roundels and on vessels, and several other Minoan elements confirm such connections. The palatial building from stratum 3 in area D at Tel Kabri excavated by Kempinski and Niemeier in Israel/Palestine from the MB IIB also consisted of an official hall system (Niemeier 1991; Oren 2002: 55-63; Niemeier and Niemeier 2002: 254-98) and clearly had an important function in the settlement. The paintings at Tell el-Dab'a, forming a great exception to Egyptian traditions, and therefore included here, were also found in a palatial setting (see for instance Bietak 1996; Jánosi 2002). Lastly, the small amount and tiny fragments of painted plaster from Hattusha have been ascribed to Temple 5, 9 and some to the palatial structure (Neve 1993; 1999; 2001; Seeher 2002: 68). In any case, it is quite clear that the painted plaster from this site comes from important settings, as it is the capital of the Hittite rulers in that period (early 13th century BC).

It should be evident from this quick sketch that what is under investigation here is a form of elite art (for other examples of elite art and their functions, see Feldman 2006: 31, figure 1, 37, figure 10, 44, figure 20, etc.), a factor which must be kept in mind throughout this study, as well as in this chapter's discussion of materials and techniques. This fact, furthermore, results in specific issues being discussed, at a theoretical level, in Chapters Two and Three and, more fully integrated, in Chapter Seven.

The current debate about the directionality of technological transfer of painted plaster is in part connected to debates about the geographical origin of the artisans and their possible mobility. As I explained fully in Chapters Two and Seven, these aspects of technological transfer and artisans' mobility are understood as part of the multiple social identities that people have, define and (re)create. The finds from Tell el-Dab'a, Tel Kabri, Qatna, Alalakh, Miletus and Hattusha, which seem to demonstrate a similar iconography to paintings found in the Aegean, need to be explained within a larger network of contacts between the regions during MM IIIA-LM IB or the later part of the Levantine Middle Bronze Age and Late Bronze Age. The date of the Hattusha material and the new dating of the Qatna paintings both to Late Bronze Age, and the redating of the Tell el-Dab'a material to well into the 18th Dynasty, show that this painting tradition, even outside the Aegean, lasted well into the Late Bronze Age, including its later part LM II/LH IIB-IIIB (see Knapp 1998: 200, on broad chronological correspondence for the cultural horizons involved). In this context, my technological analyses of the Tell el-Dab'a material are not contradicted by Bietak redating the painted plaster (Bietak 1992 versus Bietak 2000a; Cline 1998). A final brief word on chronological issues follows.

This book does not attempt to contribute to the ongoing chronological debates in any of the areas under discussion. I have, therefore, adopted the dates for the painted plaster material as provided by the excavator, or as found in the publications. All these refer to relative dates, and their interrelations are not influenced by the absolute date, no matter which one is followed (see Manning 1999: 322). So clearly, absolute dates cause the problem. Manning (1999) has pulled together all the evidence and provides a complete review in his book *A Test of Time* (1999), and the most recent update is found in Manning *et al.* (2006). This complex issue is not summarized here but I will briefly state which chronology is followed here and why.

At least three chronologies are in use in the Aegean (high, revised high and low or conventional (Manning 1999: 4-5), and as far as the eastern Mediterranean and Egypt are concerned, the picture becomes even more complex (see e.g. Dever 1997: 294, 295 figure 9.4). The fact that several chronologies are in use for each region severely complicates any discussion involving a comparative study of materials. The problem of dating is at the level of assigning absolute dates to archaeological strata and material culture styles (Manning 1999: 231) and, in some cases, to single events. The absolute date of the volcanic eruption on the island of Thera (its relative date is during mature LM IA, see Manning 1999: 17, n. 79) is the most famous and crucial event for the whole of Mediterranean 'history'. It remaines controversial despite the recent radiocarbon dating of an olive branch, including the bark, found in the ashes of the volcano eruption on Thera (see Friedrich *et al.* 2006). The date of the olive branch, in combination with Manning's (*et al.* 2006) most recent dates on a series of short-lived samples, seem to suggest that the civilizations of Crete and Cyprus were much more closely linked to the Levant (Lebanon, Syria, Palestine) and that the Minoan civilization is now linked to the 2nd Intermediate Period in Egypt rather than with the New Kingdom, stretching the Second Palace Period on Crete to a 250-year period.

For this study, the relative dates given to each plaster group remain, but scholars of several different opinions in terms of dating stand behind these (e.g. Bietak versus

Niemeier). The choice for this book falls on the high chronology based on the current scientific evidence that seems to support such dating in correlation with archaeological evidence, namely dendrochronology, $^{14}C$ dating of short-lived species (e.g. Housley *et al.* 1999; Manning 1999: 256; Manning *et al.* 2006), ice-core dating (Manning 1999: 263), and now the radio-carbon dating of the olive branch. These scientific dating results are, according to Manning (1999: 243-44), far more quantifiable and can be controlled more rigorously, with often smaller errors, than dating through art historical linkages. In 2004 (Brysbaert 2004b) I wrote: 'Even if the interpretation of the scientific data is not completely in place yet, the improvements of the scientific methods over the last 20 years show the future trend. These results will eventually support either the high or the low (conventional) chronology based on whether the Thera eruption falls at 1628 BC (high) or not.' Since the publications of Friedrich *et al.* (2006) and Manning *et al.* (2006), it seems that we are moving towards stronger support for a high chronology through scientific dating methods. When we will finally reach that stage, we may then 'let the material speak for itself'.

# 6 Analyzed to Bits: Technological and Iconographic Transfer

## Introduction

As stated in Chapter Four, a strong interdisciplinary approach to the topic of painted plaster, applied for the first time here, is the result of the combined use of scientific analysis, experimental replication studies and the application of social theory. This interdisciplinary approach is necessary in order to give very fragile and scarce material the best chance to be understood, not just the technical aspects (see Chapter Four and here below), but also their meaning when contextualized (Chapter Seven). Having outlined in Chapter Four the significance of combining macroscopic and microscopic techniques with instrumental analysis to obtain a multivariate set of results, it should not come as a surprise that the data, collected in such a manner, become far stronger in their ability to support interpretive statements. This will be demonstrated in this chapter.

In combining data from scientific analysis and experimental replication, this chapter presents and discusses the material culture remains of painted plaster at three levels: production, distribution and consumption. Moreover, it indicates the degree of interrelation of these painted plaster remains both technologically and, to a lesser extent, iconographically, and how other material remains and technologies may influence the former. Consequently, patterns of material transfer, mobility and cross-craft interaction are demonstrated. It further shows that the triptych of technology, iconography and style is clearly interwoven, forming an Aegean painting craft as a 'technological style' in its own right.

This chapter is subdivided into six sections which resemble an onion structure where each section forms a layer moving deeper into the matter. First, all macroscopic observations are presented according to a method outlined previously (Brysbaert 2002b). This is followed by a discussion of the features observed microscopically by means of a range of different microscopes. The mineralogical identification of plaster and pigments by means of XRD analysis is the next layer revealing the multiple scope of this material, and this is followed by the dissection of the blue pigments. After a discussion of chronological issues relating to blue pigments, the theoretical and hypothetical discussions of technological transfer are illustrated with supportive material evidence. Finally, two non-destructive approaches are discussed: the testing of non-destructive instrumental analyses and the usage of experimental replication studies as a non-destructive technique in the study of painted plaster.

Several of these investigations are already published or under way, so none of the data or methodological discussions will be repeated here. A summary may be provided, where relevant, which includes the necessary references for the reader who is interested in detailed analyses.

## Macroscopic Study Results

### Description of the Features Observed

Only a short description of each feature follows because a more detailed account of several of the features has been provided elsewhere (Brysbaert 2002b). Its presence at each site is indicated in Table 6.1. These macroscopic features are divided into a) surface treatment features (1-16) and b) materials composition features (17-21).

- 'Surface treatment features' refer to technical features related to the act of painting onto the surface as well as any other activity that involves surface decoration, whether this took place before, after or even during the painting.
- 'Materials composition features' relate to the pigments, the plaster and the backing support of the plaster. Some of these features relate directly or indirectly to the painting techniques. Where appropriate, an indication of the techniques is stated.

*1. Fingernail, fingertip and knuckle impressions, (possibly elbow [with cloth] impression)* (Figure 6.1). These impressions, leaving a specific mark in the damp plaster, are noticeable by projecting raking light over the surface. They are almost certainly left by the painter who worked on a damp surface painting *al fresco* (see Cameron *et al.* 1977). Fresco artists have rightly commented that such marks can be made at any time prior to the drying of the surface and that painting could have been executed after the drying (S. Hocombe 2002: pers. comm.). These artists, therefore, only see fingernail impressions *in association with* brushed-up wet plaster (leaving brush marks) as a clear sign of *al fresco* painting. Having seen fingernail marks on most occasions in close association with the painting of a small or repetitive motif, however, I believe strongly that this is proof positive of the *fresco* technique.

*2. The formation, in plaster, of tiny decorative elements over the surface prior to its over-painting* (Figure 6.2). When tiny features such as eyes or edges of cloth were painted, they were often first 'accentuated' in a thin layer (<1 mm thick) of lime plaster over the face or cloth before these were painted. I call this 'localized *fresco* painting' because exactly the effect of applying paint onto damp plaster is achieved here. In several cases, this feature is also accompanied by additional fingernail impressions.

*3. The application of lime plaster in (high) relief for a specific feature on a coloured background prior to its painting.* This feature is a larger version (up to a few cm) of the previous one, but differs from it in that it may be modelled into a specific (relief) shape such as a flower or a leaf, etc. (see Brysbaert 2002b).

4. *The rough surface onto which blue pigment is applied.* This surface often seems rougher than the areas for the adjacent pigments. I believe this is done *intentionally* by leaving it rougher in the surface flattening, or by gently pecking or pressing this area more than the surrounding areas. This would provide a better key for a larger grained pigment to the plaster surface. Alternatively, this rough surface may be caused by polishing, after painting, of a thickish blue pigment coat with largish grains, clearly demonstrating *al fresco* painting.

5. *The impression of a repetitive shape in the plaster (possibly already painted with a background colour) into which paint is applied.* These impressions are possibly formed by the use of rigid templates (see also Bietak *et al.* 2000: 88-89; and indirectly, Cameron *et al.* 1977: 169, point 22) in order to form a repetitive motif before this is expressed in paint. Such impressions would guide the painter, both in order to repeat the motif and to be certain that it fits into the dedicated space.

6. *Oblong and curved marks into the plaster surface* (Figure 6.3). These marks are formed by the actions of trowelling (see experimental replication: Brysbaert 2004a) and floating the surface with a stone or other tool. Therefore, these marks are logical only if real *fresco* work was intended. The trowelling (and the polishing if carried out before painting) has the function of flattening and smoothing the surface. It also pushes down the heavier filler and, hence, 'draws up' a higher lime level to the surface, very suitable to 'lock in' or fix the pigments upon drying. This also results in a shinier surface.

7. *String impressions* (Figure 6.4). In the top plaster layer (or also in underlying layers), an impression was formed by releasing a string, which snapped onto the surface leaving a perfect straight line that could be used as a guide. In itself it is not a sign of *fresco* painting but was never associated with *secco* painting. On Egyptian paintings, executed in *secco*, string lines as guides were formed by releasing a string dipped in red paint, leaving a red paint line on the plaster (Robins 1994: 26; Shaw 2000a: 65). The string impressions aided the 'Aegean' painter in 'organizing' his/her composition. A distinction can be made between string lines used as a guide for larger linear features and those used more 'internally' as a grid to divide a smaller surface into squares in order to paint a detailed repetitive motif (Shaw 2000a). M. Shaw suggests that entire Egyptian compositions were organized by using the grid (see also Robins 1994: 23), while the Aegeans only used the grid within a free motif to aid the internal decoration of that motif (Shaw 2003: 187). Iversen (1975: 20) also mentions the hypothesis that the grid system and the metrological canon may have been connected.

8. *Incision lines* (Figure 6.5). Lines (both straight and curved, sometimes done with a compass for spirals; Shaw 2000a) were incised with a sharp tool into the damp or dry

| *Feature** | *AL* | *MI* | *HA* | *QA* | *SA* | *DB* | *TK* | *MY* | *TI* | *TH* | *GL* | *OR* | *KN* | *PK* | *MO* | *PH* | *MP* | *Comments* |
|---|---|---|---|---|---|---|---|---|---|---|---|---|---|---|---|---|---|---|
| *1* | + | + | + | + | - | + | + | + | + | + | + | + | +/- | + | - | + | + | |
| *2* | + | - | - | - | - | + | - | ? | + | + | + | + | +/- | - | - | ? | - | |
| *3* | - | + | - | - | - | + | - | ? | + | + | + | + | +/- | - | - | + | - | |
| *4* | - | + | + | + | - | + | - | + | + | + | + | + | +/- | + | - | ? | - | PH: often flaky blue |
| *5* | - | + | - | - | - | + | - | ? | - | + | - | - | +/- | - | - | + | - | MI, TH, PH: only for lilies |
| *6* | + | + | + | + | - | + | + | + | + | + | + | + | +/- | + | - | + | + | |
| *7* | + | + | + | + | - | + | + | + | + | + | + | + | +/- | + | - | + | + | |
| *8* | + | + | - | - | - | + | + | + | + | + | + | + | +/- | + | - | + | + | |
| *9* | + | + | + | + | + | + | + | + | + | + | + | + | +/- | + | - | + | + | No blue at AL&MO |
| *10* | + | + | + | + | - | +* | + | + | + | +* | +* | + | +*/- | - | - | + | + | * On surface below paint layer |
| *11.1.* | + | - | + | - | - | + | - | - | + | - | - | - | +/- | + | - | - | + | AL: see Woolley 1955: 229 |
| *11.2.* | - | + | + | - | - | + | - | + | + | + | + | + | ?/- | - | - | + | - | TK: dark blue = black |
| *11.3.* | + | + | + | + | - | - | + | + | - | + | - | - | +/- | + | - | - | + | |
| *11.4.* | + | - | - | + | - | + | - | + | + | + | + | + | +/- | + | ? | + | + | |
| *12* | + | + | - | - | - | + | - | + | + | + | + | + | +/- | - | - | - | + | HA: brush strokes in red paint |
| *13* | - | - | - | + | - | + | - | + | + | + | + | + | +/- | - | - | + | + | |
| *14* | + | + | + | + | - | + | + | + | + | + | + | + | +/- | + | - | + | + | |
| *15* | + | + | - | + | - | + | + | ? | + | + | + | + | +/- | + | - | - | - | HA: signs of polishing |
| *16* | + | + | + | + | + | + | + | + | + | + | + | + | +/- | + | + | ? | + | SA: key holes *in situ* |
| *17* | - | + | - | - | - | + | - | + | + | + | - | - | +/- | + | - | + | - | MI,MP,OR,AL,KN,DB: crushed ceramic; HA: crushed white, translucent stone |
| *18* | + | + | + | + | +/- | + | + | + | + | + | + | + | +/+ | + | + | + | + | PK, MO, MP: mud plaster too |
| *19* | ? | + | ? | + | - | + | + | ? | + | + | - | + | +/- | + | - | + | + | AL: see Woolley 1955: 233 |
| *20* | + | + | + | + | - | + | + | + | + | + | + | + | +/- | + | - | + | + | AL: only pigment sequence |
| *21* | - | + | + | + | + | + | + | ? | + | + | + | + | +/- | + | - | + | + | SA: from straw thus larger |

**Table 6.1** Macroscopic features of the painted plaster.

*Key to Table 6.1:*

1. Fingernail impressions, fingertip and knuckle impressions, (possibly elbow (with cloth) impression)
2. The building up of tiny decorative elements in extra plaster over the overall surface prior to its over-painting
3. The application of lime plaster in (high) relief for a specific feature on a coloured background prior to its painting
4. The rough surface onto which blue pigment is applied
5. The regular impression of a same shape in the plaster (possibly already painted with a background colour) into which paint is applied
6. Oblong and curved marks into the plaster surface
7. String impressions
8. Incision lines
9. A varied but restricted colour palette
10. Traces of red or yellow paint below the actual paint surface
11. Mixed colours
12. Plaster surface disturbed by the painting process
13. Indented areas in the plaster, often covered with background colour and filled in with other pigment
14. Dragmarks on top surface or just below the painted surface
15. Oblong, sometimes swirly shallow indentations, in some cases associated with craquelures of this surface area
16. Protruding round parts in the back of the plaster, also formed in plaster
17. Crushed shell fragments as filler in the plaster matrix
18. The plaster used is lime plaster
19. The presence of older plaster fragments of different sizes in later plaster layers
20. The layered structure of painted plaster
21. Tiny oblong surface marks in white smooth plaster

plaster surface, forming different types of grooves and offering a guide to paint specific motifs onto an (un)painted background. These lines were, however, not always strictly followed with paint. They make more sense when *fresco* is planned since *secco* could have been prepared by just sketching with paint or charcoal (Cameron *et al.* 1977: 167).

*9. A varied but restricted colour palette.* The colour palette is varied but limited by a small number of pure materials. The basic colours are red, white, black, yellow and blue. Additional colours are brown, green, beige, pink, purple, grey, most of which are mixed (see below) or painted over each other (but there are exceptions).

*10. Traces of sinopie.* Under a flaky top paint surface, underdrawings or *sinopie* may become visible. These may, at the actual painting time, be covered with a thin plaster layer, the *intonaco*, or simply be overpainted. These underdrawings formed a guide for the painter as to where to execute his/her composition. In the Renaissance period, this was done for *fresco* paintings to see how to fit the work of one day (*giornata*) onto the wall. Strictly speaking, only by analogy can we say that *sinopie* is a sign of *fresco* painting because *intonaci* are not omnipresent in the paintings under study (see below). Most often invisible, this influences how frequently *sinopie* can be observed. In some cases we also see red, black or yellow drawings directly under the final paintings. In the case of the Qatna spirals, they are done before *and* after the spiral painting, and in a rather clumsy way.

*11. Mixed colours.* The issue of mixed colours is complicated by the state of preservation. Even in cross-section it is not always totally clear whether there are two separate paint layers or whether the pigment is mixed. The top layer in particular is often far from pure; possibly pigments became mixed during the painting process even if they were meant to be two separate coats.

*11.1. Green: formed by blue and yellow, either yellow over blue or mixed.* For Knossos see Cameron *et al.* 1977; mainland Greece: Brysbaert and Perdikatsis 2008; other sites: Brysbaert 2002b.

*11.2. Dark blue: formed by blue and black, either blue over black or black over blue.* Black over blue (Miletus and Tell el-Dab'a) seems less common than blue over black (Tell el-Dab'a, Hattusha, and several other sites).

*11.3. Purple: formed by red and black, or red and blue, either on top of each other or mixed.* Far less common is a sort of staining occurring at Mycenae and Gla (see below and also Brysbaert 2006: 252-60; Brysbaert and Perdikatsis 2008).

*11.4. Orange: formed by painting red over yellow or mixing them.* See the Palaikastro paintings (Building 6).

*12. Plaster surface disturbed by the painting process* (Figure 6.6). In many cases the plaster is dragged up by the paintbrush during the painting process. As a clear indication of *fresco* painting, it also shows that the painter did not wait long enough before s/he started to paint, and, as a result, the plaster became mixed with the pigment, changing its hue and staining power.

*13. Areas indented in the plaster, often covered with background colour and filled in with other pigment* (Figure 6.7). This has to be distinguished from the *incavo* technique for paintings of Knossos, Amnisos and Phaestos (Immerwahr 1990: 206, n.12), but has been called '*impasto*' (Cameron *et al.* 1977: 168). This gives no clue as to whether the infilling paint (most often white) was carried out *fresco* or *secco* but, according to Cameron *et al.* (1977), the plaster was certainly still soft at a rather late stage of the painting. Indentations into the plaster under the *impasto* areas were noted at Thebes and elsewhere.

*14. Dragmarks on top surface or just below the painted surface* (Figure 6.8). These long relief marks are caused by floating the surface in the wet plaster (see experiment 1, Brysbaert 2004b). This slightly rough surface may form ideal keying for pigments to be applied in rather thick layers, or of large grain size (blue), and for the keying of the *intonaco*.

*15. Oblong, indented polishing marks, sometimes associated with craquelures in this surface* (Figure 6.9). Polishing marks tend to follow or precede the contours of specifically painted features. The polishing, if carried out before painting, has a very similar function as trowelling and also results in a shiny surface. We cannot say with certainty, however, whether the polishing was carried out before or after the painting. If done before, it could have formed a guide for the painter as to where to paint the motif, in addition to drawing up lime (see feature 6). If done after the painting, it was a clear sign of *fresco* painting since it left an indentation in the plaster.

The craquelures may have been formed when the plaster surface was already too dry to receive polishing. Cameron's argument for polishing to bring up lime *after* painting (Cameron *et al.* 1977: 168) does not work since such lime will surely cover the paint surface with a white wash and obscure the paint (see experiments by Chryssikopoulou *et al.* 2000).

*16. Protruding elements in the back of the plaster, formed in plaster.* Often associated with left-overs of mud plaster or mudbrick, these form the plaster shape of the keying that was carried out by the plasterer into the backing support in order to fix the plaster securely to that surface. In some cases, the mark is so clear that the direction of the tool or finger may help in reconstructing the iconographic design.

*17. Crushed shell fragments as filler in the plaster matrix* (Figure 6.10). Crushed *Murex* shell fragments (both the *trunculus* and *brandaris* species) were found in the thickness of the plaster of both floor and wall plaster. I am convinced that these visible fragments are not the result of using shells as a source of calcium carbonate to make lime. I do not deny that shells may have been used as a source of calcium carbonate in lime production but they would not have been visible (for details, see Brysbaert: 2003b; 2004b; 2007b). In some cases, other crushed shells such as *Venus vericoccus* were also found.

*18. Lime plaster is used.* A simple acid test determined that all plaster was lime plaster. Although this is not a reason to refer to *fresco* painting, real frescoes can only be executed on lime plaster (Brysbaert 2004a: 11-12; see also Chapter Two, Table 2.2). If the *al fresco* technique was not intended, one can question why so much effort and fuel was invested in the production process of this plaster if gypsum or mud plaster could have served equally well for *secco* painting, as it did in Egypt during at least all Pharaonic periods.

*19. The presence of older plaster fragments in later plaster layers.* An entire range of reused and recycled materials have been noted in plaster of this period (for a complete record on this topic see Brysbaert 2003b), and forms an interesting feature in the discussion about technological transfer (see also Chapter Seven).

*20. The layered structure of painted plaster.* In section and cross-section, layered structures of plaster (and pigments) can be noted. This is another form of reuse, in which the painted plaster forms a good enough flat surface to be used as a backing support for a new decoration, whether carried out in plaster and paint, or only in paint (see e.g. Brysbaert 2000 for details about Palaikastro paintings).

*21. Tiny oblong surface marks in white smooth plaster* (Figure 6.11). I believe that these tiny short marks are caused by either tiny air bubbles that were wiped out during the smoothing process, or by chopped hair, now decayed.

## Comparative Discussion

Table 6.1 shows that not all sites have the same features but certain clear trends are observed (with the exception of Mycenae, which could not be studied in detail for practical reasons, hence the question marks for several of the features). This evidence, for instance, shows that, apart from those from Monastiraki, Tel Sakka, and early Knossos, all paintings were carried out at least partly *al fresco*.

The most recent statement on the *fresco-secco* ratio discussion suggests that the technique is a combination of both *secco* and *fresco*, at least as far as Thera is concerned (Perdikatsis *et al.* 2000). Cameron believed that the Knossian material was mostly done *al fresco* and I can now confirm this for most sites apart from the three mentioned above. In order to state that for the majority even the tiniest detail was done *al fresco*, we need to look at the painting sequence of any wall with a specific scene. Logically speaking, the wall is plastered, flattened, perhaps polished and prepared by string lines and other guiding features. The first painting phase is painting the background. Next, large areas are painted according to the composition planned. Logically, one paints from larger to smaller surfaces and the tiny details are done at the end (see also Egyptian Pharaonic paintings, Robins 1994). This means that, if fingernail impressions are associated with tiny features of repetitive motifs (Figure 6.1) or with extra plaster applications (features 2 and 3), the entire composition, without exception, was done *al fresco.* If any work was added in *al secco*, this may have been done later and may have been confined to specific areas only. Since my samples were as small as was dictated by my permits, and no analyses were carried out on iconographically reconstructible parts of any of these paintings, no final word on the ratio of *fresco* to *secco* can be said here. Moreover, in order to state that *al secco* painting was carried out, solid positive evidence needs to be in place. This means that the preservation conditions need to be such that organic binding media have been preserved after at least 3000 years of burial in soils that have been subjected to fluctuating moderate Mediterranean climates (see also Evershed *et al.* 1995: 91), intense agricultural activities (including the frequent use of natural and chemical soil fertilizers that leave a variety of organic compounds behind that often have the same chemical composition as the basic components of organic binders), animal grazing and faeces, and ploughing activities that may have cut up and mixed lots of the evidence with surrounding soils (see also Lang 1969: 5). One specific problem with organic materials (e.g. polysaccharides such as gums or plant-based sugars) is that some need water to become soluble enough to be used as a binder. Consequently, water also forms their solvent, which is potentially disastrous in damp soil. Having said that, if the actual polymer material (i.e. the binder) has cross-linked over time, water may not dissolve it anymore, but, at that stage, the analytical equipment used may have problems in producing a result that compares with a modern reference sample, used in the interpretation of the original aged sample. The ageing of organic compounds and their cross-linking form serious problems in the correct and accurate interpretation of what may be interpreted as 'original' organic compounds, despite the fact that several methods are used frequently in this regard (e.g. Gas Chromatography/Mass Spectrometry). Several scientists have warned us also to be very cautious with the analytical results of specific methods of organic remains for reasons of contamination by inorganic materials inherent to the objects we analyze, such as inorganic pigments (e.g. Newman and Serpico 2000: 482, 485; Fourier Transform-Infra Red Spectrometry). They referred specifically to the great interference that pigments and calcite had in the analysis of certain glues and the measurements of the constituent amino acids (Newman and Serpico 2000: 484). Another potential source of contamination is the reuse of the original object for a different purpose than was originally intended (Serpico and White 2000a: 422). These specialists suggest conducting a comprehensive study of modern reference samples of soils for a chemotaxonomic comparative study before analyses of organic compounds are taken for granted. A last source of contamination is human: handling materials can contribute to the distortion of the lipid, salt and suger content of the specimen analyzed, or it can add contaminating organic compounds (e.g. amino acids, etc.) to original organic materials, which can lead to entirely erroneous interpretations. Needless to say, any conservation treatment, even careful surface cleaning with solvents (and often totally invisible or undocumented), alters any analysis of organic compounds.

Back to our material analyzed for this study, Monastiraki and EM III Knossos (1) which produced earlier material than most other sites, and the Tell Sakka (2) paintings done on completely different plaster mixtures (lime with a lot of visible organic inclusions and also on mud plaster), may not have been done *al fresco,* but no organic binder has been analyzed. For both groups, different reasons can be postulated.

There is also a rather large uniformity in the presence or absence of specific features as far as the Greek mainland sites are concerned. This corresponds well with the material from Tell el-Dab'a and later Knossos, and may indicate a close relation between all these sites. This has already been stated for the mainland as far as microscopic and mineralogical identifications of materials are concerned (Brysbaert and Perdikatsis 2008). Another feature mainly visible at the mainland sites and at Qatna is the 'messy' work. It looks as if it had to be done quickly, with little care being given to following guides. Even short vertical or horizontal lines are not straight and are not of the same thickness, if that was the intention. Moreover, paint dribbles running off the surface onto other areas are not uncommon (Figure 6.12), and these also appear on Mycenaean pottery. Were the painters in such a hurry to finish the design before the wall was dry or were they less capable than their predecessors?

The painter also had quite a range of devices in order to facilitate the work: string lines, incisions, underdrawings, surface preparations and manipulations. It seems, however, when the material is compared, that there was no homogeneous use of each aid, even if they knew about it. Incised lines and frequent use of string lines are, for instance, much more evident on Crete, Tell el-Dab'a and Miletus than on the mainland sites. They existed there but were not always used, especially not the incisions. Also at Qatna, the use of strings was known, but was only noted on a handful of fragments, all from the same design (yellow and white bands). At this site, spirals were painted freehand and give a very clumsy impression (see Figure 6.9 and Figure 6.12 for examples of such messy paint work). The same can be said for most of the mainland spirals. I return to this point in Chapter Seven.

Without wanting to cover every single aspect, a last important set of features are those related to the reuse and recycling of materials (17, 19, and 20 above). This issue has been discussed before (Brysbaert 2003b) and is touched upon later in this chapter.

In general, several features, especially relating to the painting technique, seem to have occurred at all or most sites. The clear absence of certain other features on specific sites, however, is very interesting in itself. This absence of specific features, however, may have been influenced by the state of preservation of the materials. In particular, the tiny painting fragments from Hattusha and also, to some degree, those from Tel Kabri are especially problematic for macroscopic observations. Methodologically speaking, two matters are clear. First, seeing *a whole range of features* in common on several sites forms evidence for a much stronger case for proving and confirming technological factors, such as *fresco* painting. Second, the macroscopic studies of these materials have benefited from experimental work and have also inspired it, mutually enriching the outcome of both.

## Microscopic Study Results

### Description of the Features Observed

The results described in this section derive from the study of the samples and cross-sections by means of several different microscopes (see Chapter Four). This work has not been carried out yet on the material of Myrtos-Pyrgos.

Image analysis measurements (expressed in microns where 1 micron = 1/1000 mm) on photomicrographs have been studied on a selection of cross-sections to make comparative work possible. For instance, per site the pigment grain size was measured on the four most common pigments and on a minimum of two samples per colour. Per sample, five measurements were then taken for each colour. Pores, inclusions, thickness of plaster and pigment layers were measured on a minimum of 6-8 samples. The thickness measurement of each visible plaster layer was repeated at lower magnification on the section photograph of each sample (Phylakopi and Gla, however, miss these latter measurements due to a damaged data disk). Several features are summarized in Table 6.2 and all features are described below, referring to additional tables.

1. *Number of plaster layers.* Since we cannot be sure that all existing layers were preserved until the moment of sample-taking, we have to be cautious in using this figure for comparative work. This is, for instance, the case for Alalakh, Miletus and Hattusha where past conservation treatments (the former) and environmental conditions of preservation (the latter two) may have affected the results.

2. *Thickness of each plaster layer* (Table 6.3). A similar point needs to be made for this feature although accurate measurements can be taken from the completely preserved layers. Three measurements per sample were taken.

3. *Number of pigment layers.* Apart from fragments where surface erosion, long-term burial and past maltreatment have removed parts of the surface, a minimum number is given, especially since cross-sections provide extra information on otherwise invisible paint layers.

4. *Thickness of each pigment layer.* As with the previous one, the thickness of this feature has been influenced by a range of negative circumstances. In order to approach the original as closely as possible, the thickest area was always measured. Three measurements were taken per sample.

5. *Grain size of each pigment* (Table 6.4). In all cases, the grain size varied within the sample so five measurements were taken per sample. Blue, red, yellow and black pigments were measured in two samples each, as was green where present.

6. *Pore size of each main plaster layer* (Table 6.5). The pores were caused by air bubbles that became trapped either during slaking or, more likely, during the application of the

| *Feature** | *AL* | *MI* | *HA* | *QA* | *SA* | *DB* | *TK* | *MY* | *TI* | *TH* | *GL* | *OR* | *KN* | *PK* | *MO* | *PH* | *Comments* |
|---|---|---|---|---|---|---|---|---|---|---|---|---|---|---|---|---|---|
| *1* | 1-2 | 1-2 | 1-2 | 1-3 | 1 | 1-2 | 1-3 | 1-2 | 1-2 | 1-2 | 1-2 | 1-2 | 1-2/1 | 1-3 | 1-2 | 1-2 | Most often 1 |
| *3* | 1 | 1 | 1-2 | 1-2 | 1 | 1-2 | 1-2 | 1 | 1 | 1-2 | 1-3 | 1-2 | 1/1 | 1-2 | 1 | 1-2 | Most often 1 |
| *10* | + * | - | + | + | - | + | + | - | + | + | + | + | -/- | - | - | + | Only in few samples per site |
| *11* | - | - | - | - | - | - | - | + | + | + | + | + | -/- | - | - | - | |

**Table 6.2** Microscopic features observed on the studied material, excluding the measurements. * see Woolley 1955: 229.

KN: 1st mark = Ashmolean MM III-LM IA materials, 2nd mark = EM II material + = positively attested

- = not attested at all

? = not observed

*Key to Table 6.2:*

1. Number of plaster layers

3. Number of pigment layers

10. Clear presence of *intonaco*

11. Presence of a 'black *intonaco*'

| *Site* | *Minimum* | *Maximum* | *Mean (3)* |
|---|---|---|---|
| Mycenae | 0.82 | 1.12 | 0.97 |
| Thebes | 0.58 | 0.74 | 0.66 |
| Tiryns | 0.84 | 1.07 | 0.95 |
| Orchomenos | 0.94 | 1.15 | 1.04 |
| Alalakh | 0.41 | 0.75 | 0.58 |
| Tell el-Dab'a | 1.00 | 1.26 | 1.13 |
| Miletus | 0.52 | 0.83 | 0.67 |
| Palaikastro | 0.91 | 1.12 | 1.02 |
| Monastiraki | 0.79 | 1.05 | 0.92 |
| Tel Kabri * | 0.30 | 0.50 | 0.45 |
| Hattusha | 0.25 | 0.60 | 0.35 |
| Qatna | 0.80 | 1.15 | 1.00 |
| Tell Sakka ** | 0.10 | 1.30 | - |

**Table 6.3** Summary of the plaster layer thickness per site (in cm), * most samples had 2 layers, ** referring to lime plaster only.

plaster onto the backing support. The pores, caused by deteriorated organic inclusions, are always measured by their largest diameter, never by length, for consistency. Five measurements per sample were taken.

7. *Inclusion size of each main plaster layer* (Table 6.6). The identification of the inclusions has not been pursued in this study. Therefore, the inorganic ones are described by colour, while organic inclusions were also encountered. For all inclusions, several diameters, including the largest, were measured as a matter of consistency and five measurements per sample were taken. The presence of primary calcite is outlined in feature 8.

8. *Size of primary calcite inclusions.* These measurements are only possible on SEM photographs with backscattering electron signals (as is feature 9 below). Primary calcite is the uncalcined grit from limestone (or other source) that is added to the plaster as filler, most probably intentionally. This material is both physically and chemically compatible with the secondary calcite plaster matrix. This aspect, as well as feature 9 below, has only been developed in depth for the Greek mainland sites, Hattusha and Tel Kabri so far, and both are consequently discussed in a separate publication (for the Greek mainland, see Brysbaert and Perdikatsis 2008).

9. *Size of secondary enriched calcite areas.* These areas form clusters in the secondary calcite matrix and consist of calcined and slaked material. This could indicate either the presence of reused old plaster in the present matrix, or the clustering of less well-mixed material in the overall plaster matrix.

| Nbr | Blue (be) | Nbr | Red (r) | Nbr | Yellow (y) | Nbr | Black (bl) | Nbr | Green | |
|---|---|---|---|---|---|---|---|---|---|---|
| **HA16** | 44-139-85 | **HA11** | 9-18-8 | **HA3** | 56-83-70 | **HA13** | 32-156-67 | **HA4** | 47-253-105 (be) | |
| **HA5** | 73-222-130 | **HA22** | 5-150-53 | | | **HA22** | 50-268-118 | **HA4** | 17-30-24 (y) | |
| **TK4** | 64-84-73 | **TK1** | 21-61-35 | **TK3** | 16-38-25 | **TK4** | 23-52-40 | **TK5** | 51-86-68 (be) | |
| **TK2** | 37-175-92 | **TK2** | 12-37-16 | | | **TK6** | 25-121-53 | **TK5** | 29-57-41 (y) | |
| **QA16** | 9-19-13 | **QA3** | 8-16-12 | **QA1** | 7-20-13 | **QA18** | 16-50-27 | | | |
| | | **QA4** | 5-9-6 | **QA9** | 7-31-16 | **QA7** | 9-18-14 | | | |
| **SA18** | 79-130-99 | **SA3** | 7-21-13 | **SA7** | 13-58-29 | **SA2** | 13-25-18 | **SA19** | 49-106-78 (be) | |
| | | **SA1** | 6-14-9 | | | | | **SA19** | 27-60-42 (y) | |
| **KNC1** | 19-52-29 | | | **KNC3** | 7-15-11 | | | | | |
| **DB77** | 19-81-39 | **DB95** | 6-17-11 | **DB3** | 6-14-9 | **DB93** | 5-14-8 | **DB54** | 16-57-27 (be) | |
| **DB87** | 18-64-41 | **DB97** | 6-10-8 | **DB90** | 9-59-30 | **DB7** | 7-18-11 | **DB54** | 6-29-14 (y) | |
| **MI10** | 30-61-52 | **MI7** | 7-17-12 | **MI5** | 17-39-26 | **MI20** | 27-134-72 | | | MI5 =orange |
| **MI24** | 32-118-67 | **MI8** | 13-52-33 | | | **MI29** | 22-144-69 | | | |
| **MO16** | 27-46-38 | **MO1** | 9-43-27 | **MO2** | 14-33-26 | | NONE | | | |
| **MO20** | 26-99-46 | **MO7** | 18-29-10 | **MO10** | 21-43-30 | | | | | |
| **PK38** | 15-47-25 | **PK36** | 37-148-68 | **PK20** | 10-24-16 | **PK1** | 6-15-10 | **PK16** | 21-90-46 (be) | |
| | | **PK39** | 6-20-13 | | | | | | | |
| **PH22** | 32-75-51 | **PH16** | 30-107-58 | **PH19** | 10-17-14 | **PH4** | 8-46-23 | | | |
| **PH25** | 11-78-27 | **PH19** | 9-17-11 | **PH20** | 8-14-10 | **PH12** | 18-28-23 | | | |
| **THA5** | 26-35-30 | **THA7** | 4-10-7 | **THA12** | 5-21-10 | **THA5** | 12-44-26 | | | |
| **THA2** | 13-51-27 | **THA12** | 4-7-6 | | | **THA11** | 8-36-24 | | | |
| | | **THB3** | 8-45-18 | | | | | | | |
| **TI1** | 14-36-26 | **TI3** | 7-18-11 | **TI2** | 7-18-13 | **TI5** | 8-36-24 | **TI12** | 19-44-28 (be) | |
| **TI18** | 28-59-39 | **TI9** | 8-48-16 | **TI4** | 19-26-21 | **TI19** | 39-50-43 | **TI12** | 6-29-14 (y) | |
| **GLB8** | 14-135-34 | **GLB4** | 7-27-19 | **GLB13** | 3-32-12 | **GLB15** | 18-41-25 | | | |
| **GLA5** | 28-57-36 | **GLB8** | 8-24-17 | **GLA6** | 14-25-21 | **GLA5** | 22-101-51 | | | |
| | | **GLB12** | 7-36-17 | | | **GLB7** | 11-34-18 | | | |
| **MY1** | 15-29-22 | **MY4** | 13-29-20 | **MY6** | 10-18-14 | | NONE | **MY19** | 11-26-21 | MY19 =brown |
| **MY8** | 15-89-36 | **MY10** | 17-38-28 | **MY13** | 20-66-34 | | | | | |
| **ORA7** | 8-30-17 | **ORB1** | 13-37-21 | **ORA5** | 10-18-14 | **ORB2** | 17-39-28 | | | |
| | | **ORA9** | 12-24-19 | **ORA8** | 22-33-27 | **ORB3** | 16-26-22 | | | |

Table 6.4 Pigment grain size for 4 main colours and green per site (in μm): minima-maxima-mean values.

| *Site* | *Minimum* | *Maximum* | *Mean (5)* |
|---|---|---|---|
| Mycenae | 79 | 205 | 130 |
| Thebes | 45 | 166 | 90 |
| Tiryns | 57 | 213 | 117 |
| Orchomenos | 76 | 180 | 116 |
| Alalakh | 29 | 140 | 77 |
| Tell el-Dab'a | 45 | 226 | 120 |
| Miletus | 55 | 114 | 97 |
| Palaikastro | 49 | 196 | 103 |
| Monastiraki | 26 | 190 | 87 |
| Tel Kabri | 59 | 304 | 143 |
| Hattusha | 58 | 370 | 174 |
| Qatna | 26 | 147 | 62 |
| Tell Sakka | 32 | 217 | 90 |

**Table 6.5** Summary of the pore size per site (in μm).

| *Site* | *Minimum* | *Maximum* | *Mean (5)* |
|---|---|---|---|
| Mycenae | 55 | 243 | 165 |
| Thebes | 34 | 173 | 88 |
| Tiryns | 46 | 241 | 107 |
| Orchomenos | 39 | 249 | 115 |
| Gla | 34 | 364 | 173 |
| Alalakh | 21 | 110 | 55 |
| Tell el-Dab'a | 40 | 238 | 103 |
| Miletus | 43 | 142 | 107 |
| Palaikastro | 26 | 214 | 86 |
| Monastiraki | 34 | 241 | 125 |
| Tel Kabri | 43 | 361 | 192 |
| Hattusha | 46 | 421 | 161 |
| Qatna | 16 | 99 | 51 |
| Tell Sakka | 28 | 311 | 125 |

**Table 6.6** Summary of the inclusion size per site (in μm).

*10. Presence of* intonaco (Figure 6.13). Its presence is far from everywhere and thus cannot be seen as a prerequisite for *al fresco* painting (see also Perdikatsis *et al.* 2000 for Akrotiri).

*11. Presence of a 'black* intonaco' (Figure 6.14). Under several blue pigment layers, a black layer is found made up of carbon black mixed in with lime plaster. It is not always very clearly visible as a separate layer but it makes the blue paint look darker and more intense.

*12. Pigment penetration depth*
Past literature has connected the ratio 'depth of pigment penetration' and 'uneven line of demarcation' as an indication of *fresco* painting (Cameron *et al.* 1977: 163-66). The point raised by Perdikatsis *et al.* (2000) about the penetration depth related to osmotic pressure for *secco* painting, however, is a very valid one. I believe that we can only decide about pigment penetration as a sign of *fresco* when the blue pigments are considered. Their grain size is too large to be absorbed into a dry surface by osmotic pressure. Since there is no clarity on this issue, I decided not to measure this feature because, so far, only experimental replication may offer a clearer view of this phenomenon (work in progress).

## Comparative Discussion

Although several of the above-mentioned features were not ideal for comparative work (thickness of paint layers, amount of plaster and pigment layers), they had that negative function of becoming excluded as a useful parameter for such work, each for their own reasons. For instance, the number of pigment layers is seriously biased by the sampling strategy applied in this study, which did not allow sampling of multiple-coloured fragments. Moreover, their thickness varies greatly due to the level of preservation of that top surface. From a methodological point of view, however, it was necessary to carry out these measurements in order to find out if these parameters could be used or not (see Brysbaert and Perdikatsis 2008).

As I have already pointed out, all Greek mainland sites share the interesting feature of the so-called 'black *intonaco*' under almost every blue pigment layer. This was not noted for any other material in or outside the Aegean apart from the material of Ayia Irini on Kea, which is slightly earlier (Majewski and Reich 1973). It could have been a solution to reduce the amount and grain size of Egyptian blue or even riebeckite in order to achieve a deeper blue colour. The presence of a real *intonaco* here is not frequent compared to its presence in Renaissance frescoes. An *intonaco*, however, is not a vital parameter for a well-painted *fresco* (Perdikatsis *et al.* 2000), and this is confirmed by this material.

The measurement of the plaster layer thickness has shown that, between each minimum and each maximum, only small differences were detected. Three exceptions, the paintings from Alalakh, Miletus and Hattusha, were expected since environmental conditions and past conservation treatments had affected their preservation. The other exception was Thebes, but a similar difference between minima and maxima was measured as with the other groups of material, which points to a striking homogeneity in the overall material.

The measurement of grain size has proved to be a very useful parameter for comparative work (see measurements done on yellow, red and blue pigments on the Greek mainland sites: Brysbaert and Perdikatsis 2008). These measurements agree with those made at Akrotiri (Perdikatsis *et al.* 2000), and now also match closely those of pigments from several other sites (Table 6.4), especially Tell el-Dab'a and Miletus, which only have larger maxima for blue and black. The pigments of the other sites, including the earlier Monastiraki material, all have roughly similar measurements, even though some minima, maxima and difference between them are larger.

The last two parameters that were measured are the sizes of inclusions and pores. As regards the inclusions, some interesting patterns came up:

- A range of different sizes was found at all sites, and differences between minima and maxima are large, but fall roughly again in the same ranges (Table 6.6).
- The exceptions make sense when we see their results in view of the technology. Alalakh, Miletus, and Thebes provide lower numbers for both minima and maxima measurements due to a worse preservation of their plaster layers. This is a normal phenomenon because when the surface is flattened by floating and trowelling (and possibly polishing) the heavier inclusions are pushed further down to let the finer lime to come to the surface. If only this top area is preserved, there are naturally fewer and smaller inclusions. This again shows a good understanding and control over the materials and the technological actions involved.

I find the pore size measurements (Table 6.5) most surprising because there was no direct control over this aspect. Too many and large air bubbles, however, were to be avoided because they could be disruptive to the structural strength, consistency and impermeability of the plaster layer in relation to its backing support. Skilled plasterers, therefore, have always knocked up their slaked lime to get the best contact between matrix and filler (Hocombe 2001: pers.comm.; Coopman 2003: pers. comm.; Gibbons 1995: 22-25). That voids were in direct relation to both the lime matrix and the added filler is clear in the discussion about void ratios (Gibbons 1995: 15-16), which makes the presence of many large pores unwanted (see Jones 1999: 153). Experimentally, it was clear that a much better coherence, and hence structural soundness, was obtained when the plaster was tightly applied onto the backing support. Pressing with the trowel and floating over each layer had the additional function of removing air bubbles. Moreover, modern fresco artists suggest popping air bubbles with a needle during the plastering procedure (Hocombe 1999: 11). The pore size for all of the material studied falls in the same ranges (Table 6.5), apart from larger maxima for Hattusha and Tel Kabri.

The SEM study demonstrated that all mainland samples showed a certain amount of primary calcite within the secondary calcite matrix. This was confined to specific areas, most often a certain distance away from the paint layer that was applied to the top coat of secondary calcite, and this is very logical after the plaster was floated (see inclusion discussion). While there was some variability in the size of the primary calcite grains, there was no large deviation between the values of the different mainland sites. This aspect will be followed up for the other sites too and some work has already been carried out for material from Hattusha and Tel Kabri.

## Mineralogical Identification of Plaster and Pigments by Means of XRD

The qualitative and semi-quantitative analyses of pigments and plaster by XRD provide a mineral identification of the materials. The purpose was to analyse material from a range of sites in order to be able to compare results with the main goal of looking at technological transfer. Since these XRD results indirectly relate to a chemical composition, elemental analysis was carried out on those samples whose XRD results were not satisfactory (see p. 134 on blue pigments).

More than 600 XRD measurements have been conducted, some of which have been or are to be published (Brysbaert and Perdikatsis 2008; Brysbaert 2007c; 2008). Therefore, the results are only summarized here, divided into plaster (1) and pigments (2), so that they become useful for comparative work. A chronological distinction enclosed with this material is discussed in relation to technological transfer in Chapter Seven.

### Plaster

The XRD plaster analysis of the samples from all sites gave a rather homogeneous picture. All plaster contained calcite as the main mineral. In most cases a small amount of quartz (referring to sand) and in few cases a very small amount of feldspars (Miletus, Mycenae, Tiryns, Monastiraki) were identified. Expanding clays, such as montmorillonite, have also been seen (Knossos, Palaikastro, Tiryns, Mycenae, Qatna, Hattusha, Phylakopi [Figure 6.15]), but may be the result of contamination. In certain cases, aragonite was detected as well. Specific diffractograms of quartz and aragonite, however, overlap substantially and confusion is thus possible. Aragonite can refer to a variety of origins (lime source, weathering product), but was nowhere present in large amounts in association with the plaster. Dolomite was manifest in combination with calcite in few samples (Palaikastro [Figure 6.16], Thebes, Tell el-Dab'a, Monastiraki, Phylakopi), and in all samples from Tell Sakka, and may refer to the use of local limestone sources containing magnesium. In general, the plaster always seems to contain calcite as its primary component (Figure 6.17) (e.g. Monastiraki, Tel Kabri, etc.) with a small to moderate amount of quartz in most cases (e.g. Hattusha, Qatna, etc.). These results fit with those previously published from sites such as Knossos (Cameron *et al.* 1977), Ayia Arini (Perdikatsis 1998), Mycenae (Profi *et al.* 1976) and Akrotiri (Perdikatsis *et al.* 2000).

This rather pure lime plaster is a necessary prerequisite for *fresco* painting (see Chapter Two, Table 2.2). On the one hand, lime as a raw material source for plaster cannot be a decisive parameter in discussing technological transfer in this region because it is hardly likely that limestone, locally present in all regions of this study, was exploited for export over long distances during the Bronze Age. Moreover, sourcing limestone back to its provenance may be difficult once it has undergone calcining and slaking. Lime plaster as a material for *fresco* painting, on the other hand, is certainly a parameter that can be taken into account for allowing the possibility of painting *al fresco* and subsequent discussions about technological transfer. Producing lime plaster requires far more effort, both in labour and in resources (fuel) than making gypsum plaster, so there has to be a good reason to produce lime plaster.

## Pigments (Table 6.7a)

Most of the XRD results of the pigments from all sites show a certain amount of calcite present in the diffractogram. This may have different origins:

- Lime was mixed with the pigment to tone down its hue
- Lime was used as a binder in the painting
- Calcite refers to the carbonation 'skin' that necessarily forms at the top of each *fresco* painting because carbonation goes from outside to inside (see experiments in Brysbaert 2004b)
- Where calcite occurs together with quartz in pigment analysis, part of the plaster layer must have been analyzed too (paint layer too thin) or there is contamination of the plaster on the pigment surface

I believe, in line with features discussed previously, that the carbonization effect plays the main role in this.

*Red, Orange, Yellow, Brown*

All analyzed red, yellow, orange and brown colours proved to be ochre-based, either haematite or goethite/limonite. In two places, Miletus and Phylakopi, goethite for yellow was found together with chlorite. At Miletus, red/purple consisted of haematite with siderite (iron carbonate, $FeCO_3$) (next to riebeckite, see below), and at Phylakopi haematite was found together with diopside and microcline which can be bright red (Figure 6.18); both minerals occur in association with igneous rocks (Read 1971: 421-22, 384). At Hattusha, finally, a red mineral coming from some pigment lumps was detected but not yet identified. This material is currently undergoing further investigation. These results may be through the use of local materials for these common pigments. In a few cases, yellow pigments were not clearly identified by XRD, possibly because its raw material was not crystalline (i.e. limonite); the same phenomenon was found at Knossos (Cameron *et al.* 1977: 143). Haematite was also the colouring component for most of the brown except at Gla where it was combined with chlorite. Apart from the red/purple sample from Miletus, consisting of haematite with siderite, the results are

| site | Red<br>Pink | Blue | Green | Yellow<br>Orange | Brown | Black<br>White |
|---|---|---|---|---|---|---|
| KN | Hematite<br>- | Glaucophane<br>?(Ashmol) | - | ?<br>- | - | -<br>Calcite |
| PK | Hematite<br>- | Riebeckite | Celadonite+chlorite | Goethite<br>Hematite | - | -<br>Calcite |
| MO | Hematite<br>- | - | - | Goethite<br>Hematite | - | -<br>Calcite |
| PH | Hematite<br>Hematite+diopside +microcline<br>Hematite+Calcite | Riebeckite<br>Cuprorivaite | - | Goethite<br>Goethite+chlorite<br>?<br>- | - | Pyrolusite<br>Hematite<br>Calcite |
| MY | Hematite<br>Hematite+Calcite | Cuprorivaite<br>Amphiboles? | - | Goethite<br>Goethite | Hematite | -<br>Calcite |
| TI | Hematite<br>- | Cuprorivaite<br>Amphiboles? | Cuprorivaite+- | Goethite<br>- | - | -<br>Calcite |
| TH | Hematite<br>Hematite+Calcite | Cuprorivaite<br>Amphiboles? | - | Goethite<br>- | - | Hematite<br>Calcite |
| GL | Hematite<br>- | Cuprorivaite | - | Goethite<br>- | Hematite+chlorite | -<br>Calcite |
| OR | Hematite<br>- | Cuprorivaite | - | Goethite<br>Goethite | - | Pyrolusite<br>?<br>Calcite |
| AL | Hematite | - | - | -<br>Hematite | - | -<br>Calcite |
| MI | Hematite<br>Hematite+Siderite<br>Hematite+Calcite | Riebeckite | - | Goethite+chlorite<br>- | - | ?<br>Hematite+ manganite<br>Calcite |
| DB | Hematite<br>- | Cuprorivaite | Cuprorivaite+- | Goethite<br>- | Hematite | -<br>Calcite |
| HA | Hematite<br>- | Cuprorivaite<br>Azurite | Cuprorivaite+- | Goethite<br>- | - | -<br>Calcite |
| QA | Hematite<br>Hematite | Cuprorivaite (+?) | Cuprorivaite+- | Goethite (+hematite), ?<br>-Hematite | Hematite | Hematite<br>Calcite |
| SA | Hematite<br>- | Cuprorivaite | Cuprorivaite+- | Goethite<br>- | - | -<br>Calcite dolomite |
| TK | Hematite<br>- | Cuprorivaite | - | -Goethite<br>- | - | -<br>Calcite |

Table 6.7a A summary overview of the XRD results of the colouring pigment minerals in the Aegean and Eastern Mediterranean (purple in separate table).

Figure 4.1 Selection of painted plaster samples from several sites.

Figure 4.2 Blue pigment lump from Tiryns, cat.no. 2060, unknown context.

Figure 6.1 Fingernail impressions in the painted plaster (Alalakh).

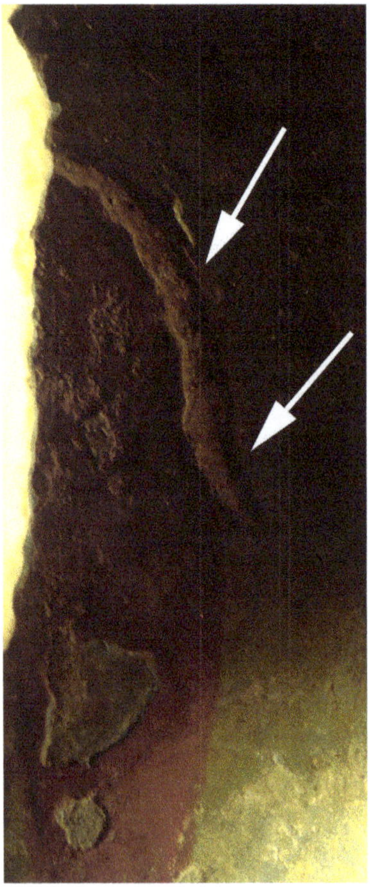

Figure 6.2 Details built up in plaster before it was overpainted (Knossos).

Figure 6.3 Tool marks from trowelling (Tiryns).

Figure 6.4 String impressions from cord snapping in damp plaster (Alalakh).

Figure 6.5 a & b. Incision lines in damp plaster (a: Tell El-Dab'a; b: Qatna: for the turtleback design).

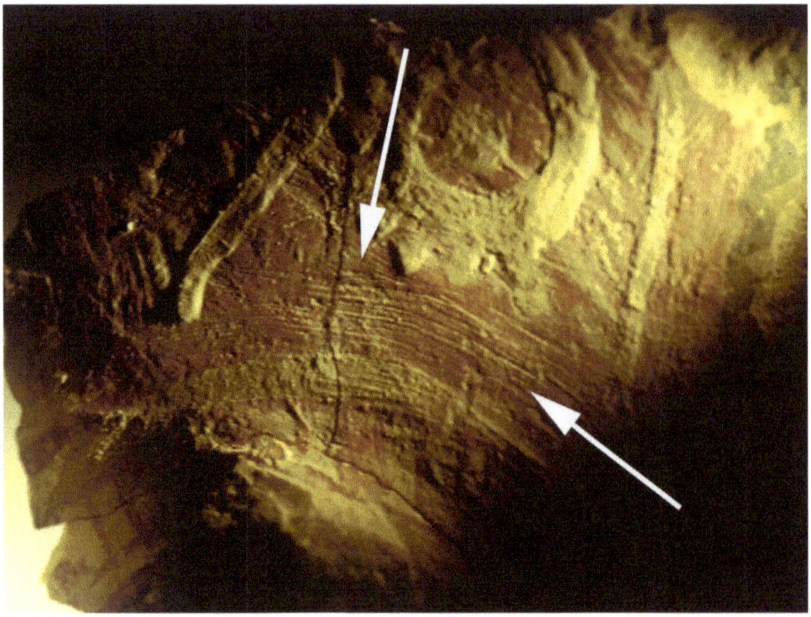

Figure 6.6 Plaster surface disturbed by the painting process (Knossos).

Figure 6.7 Traces of *impasto* applications leaving an indentation in the surface (Thebes).

Figure 6.8 Areas with drag marks from the plaster float (Tiryns).

Figure 6.10 Inclusions of crushed *Murex* shell (Miletus).

Figure 6.9 Polishing grooves left in the plaster surface (Qatna).

**Figure 6.11** Negatives of decayed hair inclusions or from floated tiny air bubbles (Miletus).

**Figure 6.13** Clear example of *intonaco* (Tell el-Dab'a).

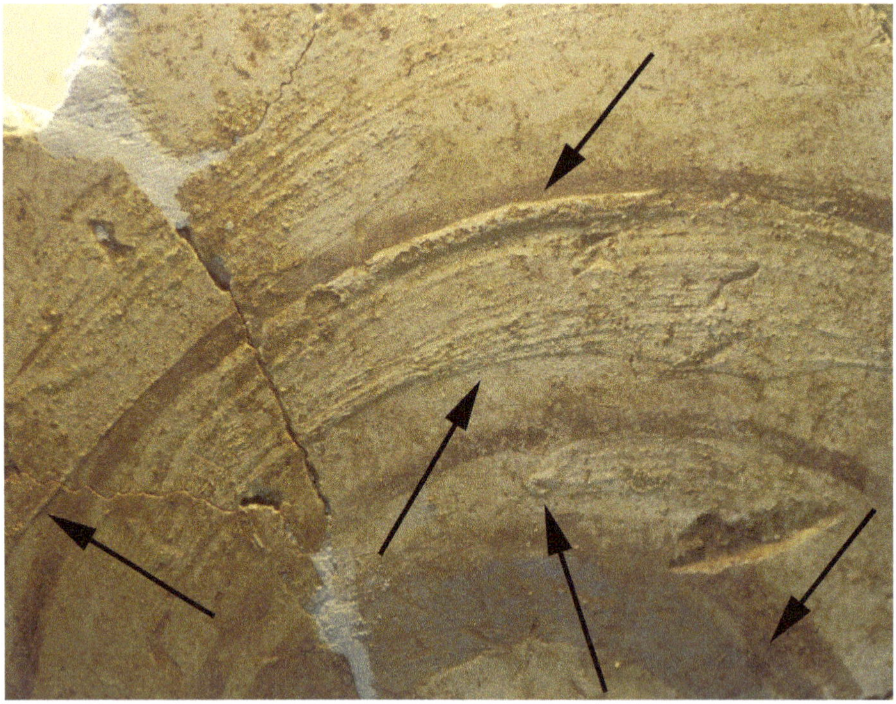

**Figure 6.12** Messy painting: paint dribbles and untidy outlines from the mainland (Orchomenos).

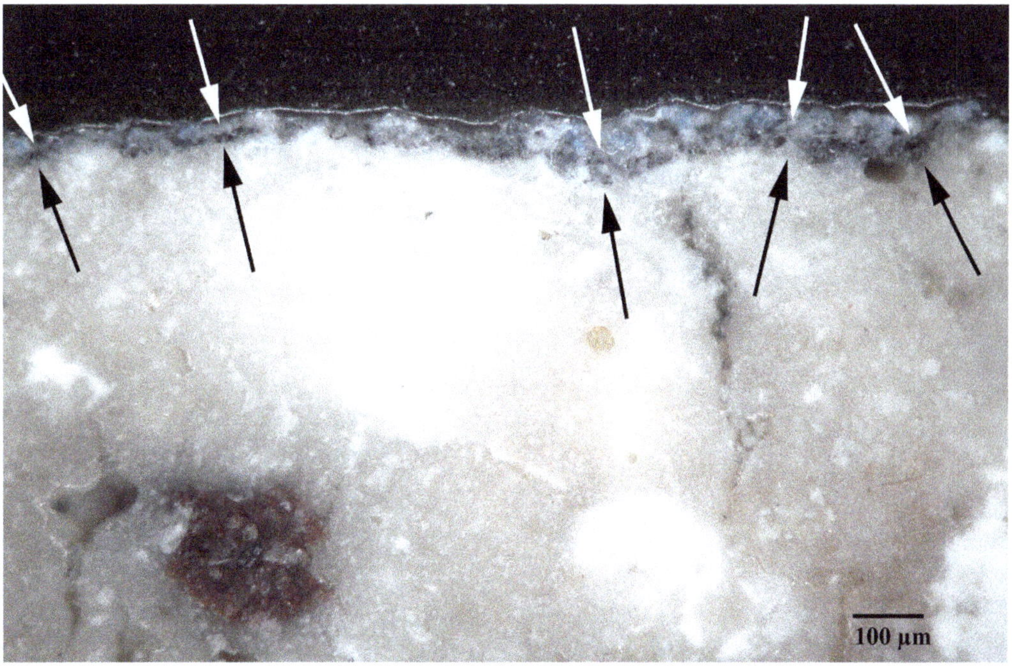


Figure 6.14 'Black *intonaco*' under the blue paint layer on the Mycenaean mainland sites (Orchomenos).

Figure 7.3 Five-year-old boy grinding coloured minerals into pigments.

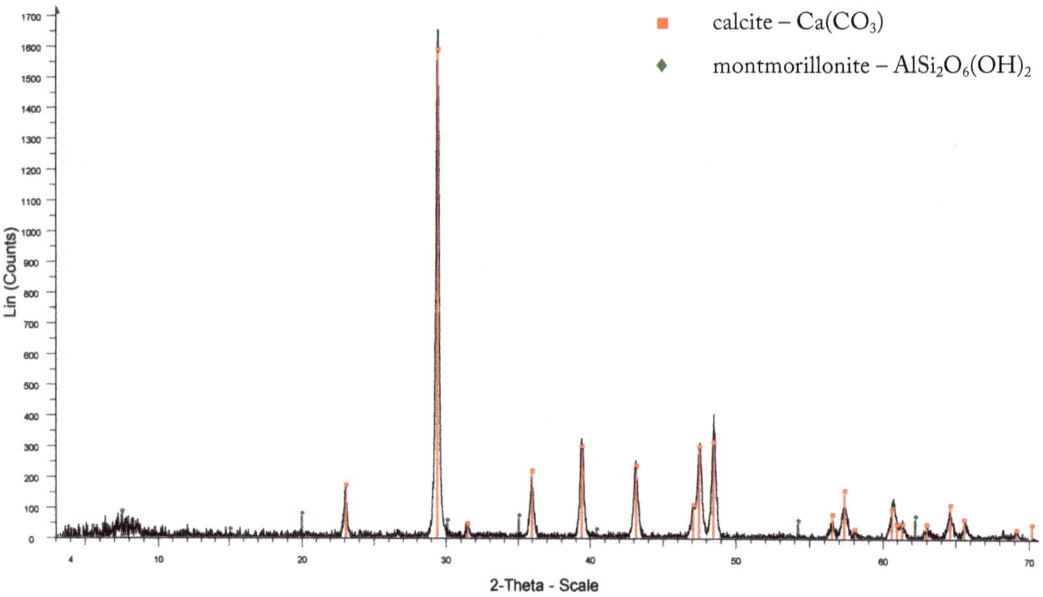


Figure 6.15 X-ray diffractogram of PH 7A: with montmorillonite.

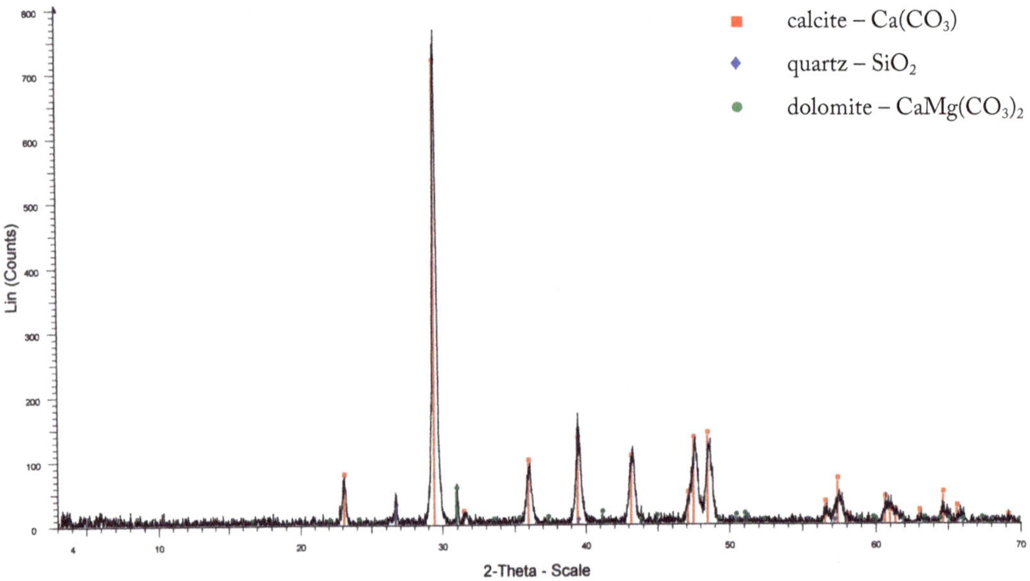


Figure 6.16 X-ray diffractogram of PK 60A: with dolomite.

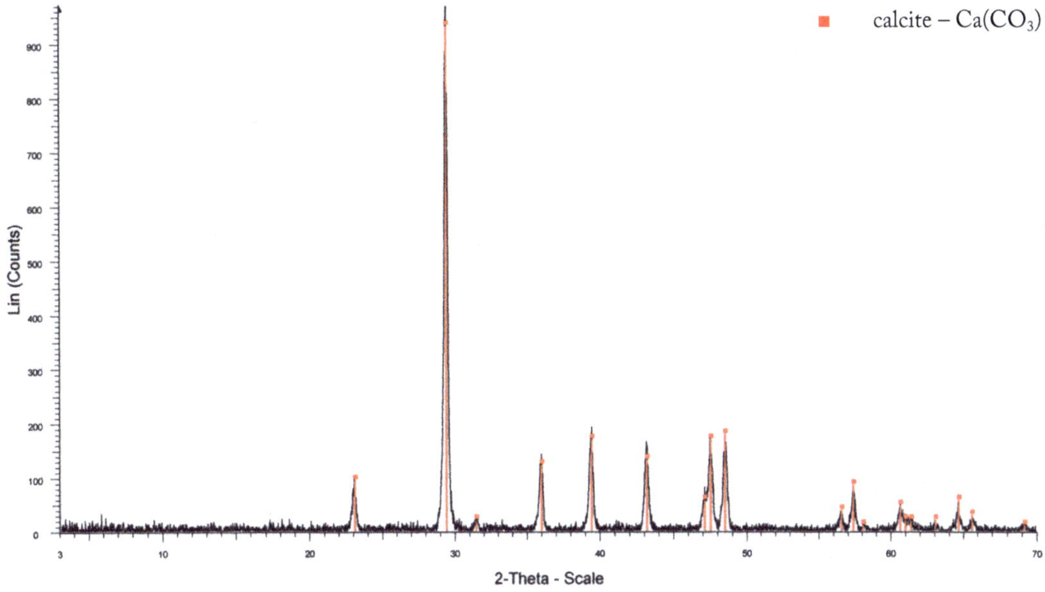


Figure 6.17 X-ray diffractogram of MO 14A: with calcite.

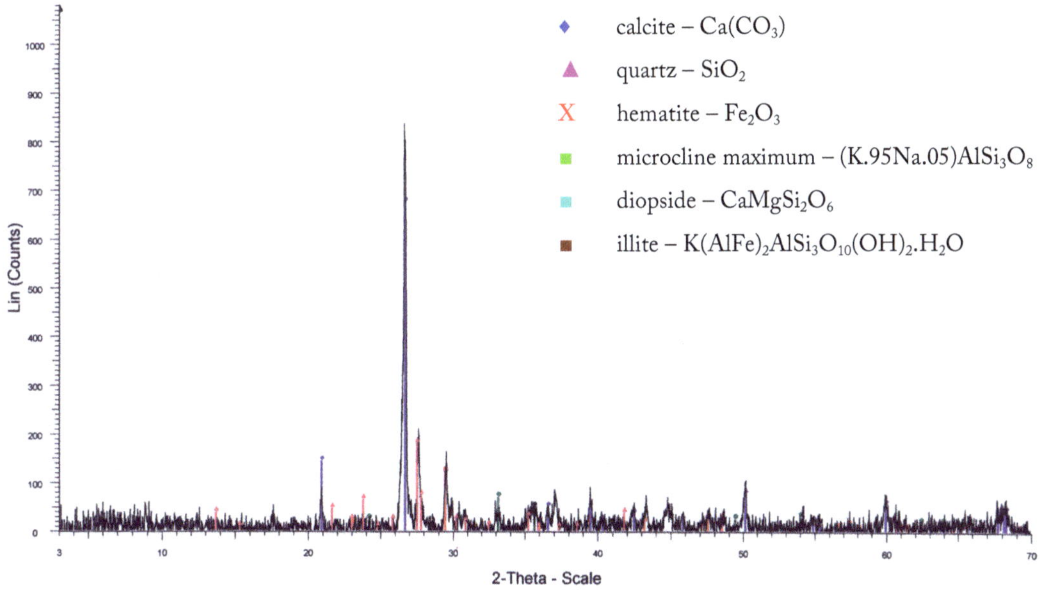


Figure 6.18 X-ray diffractogram of PH 14R: red.

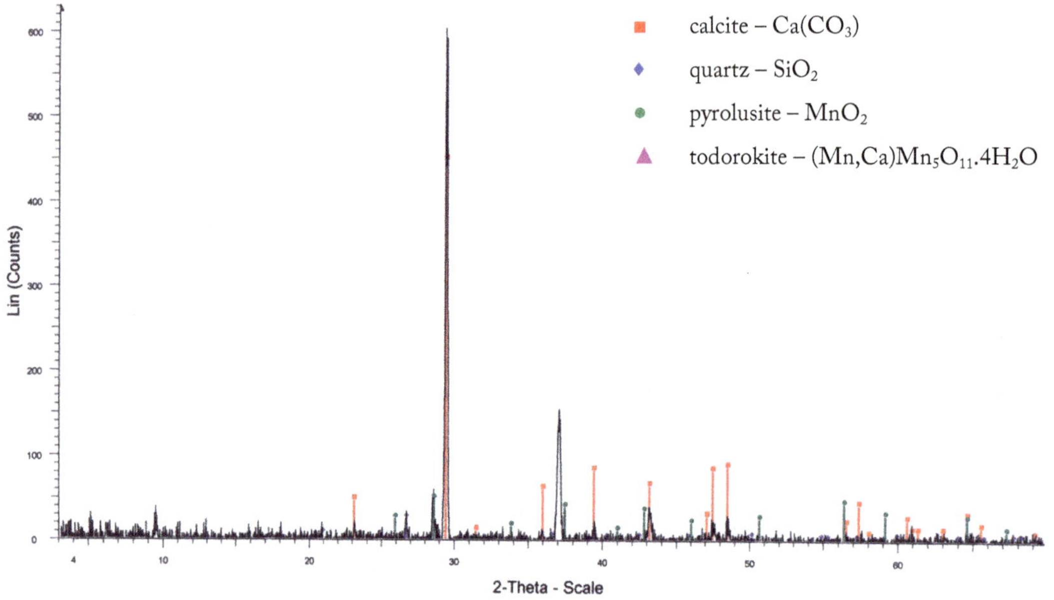


Figure 6.19 X-ray diffractogram of ORA 6BL: black.

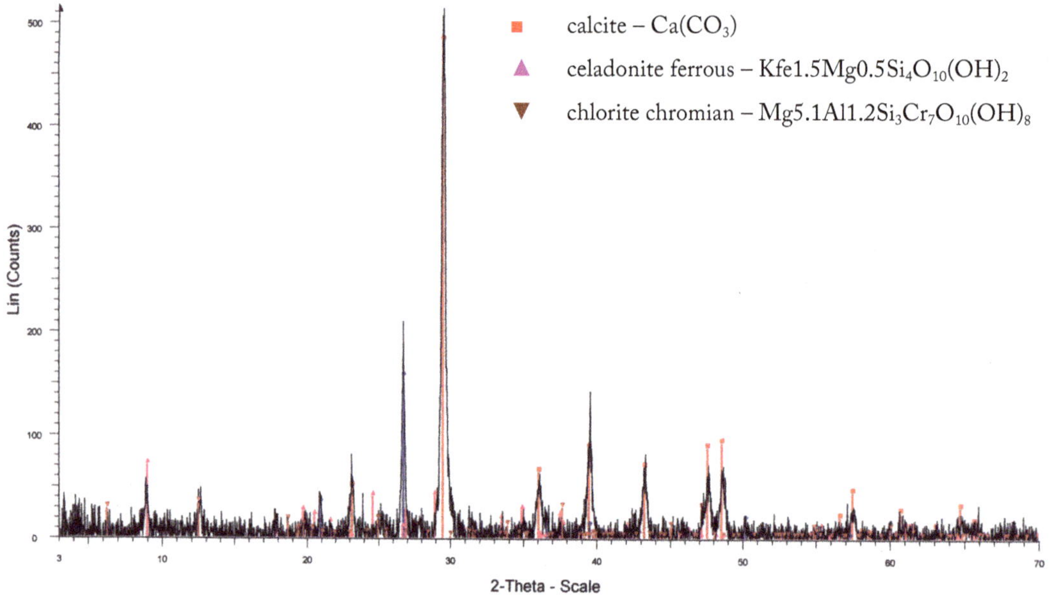


Figure 6.20 X-ray diffractogram of PK 120GR: green.

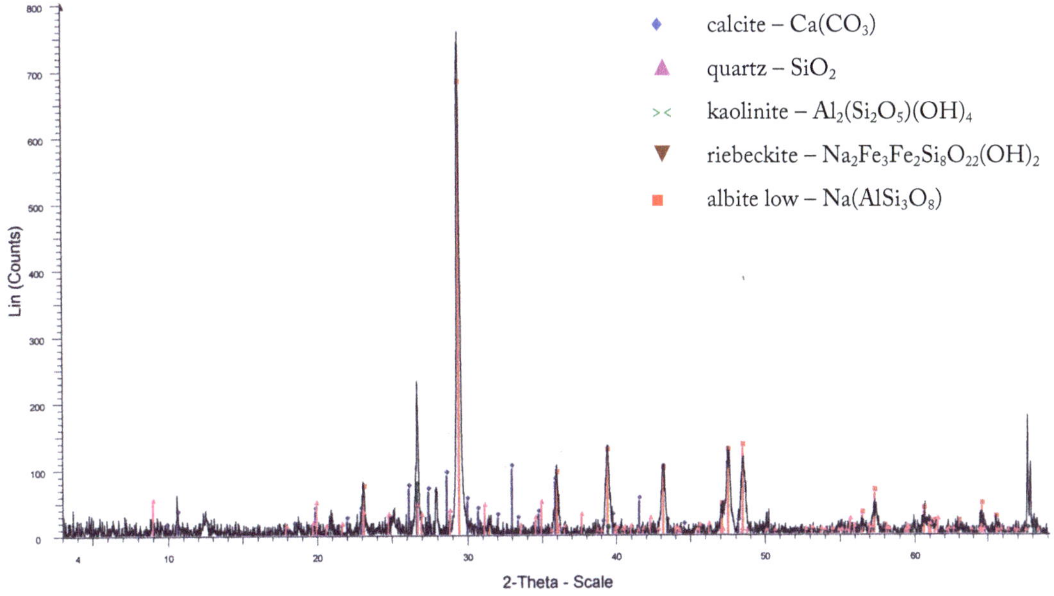


**Figure 6.21** X-ray diffractogram of MI 10BE: blue.

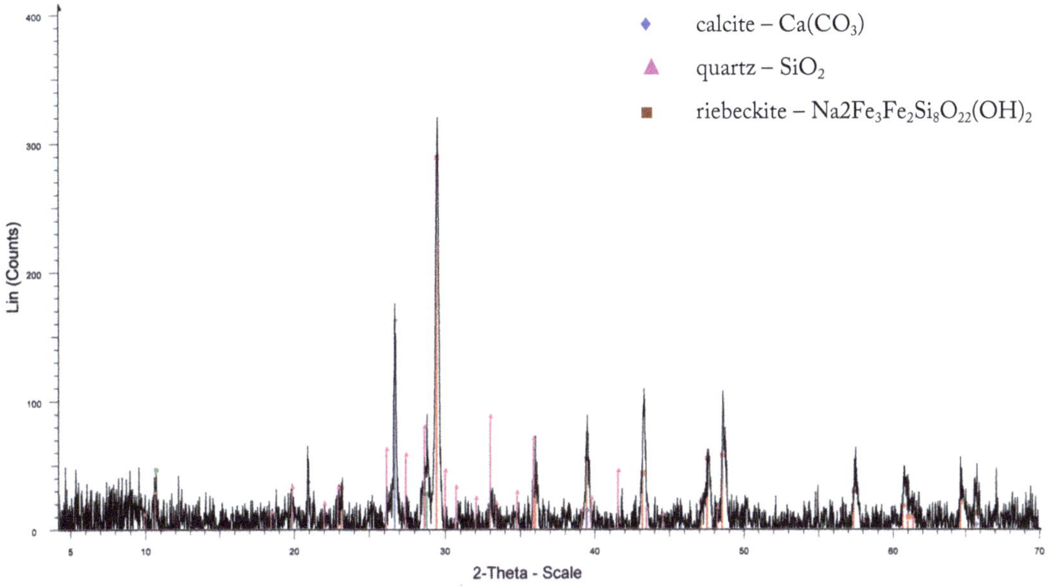


**Figure 6.22** X-ray diffractogram of PK 26BE: blue.

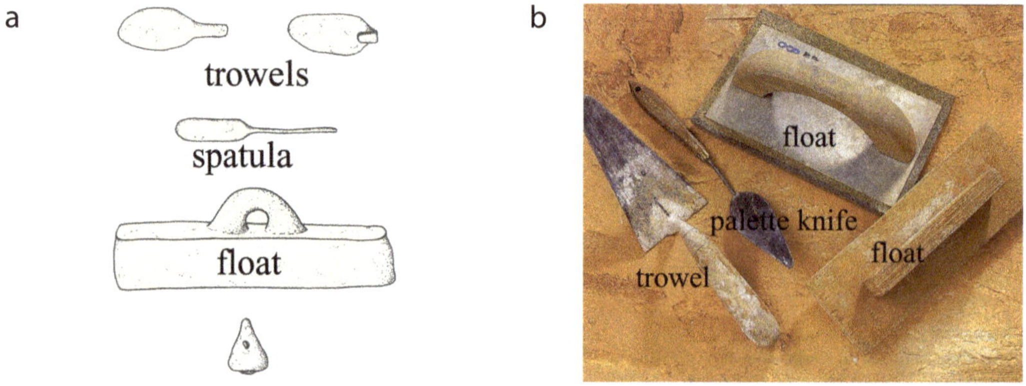


Figure 7.4 Fresco tools and materials: (a) Minoan (after Evely 1999: 155), and (b) modern (after Hocombe 1999: 12, 19).

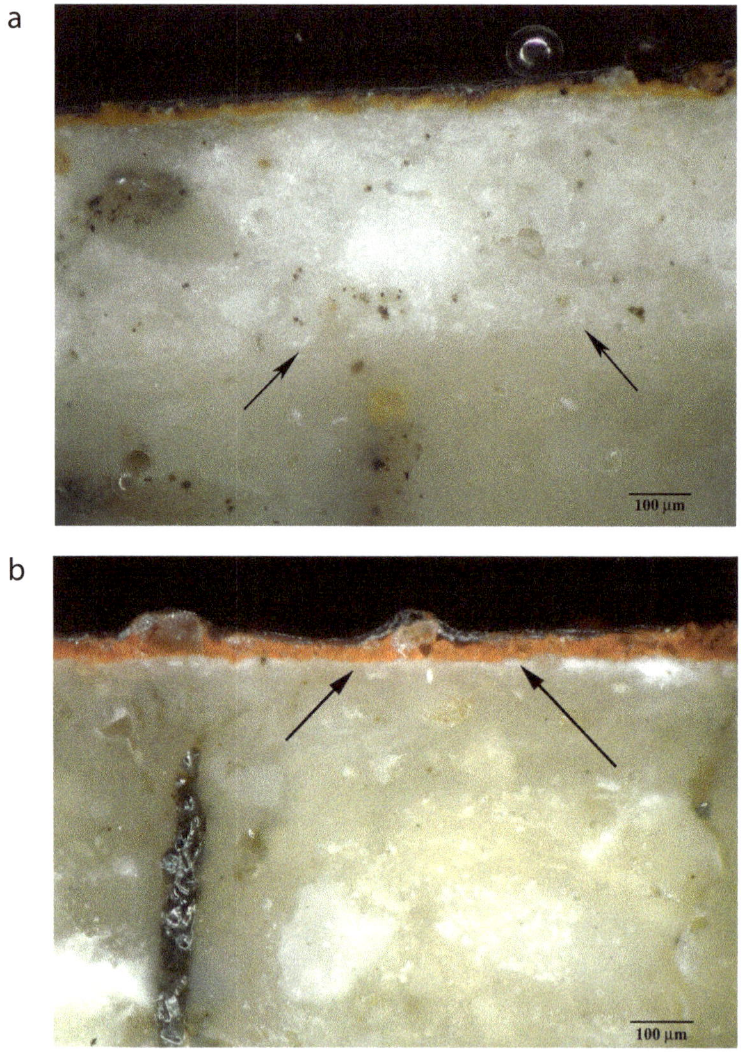


Figure 7.5 Cross-sections of painted plaster from Thebes (a) with *intonaco* and (b) without *intonaco*.

completely in accord with the known materials for the Aegean Bronze Age, even in the earlier periods in Bronze Age Crete (see for instance Cameron *et al.* 1977; Jones 1999; 2004; Perdikatsis 1998; Perdikatsis *et al.* 2000; Dandrau 1999; Brysbaert and Perdikatsis 2008). Ochres can be found almost anywhere in the world, hence their low indication level for transfer of materials. Transfer may be more apparent on a trace level study but the question is whether it is worthwhile going that far if we know that these pigments were omnipresent.

*Black and White*

Most of the black pigments were carbon, based on what was observed under the microscope. A few mineral-based black pigments were found, however: pyrolusite (manganese dioxide, $MnO_2$) at Phylakopi and Orchomenos (Figure 6.19), haematite at Phylakopi, Thebes, and Qatna, and at Miletus, in one sample, a combination of black haematite and manganite (hydrated manganese oxide, $MnO(OH)$). Manganite has never been found before and pyrolusite is not very common either (see Jones' report in Renfrew 2007 for Phylakopi). Haematite black has been encountered at Knossos (Cameron *et al.* 1977), and specific clay minerals were found at Akrotiri (Perdikatsis *et al.* 2000).

White, analyzed by XRD, gave pure calcite (see also Denninger 1974) or calcite with quartz. Since white areas on these paintings were always areas left unpainted, white corresponds with the composition of the actual plaster (see above). Neither gypsum nor kaolinite was found in any of the samples investigated for white. I need to remind the reader, however, that no white areas of detailed decorative paintings were analyzed to avoid sampling from reconstructible iconographic themes and motifs.

*Green*

Green samples were found at Tell el-Dab'a, Palaikastro, Tiryns, Hattusha, Qatna, Tell Sakka, and at Tel Kabri. In the past, Heaton (1912: 216) had identified green malachite powder from the 'figure-of-eight shield' scene at Tiryns. This has been the only malachite green on Aegean Bronze Age painted plaster so far published. Two more samples from both Tiryns and Tell el-Dab'a, one from Hattusha, Qatna, Tell Sakka and Tel Kabri have now been examined and analyzed. Microscopy clearly indicated two different materials forming green, most often Egyptian blue and yellow grains (no identification by XRD). These two colours are either mixed into one layer or are applied as two separate layers (the latter is more common). Microscopically, this follows the tradition as we see it at Knossos and several other places, such as Ayia Triada (Militello 1998). A yellow sample from Mycenae, however, showed limonite, a non-crystalline yellow ochre (also identified at Knossos and Akrotiri), which was identified by MRS (Brysbaert and Vandenabeele 2004: 690; 2006), and it may well be that most unidentified yellow samples by XRD are thus limonite. A grey-green found at Alalakh by Woolley (1955: 229) could not be identified but seemed to have been a mixture of yellow and blue.

In light of a recent article published by El Goresy (2000: 63) on Egyptian pigments from the Old to the New Kingdom, malachite (as a pigment) has been characterized as a

| *Sample* | *CROSS SECTION, (RLM)* | *XRD (and other)* |
|---|---|---|
| KNC 1 | Red x black particles | - |
| KNC 2 | Blue particles | Glaucophane |
| PK 48 | Red x black particles | - |
| PK 64 | Red x black particles | - |
| PK 80 | Red x black particles | Hematite, calcite, quartz |
| PK 94 | Dark red particles | Hematite, calcite, quartz, dolomite |
| PH 27 | - | Hematite, calcite, quartz |
| THA 7 | Red x black particles | Hematite, calcite, quartz, dolomite |
| THB 3 | Red particles | Hematite, calcite, quartz (SEM: Si, Fe, Al) |
| GLA 1 | Red x dark (black?) particles (PLM) | Hematite, calcite, chlorite |
| GLB 11 | No particles, more staining effect | Hematite, calcite, quartz |
| GLB 12 | Few tiny purple particles, more staining effect | Calcite, aragonite, quartz (SEM: Si, Ca, Fe, Al, Mg) |
| GLB 13 | Tiny purple particles, more staining effect | (MRS: Lapis Lazuli, hematite, organic substance) |
| GLB 14 | No particles, more staining effect | See GLB 13 – from same scene |
| GLB 15 | No particles, more staining effect | See GLB 13 – from same scene |
| GLB 16 | No particles, more staining effect | Calcite, aragonite, quartz (and see GLB 13) |
| MY 14 | No particles, pink staining in top plaster layer | Hematite, calcite |
| MY 18 | Red particles | Hematite, calcite, aragonite, gypsum |
| TI 9 | Dark red particles | Hematite, calcite, aragonite |
| AL* | Red and blue | Indigo (MRS) |
| MI 15 | Dark red x fine black particles | Hematite, calcite, quartz |
| MI 18 | Dark red x fine black particles | Hematite, calcite, quartz, (glaucophane ?) |
| MI 22 | Dark red x fine black particles | Calcite, quartz |
| MI 31 | - | Hematite, calcite, quartz, (glaucophane ?) |
| MI 12 | - | Hematite, calcite, quartz, talc, siderite, chlorite, kaolinite, riebeckite |
| QA 4 | Red particles | Hematite, calcite, quartz, ? |
| QA 23 | Red particles | Hematite, calcite, quartz, talc |
| QA 27 | Red particles | Hematite, calcite, quartz, talc |
| SA 1 | Red particles | Hematite, calcite, dolomite, quartz |
| TK 2 | Red x blue particles | Hematite, cuprorivaite, calcite |
| TK 3 | Red particles | Calcite, ? |
| HA 26 | Red x blue particles | Calcite, ? |

**Table 6.7b** Combined analysis (XRD, SEM, RLM, MRS, LIBS) of macroscopically identified purple and purplish samples (no purple at Monastiraki, Orchomenos, and Tell el-Dab'a). *Alalakh as described in Bingöl 1997: 24.

possible corrosion product due to devitrification and subsequent weathering of the blue glass phase in Egyptian blue. In contrast, it seems that malachite, as a natural mineral, was used as a pigment at some point later in antiquity, since Vitruvius mentioned it as being found in Macedonia near copper mines (Kakoulli *et al.* 2001: 265, 271). This means that care needs to be taken before malachite is determined as a pigment. XRD analysis should be able to distinguish clearly between natural malachite and a copper corrosion product as two different mineralogical phases with the same name. From this point of view, it would be useful indeed to re-examine the Tiryns sample too.

Regarding several of the green samples from Palaikastro, two formed a mixture of celadonite (commonly called 'green earth', Grissom 1986) and chlorite, which occur together geologically. In my opinion, this reflects a local source, and its use as green pigment has not yet been identified elsewhere in the Bronze Age. Chlorite may give yellow to greenish grains but its combination with celadonite in the two samples is unique (Figure 6.20). Two other green samples from Hattusha have not yet been identified mineralogically, even though spectra were produced, and further work is being undertaken at present. One indicates the possibility for a green clay source but this needs further confirmation.

*Purple (Table 6.7b and c)*

Purple has appeared on many sites and has been identified in various ways. The range of different observations shows how 'purple' can be achieved through a variety of original materials. The synthesis of these observations and others relating to the use of *murex* in the painted plaster craft are partially presented in Brysbaert (2007b: figure 1), but completely in Tables 6.7b and 6.7c. Purple is often achieved by the use of different intensities of haematite. Haematite has also been combined with Egyptian blue (Knossos: Cameron *et al.* 1977, Hattusha, Tel Kabri, Alalakh) or with riebeckite (Miletus). In most cases, however, we see haematite combined with fine black particles, not detectable by XRD, but recognizable under the microscope as organic (carbon).

A very enigmatic purple colouring material was published by Iakovides (2001: 237) in his second volume on Gla and described as a sort of magenta, probably originating from madder, while French (2002: 122) recently referred to it as a 'yet unanalyzed purple'. Samples were taken, examined and analyzed. XRD analysis indicated the presence of both calcite and a rather large amount of aragonite, possibly pointing to the

| *Pure material* | *Organic (*murex*?)* |
|---|---|
| | Inorganic (hematite) |
| Mixture of materials | All inorganic (hematite+EB/amphibole) |
| | Organic + inorganic: |
| | 1-hematite + charcoal black |
| | 2-hematite, lapis lazuli + organic |

Table 6.7c Ways of obtaining purple in Aegean and eastern Mediterranean painted plaster.

use of *murex* purple (Brysbaert and Perdikatsis 2008; Brysbaert 2006; 2007b). MRS has produced a clear spectrum in the organic range (suggested to identify indigo [P. Vandenabeele 2007: pers.comm]), but so far no reference can be matched with it to provide secure identification. Inorganic MRS spectra of that same sample, however, identified both haematite and *lapis lazuli*, the oldest positively identified presence of this latter on paintings (Brysbaert 2006). Further work, including gas chromatography with mass spectrometry, has been carried out at the University of Leicester, Department of Chemistry, but no positive results were obtained.

## The Blue Pigment Issue

Most blue samples were identified as cuprorivaite (mineral name for Egyptian blue, $CuCaSi_4O_{11}$) which had already been in use since the MM IIA, as analyzed samples from Knossos testify. At Mycenae, Tiryns, Thebes, and Phylakopi, however, riebeckite, a blue amphibole (Leake 1978), may possibly have been used, even though XRD did not positively identify this mineral. This tentative identification is based on: (1) the presence of iron in some samples (both riebeckite and glaucophane are iron-based, see below), (2) no positive identification of Egyptian blue or Cu, and (3) accessory minerals pointing to amphiboles (see below). This is the first time that this iron-based blue pigment had been found outside Crete, Akrotiri and Kea. Accessory minerals to amphiboles (talc and chlorite) in metamorphic rocks (Perdikatsis *et al.* 2000), have also been identified in some blue samples from Phylakopi, Palaikastro and Miletus, which form additional confirmation for the identification of amphiboles.

At Miletus, riebeckite has been identified for the first time outside the Aegean, which forms a great indicator for material transfer (Figure 6.21), and the presence of talc as an accessory mineral to amphiboles has now also been found at Qatna (Syria) and Hattusha (Turkey). Riebeckite has been found previously at Akrotiri (Perdikatsis *et al.* 2000: 111), Ayia Irini (Perdikatsis 1998: 107), Knossos (Cameron *et al.* 1977), Ayia Triada (Militello 1998: 381) and several other sites on Crete. It has also been identified for the first time at Palaikastro on Crete (Figure 6.22). At both Knossos and Akrotiri, amphibole blue and Egyptian blue were mixed together in order to form a rather dark blue. This combination is now assumed to have been discovered also for the first time at Phylakopi where both talc (XRD) and Cu (SEM) were found. In addition, pure Egyptian blue was used at Phylakopi, confirming the suggestion made by Jones (1978). Other identifications for blue pigments within this study form the newly discovered *lapis lazuli* at Gla (Brysbaert 2006: 252-66) and azurite at Hattusha, Turkey. The other examples form either pure Egyptian blue, riebeckite or a non-identified blue. The first use of amphibole blues seem to occur as early as MM IIIA at Palaikastro (pottery date) and seem to continue until late LH IIIA/early LH IIIB1 (Mycenae, Thebes, Tiryns). These dates correspond with those for Knossian samples (Cameron *et al.* 1977: samples 22 and 25).

Emanating from the above, the blue pigments and their composition may be crucial in relation to technological transfer, whatever its form. We will now look at the Egyptian blue and the amphiboles in more detail in order to extract the maximum information about their composition and what this, consequently, can tell us about technological transfer.

The composition of the copper element in Egyptian blue has been understood as an important chronological indicator in relation to its use in well-dated contexts, whereby arsenical copper preceded bronze (copper alloyed with tin) which, in turn, preceded copper alloyed with lead on Egyptian painted plaster analyzed by El Goresy (2000). I wanted to see if a similar trend was detectable as far as Aegean and eastern Mediterranean uses of Egyptian blue were concerned. I further intended to find out whether the analysis of the elemental compositions could determine whether the tin, arsenic or lead was naturally present or had been intentionally added to the copper. Bronze is an alloy to which at least 2% tin (but up to 10% or more is possible) is added (Cronyn 1992: 213). Copper-tin alloys with less tin exist, but are, technically speaking, not bronze.

The results produced by both EPMA, and a test done by ED-XRF on blue pigment grains, indicate that tin occurs more widely than originally found (Knossos, Akrotiri and Mycenae), and arsenic has now been detected too. The first evidence for the presence of tin in the copper phase of Egyptian blue at Orchomenos and Phylakopi is now accounted for as well. One sample from Tell el-Dab'a shows tin, and we know from Seeber (2000: 102: discussion) that tin was found there. Tin was also detected for the first time at Thebes (THA 2) (Brysbaert 2008). This indicates the presence of tin in all sites on the mainland, apart from Gla. The analytical results gave values of below 0.20% for tin and 0.26% for arsenic (apart from 1.32% at Tell el-Dab'a) (Table 6.8). The detection of arsenic in the blue pigment from Gla was new, together with its first detection at both Mycenae and Tell el-Dab'a.

These very low values for both tin and arsenic are to be expected considering that the composition of the entire pigment was examined, with the copper component usually being around 20% when fully quantitative.

With a minimum of 0.01% weight (wt) for tin in copper, however, the tin values here are large enough to suggest that recycled 'bronze' scraps were used as the source for the

| *Sample* | *% tin* | *% arsenic* |
|---|---|---|
| PH 23 | 0.02-0.06% | - |
| MY 1 | 0.03-0.97% | 0.39 % |
| MY 8 | 0.01-0.67% | - |
| ORA 7 | 0.02-0.08% | - |
| DB 9 | - | 1.32% |
| DB 46 | 0.01% | - |

**Table 6.8** Normalized tin and arsenic percentages in Egyptian blue.

copper in Egyptian blue (Th. Rehren 2005: pers. comm). We can be less sure for the low values of arsenic (apart from the Tell el-Dab'a sample), which may thus also indicate a natural presence of the element in the original copper ore. These measurements put an interesting perspective on the article by El Goresy who convincingly associates the presence of arsenic, tin and lead in the copper compound of Egyptian blue with the dates of the use of these pigments in Egypt (El Goresy 2000: 64-68). Based upon the results we have today, I feel it may not be possible to transfer his statement to the Aegean and eastern Mediterranean paintings.

Chronological deductions *have* been made for the Aegean, based on the differential pigment use between pure Egyptian blue, Egyptian blue mixed with amphiboles and pure amphiboles (Philippakis *et al.* 1976), but this was rejected by Dandrau (1999). Therefore, the present and published evidence was examined to find out if any chronological conclusion may be made for the Aegean and its sphere of influence, based on the presence of the traces in the copper phase of the Egyptian blue, and its relation with amphibole blues.

Table 6.9 puts together all the blue pigments analyzed in relation to their date where known. As already stated (Brysbaert and Perdikatsis 2008), it is of interest that the examples of possible amphiboles on the mainland come mainly from the earlier contexts at Tiryns, Mycenae and, a little later, at Thebes. Gla and Orchomenos (and possibly Alalakh, see Woolley 1995: 233, 234) only show Egyptian blue. The sites producing MM IIIB-LM IA material (Palaikastro, Phylakopi, Miletus) all seem to have employed mainly amphiboles and a little Egyptian blue at Phylakopi, while no amphiboles were found or could positively be identified at Tell el-Dab'a, Tell Sakka, Qatna, Hattusha and Tel Kabri. This makes sense for all apart from the situation at Tel Kabri where one could have expected the presence of amphiboles due to its earlier date; Tell Sakka fits, in any case, into a different painting tradition (see Chapter Seven).

In terms of dating, the results appear to show that no amphiboles were in use towards the end of the Late Bronze Age. All of the fragments found in the final destruction levels of the mainland palaces contain pure Egyptian blue, and the same can be suggested for the materials from Hattusha and Qatna even though talc appeared in a blue at Hattusha and in a purple from Qatna. Egyptian blue seems to have existed both during the entire period covered within this study and much earlier (Cameron *et al.* 1977). The use of amphiboles can be chronologically outlined now. The earliest possible amphibole sample from this study comes from Palaikastro from MM IIIA late (PK 38). In this context, it is useful to point out the fact that the copper of the Cycladic islands of Siphnos and Kythnos had been exploited since the EM period (Davis 2001: 24-26; Stos-Gale 2001: 198-200), and that ferric amphiboles (riebeckite and/or glaucophane) are geologically present on both islands (see mineral rock collection in the Wiener Laboratory, Athens).

The question whether these minerals were discovered during copper mining expeditions is permissible, because, on the one hand, quite a few of the minerals used as pigments are related to metal ores (e.g. ochres or iron oxides, malachite, azurite). On

| *Sample* | *Colour* | *Identification* | *Date* |
|---|---|---|---|
| MY 1 | Blue | Cuprorivaite | LH IIIA2 |
| MY 8 | Blue | Cuprorivaite? | LH IIIA2 |
| MY 2 | Blue | ? | LH IIIA2 |
| MY 11 | Blue | ? | LH IIB/IIIA1 or LH IIIA2 |
| MY 7 | Blue | Cuprorivaite | LH IIIA2 |
| MY 5 | Blue | Amphibole? | LH IIIA2 |
| MY 3 | Blue | Amphibole? | LH IIIA2 |
| ORA 7 | Blue | Cuprorivaite? | LH IIIB |
| ORA 10 | Blue | Cuprorivaite | LH IIIB |
| GLA 4 | Blue | Cuprorivaite | Early LH IIIB1 |
| GLA 5 | Blue | Cuprorivaite | Early LH IIIB1 |
| GLB 2 | Blue | Cuprorivaite | Early LH IIIB1 |
| GLB 1 | Blue | Cuprorivaite | Early LH IIIB1 |
| GLB 4 | Blue | Cuprorivaite? | Early LH IIIB1 |
| GLB 6 | Blue | Cu, As | Early LH IIIB1 |
| GLB 3 | Blue | Cuprorivaite | Early LH IIIB1 |
| GLB 10 | Blue | Cuprorivaite | Early LH IIIB1 |
| THA 12 | Blue | Cuprorivaite | LH IIIA |
| THA 10 | Blue | Cuprorivaite | LH IIIA |
| THA 2 | Blue | Cu, Mn, Fe, Sn | LH IIIA |
| THB 1 | Blue | Amphibole? | LH IIIB1 |
| THC 8 | Blue | Cuprorivaite | LH IIIB |
| TI 1 | Blue | Cuprorivaite | LH IIIA2/earlier |
| TI 5 | Blue | Cuprorivaite | LH IIIA2/earlier |
| TI 7 | Blue (lump) | Cuprorivaite | - |
| TI 8 | Blue (lump) | Cuprorivaite | - |
| TI 13 | Blue | Cuprorivaite? | Before LH IIIB2 |
| TI 14 | Blue | Cuprorivaite | LH IIIA |
| TI 18 | Blue | Cuprorivaite | LH IIIB |
| MI 2 | Blue | Talc → amphibole? | LM IA |
| MI 4 | Blue | Amphibole? | LM IA |
| MI 10 | Blue | Riebeckite | LM IA |
| MI 20 | Blue | Amphibole? | LM IA |
| MI 22 | Blue | Talc → amphibole?, calcium copper oxide | LM IA |
| MI 23 | Blue | Amphibole? | LM IA |
| MI 24 | Blue | Talc, chlorite → amphibole? | LM IA |
| MI 25 | Blue | Talc, clinoclore → amphibole? | LM IA |
| MI 26 | Blue | Talc, clinoclore → amphibole? | LM IA |
| MI 27 | Blue | ? | LM IA |

| *Sample* | *Colour* | *Identification* | *Date* |
|---|---|---|---|
| MY 1 | Blue | Cuprorivaite | LH IIIA2 |
| MI 2 | Blue | Talc → amphibole?, Fe, Ni, Pb | LM IA |
| PH 5 | Blue | Talc → amphibole? | LM IA |
| PH 22 | Blue | Fe, Ti, Mn, Pb, K, riebeckite | LM IA |
| PH 23 | Blue | Cuprorivaite | LM IA |
| PH 25 | Blue | Clinoclore, riebeckite | LM IA |
| PK 13 | Blue | Fe, Amphibole? | MM IIIB-LM IA |
| PK 14 | Blue | Fe, Amphibole? | MM IIIB-LM IA |
| PK 15 | Blue | Fe, Amphibole? | MM IIIB-LM IA |
| PK 26 | Grey-Blue | Fe, riebeckite | MM IIIB-LM IA |
| PK 62 | Blue | Fe, Amphibole? | MM IIIB-LM IA |
| PK 38 | Blue | ? | MM IIIA late |
| PK 46 | Blue | Amphibole? | MM IIIB |
| PK 58 | Blue | ? | MM III |
| PK 74 | Blue | Amphibole? | ? |
| DB 46 | Blue | Cuprorivaite | 18th dynasty |
| DB 9 | Blue | Cuprorivaite | 18th dynasty |
| KNC 2 | Blue | Riebeckite? | MM III-LM IA?? |
| HA 4 | Blue | Cu, Si, Ca | LB (14th Century BC) |
| HA 5 | Blue | Cu, Si, Ca | LB (14th Century BC) |
| HA 6 | Blue | Cu, Si, Ca, cuprorivaite +? | LB (14th Century BC) |
| HA 8 | Blue | Talc + ?? → amphibole? | LB (14th Century BC) |
| HA 9 | Blue | Azurite + ? | LB (14th Century BC) |
| HA 10 | Blue | Azurite + ? | LB (14th Century BC) |
| TK 2 | Red + blue | Haematite + cuprorivaite | MB IIB |
| TK 4 | Blue | Cu, Ca, Si, cuprorivaite | MB IIB |
| TK 5 | Blue | Cu, Ca, Si, cuprorivaite | MB IIB |
| TK 13 | Blue | Cuprorivaite | MB IIB |
| QA 16 | Blue | Cuprorivaite | LB (14th Century BC) |
| QA 7 | Blue | Cuprorivaite + ? | LB (14th Century BC) |
| SA 19 | Blue | Cuprorivaite | MB IIB |
| SA 16 | Blue | Cuprorivaite | MB IIB |
| AL * | Blue | Silicate of copper and soda | MB IIB |

**Table 6.9** Identification (XRD—colouring element only, SEM) and dates of all blue pigments of this study (* see Woolley 1955: 233, 234).

the other hand, we can see that by the time the amphiboles started to be in use as blue pigment (MM III), copper mining activities in the Cyclades had decreased drastically in comparison with the First Palace Period (Stos-Gale 2001: fig. 10.2). This discussion has made it clear that the blue pigment issue continues to challenge us for the period under study, and further work, therefore, has already been planned.

## Non-Destructive Approaches 1: Instrumental Analyses

One of the main considerations that features throughout the book is the methodological issue of a non-destructive analytical approach to archaeological material. The practical outcome of this working methodology resulted, first, in the testing of two non-destructive instrumental techniques, MRS (Brysbaert and Vandenabeele 2004; 2006) and LIBS (Brysbaert *et al.* 2006). Briefly summarized, MRS provides a molecular identification of organic and inorganic materials, and LIBS provides an elemental analysis of inorganic materials, both of them being non- or micro-destructive. Neither technique has previously been employed on Bronze Age painted plaster from the eastern Mediterranean.

One of the outcomes of this work, in relation to more traditional techniques requiring a physical sample of a specific size, was that the results of these non-destructive techniques compared very closely with those of more traditional techniques. Consequently, this can then lead to a sampling strategy reducing the number and size of the samples for future work. Another surprising outcome of the MRS application was the first identification of the use of indigo as an organic pigment on one of the Thebes fragments (Brysbaert and Vandenabeele 2004; 2006), as well as at Alalakh. The use of this instrument, therefore, shows its strong potential in contributing towards our understanding of a wider use of pigments in the eastern Mediterranean Bronze Age, including those of organic origin.

These practical outcomes are crucial, firstly, in the consideration of the fragile nature of archaeological remains and hence our responsibilities towards their preservation for future generations. Where we are able to continue our investigations and studies of archaeological material with the aim of communicating their outcomes to the public which 'owns' its heritage, non-destructive analyses facilitate such study and encourage us to work closely with researchers from the sciences in order to improve the instrumentation according to our needs. So far, it is not entirely possible (*yet*) to escape sampling, by employing only MRS and LIBS, but taking fewer samples as a result of their capacities is certainly an important move in the right direction, and coming to this vital conclusion was the initial aim of setting up these pilot projects. Second, our understanding of the nature of the materials used in producing these paintings becomes enriched by being able to diversify our application of the range of instrumental analysis. This trend will develop exponentially in the future for the benefit of both the materials and our understanding of them.

## Non-Destructive Approaches 2: Experimental Replication

### Introduction

In the context of the interdisciplinary methodology employed for this study, four small-scale experiments and a conservation project were conducted; four papers describing this work have been published partly or in full (Brysbaert 2003a; 2003b; 2004a; 2007b). None of the published information will, therefore, be repeated here, but the importance of the studies within their archaeological context, both intrinsic and methodological, is summarized.

Each specific experiment was carried out, firstly, to answer specific questions that are briefly summarized below. These experiments were thus aimed at completing our knowledge and understanding of the painted plaster technology for both the experimenter and in general. Second, carrying out experimental replication 'puts people back into the picture' and hence this may throw light upon certain social aspects of technological actions (see Chapter Four).

- *Dolphin fresco and the Tripartite shrine fresco* (Brysbaert 2004a), to investigate and follow each step involved in building up a plaster support that will take a figurative design painted *al fresco*, to design the motif and to follow the sequence of colours applied by painting a simple design rather than just coloured bands. Time was monitored for each step. For the tripartite shrine *fresco* only, to test how a *giornata* was obtained, to paint a more complex design and monitor time.
- *Sgraffito* (Brysbaert 2003a), to try to understand how this technique works and which method(s) of those discussed in the literature was/were the most appropriate one(s) for success.
- *Pigment experiment* (Brysbaert 2004a), to go into the field and procure raw materials, based on human perception modes (seeing, smelling, touching) rather than scientific knowledge, to investigate the steps involved in manipulating raw materials into usable pigments and to apply them *al secco* by means of a range of organic binders.
- *Shell experiment* (Brysbaert 2003a, 2007b), (1) to carry out, observe and study every step from manipulating raw materials to the finished product including calcining (time + temperature), slaking (time + temperature), applying, curing time and phases (time + temperature), and (2) to test, by experiment and analysis, how far we can trace the use of shell as a raw material for lime production as its end product, and to test the results obtained against the archaeological record and available analytical data.
- *Painting on gypsum*, to see if the possibility exists to paint on wet gypsum plaster, to observe whether this became permanent, how much filler one can mix with it to influence its drying rate, and to test its strength upon drying.

The original account of all experiments (Brysbaert 2004b: Chapter Seven) was written in 'active sentences' in order to demonstrate the presence of the *actor* (whether an artist, craftsperson, technician or experimenter) in the technical production and inherent decision-making. The observer in the experiments was also the actor, and the act of observing is understood as an active role in the experiment (see also Gooding 1990: 15-16). This active way of describing the experiments was, therefore, executed on purpose to demonstrate that *people* are very much part of (1) any past production process of which archaeologists usually only find the material remains, and (2) of the experimental process itself.

Each experiment was divided into four subsections in order to promote a clear and logical outline:

- Design: identification of parameters and variables, each experimental step, level of control over variables
- Materials used: both qualitative and quantitative
- Outline of the work procedure: outline of every step
- Evaluation and analysis of the results: in view of the initial research question

Only the gypsum painting project is discussed in more detail here, since this shows why gypsum plaster cannot take *al fresco* painting, linking *al fresco* with lime plaster only.

## Gypsum Plastering and Painting

Most plaster that was found to be painted in the Aegean and eastern Mediterranean Bronze Age was lime-based ($CaCO_3$ or calcium carbonate). As far as contemporary Egypt is concerned, however, most painted plaster was gypsum-based ($CaSO_4.2H_2O$ or calcium sulphate dihydrate), although the gypsum was not always pure. We cannot say much with certainty about the Levant and Anatolia until now (Moorey 1994), due to the lack of published analytical results. Of concern here is how the chemical (and mineralogical) basis of plaster has a profound influence on the painting technique.

This work aims to confirm or reject the view that gypsum plaster cannot provide the backing support for painting *al fresco*. This goes very much against what is stated erroneously by Coburn *et al.* (1989: 2), but it confirms what has already been established through several analytical projects both on Crete (see for example Cameron *et al.* 1977), and all over the Aegean (Perdikatsis *et al.* 2000 for Akrotiri; Brysbaert and Perdikatsis 2008), and now also in the eastern Mediterranean (Seeber 2000: for Tell el-Dab'a; Brysbaert 2002b: for Miletus, Tell el-Dab'a [see also Brysbaert 2007c] and Woolley 1955: for Alalakh). By explicitly making this point once more, and by comparing the findings obtained with the archaeological remains of the eastern Mediterranean and Aegean paintings, we may be able to narrow down the possibilities for the place of origin for the first *al fresco* paintings ever carried out and so frequently and skilfully employed during the late Middle and Late Bronze Ages. Consequently, we can then see how the technique might have travelled, and in which possible direction(s), and how this corresponds with what we know about iconographic motifs and theme transfer. This is crucial in our attempts to understand not only the pure technical aspects of the craft, but also many social and organizational skills that were required to 'set it all up and do it' since both plaster types (gypsum and lime) imply a very different technology, starting from the procurement of raw materials to the final stages of applying it onto a surface and receiving the paint. No actual design took place since this was not planned as part of an intended experiment. This practical work has, however, involved enough experimentation and it has implications for archaeological questions.

*Materials Used*

Having had the chance to work with plaster of Paris (basically gypsum plaster that comes in powder form as calcined raw gypsum or hemihydrate: $CaSO_4.1/2H_2O$) as a gap-filling and moulding material for over ten years, I have been able to make a few observations that are intrinsic to the qualitative (chemical and mineralogical) composition of the material. These exclude *al fresco* painting on this material completely because of: (1) drying, setting and curing time, and hence, (2) its behaviour when mixed with a foreign filler (Coburn *et al.* 1989: 9-10, 16). These observations are described below. Other materials were (not including those specifically used for gapfilling): rubber mixing bowls, tongue pressers, artists' stainless steel painting trowels, red and yellow ochre pigments.

*Outline of the Work Procedure*

1. The best way to 'slake' plaster of Paris is to put a certain amount of water in a mixing bowl and slowly add plaster of Paris, making sure that there is always a certain amount of water standing above the plaster of Paris. Slaking is an exothermic reaction, which gives a temperature of 35-40 °C.

When the mixture is used for gap-filling, only gentle stirring (or no stirring at all) is required, in order to prevent air bubbles getting trapped in the mass, especially if one deals with large amounts. The ratio of water evaporation depends on the ambient temperature in the room but also on the amount of stirring. The harder the mixture is stirred, the faster the water evaporates, the faster the mixture dries out and setting starts. If work needs to be speedily executed, it is better not to stir (fast), so allowing the material to remain workable longer. In any case, the setting of the plaster starts at the surface and goes inwards so thicker areas take longer to set and cure completely. *Without any added filler* which would require stirring, in a Mediterranean spring to summer climate with temperatures of 22-28 °C, and a relative humidity of 60-70%, this material is no longer workable after a maximum of 3-4 hours, even if still slightly wet.

Knowing that *fresco* paintings are carried out on a damp but firm surface of a few millimetres to maximum 1 centimetre thick, and may take several hours up to days to finish, it seems obvious that this technique cannot take place on wet gypsum plaster. The surface is dry in far less than 4 hours. I have tried to apply ochre pigments, suspended in water, onto such a dry gypsum surface and they come off immediately upon the slightest touch. When applied when wet, they get absorbed quickly by the fast drying plaster and do not leave a vivid, but a rather dull or pastel colour impression. This stands in stark contrast to employing the same colours with binder to a dry surface, which leaves a much brighter coverage. The difference in colour intensity between *fresco* painting on gypsum and on lime may also relate to different opacity levels formed by the different rates of crystal growth upon curing, and in reaction to the different forms of light-surface interaction due to the different crystal-forming structures (see Crawford 1973: 159-66).

2. For a different purpose than that of retarding the drying rate of the plaster of Paris, I mixed a powder colourant of red and yellow ochre into two different bowls of plaster of Paris and tried to apply these after the slaking as gap-filling material. I added about 10-15% of this material because less than that did not give much colour intensity, which was the effect I was aiming for. To my great surprise it took *days* before these mixtures were dry, and once dry they were so weak that I could pulverize them between my fingers without exerting any pressure. This was especially the case for the mixtures containing 10% or more pigmented material. Thus, it became clear that trying to retard the drying process of plaster of Paris was not an easy process.

*Evaluation and Analysis of Results*

The method of retarding successfully the setting of gypsum-based plaster, by using less than 10% of a foreign additive often of organic origin (Coburn *et al.* 1989: 9-10, 16), is employed for construction purposes, but it is in no case related to painting *al fresco*. In fact, adding a foreign material to the gypsum-based plaster makes it expand during setting (which it does naturally) to the point that it weakens (except for ash plaster: Coburn *et al.* 1989: 8).

In contrast to this, gypsum-based plaster with no foreign additives sets and cures well to a strong material (but less strong than lime plaster), and can thus form a base for painting *al secco*, as was used throughout the Egyptian Pharaonic Period (McCarthy 2001: 18) and later. Lee and Quirke (2000: 117-18) mentioned gypsum-based plaster and also one case of lime plaster, but this was painted *al secco*; while Newman and Halpine (2001: 22-32) mentioned organic binders for all periods, but did not refer to gypsum-based plaster. At Amarna, Weatherhead (2001: 53) analyzed plastered floors made up of a mixture of lime and gypsum plaster, but no absolute quantities were given.

As stated before, grids on Egyptian walls were clearly done on a dry wall and the string was dipped in red ochre paint and then allowed to snap against the surface to form a perfect guiding line (Lee and Quirke 2000: 118; Robins 1994: 182-200; 2001: 60); while in Aegean art, the string is snapped clearly into the still soft lime plaster leaving no colour but an indentation impression (personal observations, see also Shaw 2000a). The only clear and complete exception to all these features is the painted plaster material from Tell el-Dab'a in Egypt, which therefore forms an interesting case within this research.

## Conclusions

I aimed to clarify why experimental replication has a crucial place in a technological study of this nature, since it is a logical continuation of the technological and experimental work that has been done in the past and it adds new directions.

1. As a non-destructive method, experimental archaeology is complementary to analytical work within the framework of a much-needed interdisciplinary approach to studying the technological production modes within pre-industrial societies. As such, the value of experimental archaeology (and ethnoarchaeology) as proper disciplines, side-by-side with archaeological sciences, may help to approach technological questions of Coles' (1979: 42-43) level four of experimental archaeology, 'how' and 'why' (see also Mathieu 2002: 2), while being very cautious when using the analogies. Once it has been fully recognized that these disciplines have matured and are fully part of interdisciplinary research, material science specialists may finally be able to quit the position of being forced to only write 'appendices' for archaeological reports (see also Skibo 1992: 4; Jones 2002).

2. By carrying out experimental work I tried to understand better the past aspects of technology and craftsmanship of plastering and painting through creating an artificial system whereby I tested the techniques involved myself (see also Brysbaert 2003a; 2003b; 2004a). Based on existing knowledge of the archaeological evidence, experiments and replication have serious limitations but also great advantages. They evoke and recreate the relationship between human behaviour and material culture (Coles 1979). We have to see this relationship in its present context, however. Knowing that many experimental findings have served as the basis for archaeological knowledge about certain prehistoric techniques (such as the stone tool technology, see Skibo 2000: 200), I believe that we can continue to appreciate this method of studying ancient technologies.

In small-scale experiments, the actor and observer are often one and the same person (Brysbaert 2003a). In the artificial system that the experiment represents, the actor and his/her actions create a *contemporary* relationship between human behaviour and material culture. This is of great importance when we consider that observing this relationship in its correct timeframe has revealed and established our knowledge about entirely lost past technologies. In the context of material studies, this approach also adds to the insight that techniques are in essence social productions (Lemonnier 1993: 3), and it shows how people (actors) may have transformed or manipulated raw materials into artifacts or usable products.

From this point of view we can identify experimental work as an agency-centred approach which, when viewed in its right context (i.e. contemporary reproduction), can help the archaeologist to understand technological aspects more fully (Ingold 1999: xi) and so see technology in its social context. Therefore, careful planning and accurate documentation of each step of the process is essential so that these steps can be reconstructed for more detailed observation and analysis after the experiment is over. Good documentation also forms the basis for refining the experiments. Strict monitoring and documentation, furthermore, provide the ingredients to test the experiments because the results of every published experiment, whether successful or not, should be reproducible.

When done on a small or reduced scale (see Coles 1979: 41-44) experiments do not have to be expensive or labour-intensive, so long as the experimenter realizes what limitations such experiments have. Finally, experimental work can be carried out by most people who have, on the one hand, the necessary understanding of and background in the material under study and, on the other hand, a good level of dexterity, or they can collaborate with those who have the necessary skills.

In a field such as painted plaster in the Bronze Age Aegean, where calling iconographical studies 'numerous' is a serious understatement, it is almost shocking to see the serious gaps in the study of its technology via experimentation, although much has already been achieved. I can only state very briefly why I think this is the case:

- not glamorous enough as a topic (in comparison to making pottery by hand or on the wheel, or metallurgical experiments).
- the techniques involved (or at least most of them) still exist so they can be studied in ethnographic contexts.
- the technical literature on painting *al fresco* is mainly based on what was done during the Renaissance period in Italy. This literature has been taken as *the* (only correct) literature on this topic, and as a sufficiently explanatory guide for any wall painting study, so the need to experiment did not emerge.
- the possible consideration that painted plaster techniques only had minor impact on its society, although the spread of these wall paintings in the Aegean Bronze Age context, combined with the application of social theory literature applied in this book, show a different picture.
- on the one hand, many of the original raw materials may not have been considered very valuable, but that may not have applied to Egyptian Blue, *murex* purple or *lapis lazuli*. On the other hand, value, not easily traceable in archaeological remains, may have been related to labour intensity of the production of paintings, or to the labour transfer and mobility of artisans, and others.

The fact that all these experiments go back either to technologically unanswered questions (and there are many more!) or to specific uncertainties related to materials used within the craft has, as its consequence, the realization that several aspects of these individual experiments brought about a range of different values. Of these, the educational value became quickly apparent, both for the experimenter and for the reader. Some also have specific scientific value, in that they test hypotheses or, rather, uninformed statements found in the archaeological literature (see shell experiments, procurement of raw materials, real ochres). Again, other experiments showed that certain techniques seem to be favoured over others in order to obtain a specifically desired effect (*sgraffito*). A last group of experiments resulted in the experimenter being able to formulate a set of criteria, in combination with comparative analytical work on actual archaeological material, on the ways that specific techniques can be detected within the material (*fresco* in dolphin and tripartite shrine).

None of this painting and plastering craftwork should be taken for granted or as an easy way of decoration. Much combined artistic, designer and organizational skill preceded the outcome of these painted programmes found in elite contexts. Our concepts of economic value may not be transferable, but in this context we may want to refer to the 'marketing value of skills'. This aspect is further explored in Chapter Seven.

# 7 Considering Material Culture and Social Identities

## Introduction

This chapter demonstrates how and why the material remains of painted plaster aid in shaping the complex social relationships and shifting social identities of people. It suggests that the human actor plays a central role as the connector between all relationships and any significant aspect of painted plaster technology, irrespective of its level or scale.

Chapter Seven brings together the data and results under discussion and looks at how these data may be interpreted in order to answer the questions posed, by means of applying the interdisciplinary approach as outlined in Chapter Four. This chapter, therefore, is divided into two subdivided parts. The first part answers the first research question or issue (see Chapter One), which looks at aspects of technological transfer between paintings found in the Aegean and those found beyond. Furthermore, it looks into the relation between the technological and iconographic evidence of these paintings. The second part takes the data to the social level of the second question (see Chapter One), which investigates the possible forces behind technological transfer and whether the archaeological evidence can tell us something about human agency and social identities within the sphere of craft specialization in the Late Bronze Age eastern Mediterranean.

While investigating the technological aspects of the transfer of the paintings and setting this investigation in its archaeological context, I deal with the issues of technological transfer and discuss whether the archaeological evidence may support and illustrate the actual direction(s) of this transfer. Then I take this a step further by integrating the painting technology into a broader framework, including its iconography. This suggested combination, as a way of seeing the entire craft, illustrates how this Aegean painting craft can be determined as a 'technological style' in its own right. As such, its technology and iconography became structurally interwoven over time as the craft developed. The second part of the chapter is subdivided into six sections, each unravelling the six relationships that have been determined and discussed in Chapter Two. Each of these relationships is discussed in detail in the light of the data presented.

## Technology and Social Agency: Archaeological Context

### Intra- and Inter-Site, and Regional Connections: the Material Evidence

As stated above, there are two opinions concerning the direction of the transfer of iconographic themes and motifs of Bronze Age painted plaster. There are those who believe that the painting tradition went from west to east, based on early figurative painted plaster found at Knossos and Phaestos and on much earlier material from Vassiliki, Myrtos Fournou-Korifi, Knossos and other EM sites. The first string impressions were known from MM IIA Knossos (Cameron 1975: 16). Much earlier painted plaster, however, has been attested at places such as Neolithic Çatal Hüyük (Mellaart 1967), and Aceramic Neolithic Hacılar (Mellaart 1970: 20) in Turkey, Middle Bronze Age Mari in Syria (Parrot 1958; see also Moortgat 1969: 69-84), and Chalcolithic Teleilat Ghassul in Jordan (Hood 1978: 48). None of this earlier material, however, can be positively identified as *al fresco* paintings. Scholars who suggest an alternative, namely that the tradition may have gone from east to west, argue that most innovations seem to travelled in that direction, a move that clearly implies the supremacy of the more advanced cultures of the Near East and Egypt in comparison to the 'secondary' cultures of the Aegean and Cyprus (for critique on arguments in the same vein, see Chapman 2005: 78, 80). Sherratt (1994: 238) stated that the west-east argument supports the view of Aegeocentric archaeologists who believe that most material cultural phenomena in the Aegean were indigenous innovative craft productions that subsequently moved to the east.

Taking the evidence and these arguments together, the second opinion looks more likely now that more studies related to iconographic comparison deal with both the Aegean and the Near East (Collon 1994; 2000; Crowley 1989; and others). Winter (2000: 745) and others have argued that the transfer of motifs and themes from one wall painting to another may not necessarily have gone directly but may have travelled via other media such as pottery. Textiles and seals may also have been partly responsible for such a transfer.

I believe, however, that the use of iconography alone as a method of discussing transfer of the painting tradition does not provide adequate evidence. First, several themes and motifs may have been 'invented' in one area, travelled to another and eventually returned in a different form back to the region of origin. Second, since we deal with fragmentary remains, our opinions may continue to change when new evidence appears. Third, the use of single motifs does not suffice to make solid statements, nor does it work to compare motifs applied onto a single medium (Winter 2000: 746). For instance, the early use of dados and spirals from Mari has been used as an argument for transfer from east to west. Other motifs such as the half-rosette (Bietak and Marinatos 2000: 42) have been used to make a statement about transfer from west to east. Equally, to argue that the spread of red ochre use or lime plaster is a criterion for technological transfer from one direction to another is simplistic. The point I want to make is that combining several parameters in research makes any statement stronger (see Chapter

Four). For instance, at Qatna we see the *combination* of a dado, a rocky landscape, date palm trees and spirals, all painted *al fresco* for which we have several strong indications. This is a strong case for Aegean motif combination *and* technology. Such a multivariate approach of iconography and technology may shed more light on the realities of transfer on any media, and of any theme, motif or technique as advocated in Chapter Four (Brysbaert 2004b).

In iconography, moreover, themes and motifs are free to move in all directions. They change, adapt to and become adopted by, possibly under the influence of personal taste, the media onto which they are applied, the artisans involved and socio-cultural preferences and limitations. Furthermore, Feldman (2003) argues that images such as attack scenes involving lions (metaphorically representing the ruler or military prowess, Feldman 2006: 76), or those depicting elite people on prestige objects and materials, are very much the constructs of elite patrons and artists with the intention of conveying specific meanings. This also means that politico-ideational factors may influence such iconography. I argue that these factors may play an equal role for technological aspects (see below) and that using technology to study transfer may be a more secure approach. Technology, up to a point, is intricately connected to the medium and, therefore, techniques cannot travel *that* easily via other media (*contra* Cameron 1975: 34). This is quite clear in adapting skills that transfer certain technological aspects onto another medium, for instance the depiction of flowers from miniature (on an ivory seal for instance) to large scale (on damp plaster). One technique involves, most likely, one person carving with very fine tools in soft layered material (ivory), while the other technique involves a group of people being able to paint with pigments suspended in water onto vertical still-damp plaster. All these differences (having made it simplistic) are far from easy to transfer without the necessary skills, observation periods, practice, knowledge, experimenting and feeling about the materials, and tools and equipment involved. As technology is thus more connected to its medium, technological transfer is more stable and it *has* to be, because it forms the backbone for the painting (or any art work for that matter) to be carried out successfully. Because of the interwoven nature of technology with its medium, its development, improvements and changes may be traced more easily. Moreover, the technology of painted plaster is immobile, executed on architecture; it is fixed to its locale. This book, therefore, integrates and combines iconography, style and technology as criteria to study transfer. The relationships between the different materials of painted plaster discussed here are based upon the macroscopic and microscopic observations made, combined with the results obtained by instrumental analysis, while also taking into account the experimental results that have been discussed in detail elsewhere (Brysbaert 2003a; 2003b; 2004a; 2007b).

1. *Macroscopically detected features* showed three main groups (Table 7.1). The difference between Groups 2 and 3 lies in features 2, 3, 5, 11, 13 and 15 which are less common in Group 2 (for the features and their description, see Chapter Six, Table 6.1).

| *Group 1* | *Group 2* | *Group 3* |
|---|---|---|
| EM Knossos | Palaikastro | Mainland |
| Monastiraki | Myrtos-Pyrgos | Tell el-Dab'a |
| Tell Sakka | Phylakopi | LM Knossos |
| | Miletus | |
| | Alalakh | |
| | Qatna | |
| | Tel Kabri | |
| | Hattusha | |
| **No *al fresco*** | ***Al fresco*** | ***Al fresco*** |
| | Less micro surface manipulations and technical aids | More colour and surface sophistication |

**Table 7.1** Three groups based on macroscopically observed differences in the painted plaster.

There was more colour sophistication (e.g. mixing) in later periods, represented by Group 3, and also a higher level of manipulation of detailed plaster and paint work. We could, therefore, conclude that the earlier material (Group 2, with the exception of Qatna and Hattusha), seems to demonstrate less sophistication than the later material of group 3. As outlined earlier, the study of the material from Hattusha was limited due to its state of preservation and, thus, could potentially fit in the wrong group, but it was only possible to work with positive evidence in this study.

Features 17 and 19 deserve a closer look. In the light of earlier work (Brysbaert 2003b on plaster dumps with potentially symbolically charged meaning, see also Bennet 2004: 99-100), the level of reuse and recycling indicated that the artisans understood very well the compatibility of the materials they used.

2. *Microscopic features, not involving measurements, (i.e. 1, 3, 10, 11)*, form the following two groups (Table 7.2). No *intonaco* from EM Knossos could be seen, yet Cameron *et al.* (1977: 167) reported on the existence of *intonaco* use at Knossos. Furthermore, *intonaco* is no longer considered a criterion for *fresco* work in Aegean Bronze Age paintings (Perdikatsis *et al.* 2000: 115). The interesting presence of the so-called 'black *intonaco*' (see Chapter Six) on the mainland and at Kea (Majewski 1973) may indicate a direct link between the Cycladic painting tradition (Immerwahr 1990: 103) and what happened on the mainland afterwards.

The two groups here represent chronological or geographical differences or a combination of both. We cannot state when the use of *intonaco* started but it was in use sporadically by LM IA (Phylakopi). The 'black *intonaco*' was in use later (LM IB on Kea) and continued until the end of LM III on the mainland. The number of plaster layers may indicate that on most sites reuse of the same wall surface for overplastering (and overpainting) was considered (see Brysbaert 2003b: Pl. XLII; but *contra* Cameron 1975: 291).

| *Group 1* | *Group 2* |
|---|---|
| Tell Sakka | Mainland |
| Miletus | Tell el-Dab'a |
| Palaikastro | Phylakopi |
| Monastiraki | Qatna |
| EM Knossos | Hattusha |
| | Tel Kabri |
| | Alalakh |
| **No intonaco at all** | **Some intonaco** |
| **Mostly 1, sometimes more plaster layers** | **Mostly 1, sometimes more plaster layers** |

Table 7.2 Two groups based on microscopically observed differences in the painted plaster.

3. *The plaster thickness measurements* show that there may be an ideal thickness with no greater deviation than 2 mm, but with certain local differences (Tell el-Dab'a, Thebes). There may have been a common understanding of how thick a given plaster layer had to be in order for it to provide enough separation from the backing support and enough 'body' to be sufficiently slow drying so that painting *al fresco* could take place (see Laurie 1988: 194). The evidence from Hattusha again is less conclusive. Its preservation often left only a few millimeters thickness of plaster, which was not conducive to comparative work. In any case, we can state what the mimima were for Hattusha, because a few fragments illustrated a thickness of plaster closer to that of other sites. The situation at Tell Sakka was very different and varied. Lime plaster often formed a two millimeter layer on top of the mud plaster or mudbrick, or paint was applied directly to a coat of mud plaster, while the other fragments had the usual thickness of around 0.8-1.2 cm.

4. *Pore and inclusion size* are treated together here, because there appears to be a relationship in the nature of both features. Both have a rather large spread of sizes on a micron level, but still fall within a certain range. These measurements, combined with the composition of the plaster, seem to have formed an 'ideal' mixture with an 'ideal' consistency, in use from the Second Palace Period (MM IIIA-LM IB) onwards. This agrees with the findings of Cameron *et al.* (1977: 148) who saw two peaks in the purity of the plaster, possibly related each time to the rise of palaces on Crete (MM IB-II and again MM IIIA-LM IB). This mixture looks as if it was completely standardized by the time the painted plaster tradition reached the mainland during LH IIB-IIIB as regards the more consistent sizes for both pores and inclusions (see Tables 6.5 and 6.6).

5. *Two groups could be distinguished, based on the pigment grain size* (Table 7.3), which forms a very useful indicator for connections (see Brysbaert and Perdikatsis 2008). Interestingly, the pigment grain size seemed to differ between the sites even though the technique of execution was the same.

| *Group 1* | *Group 2* |
|---|---|
| Palaikastro | Mainland |
| Miletus | Tell el-Dab'a |
| Monastiraki | Phylakopi |
| Hattusha | Qatna |
| Tel Kabri | |
| | |
| Larger differences between small-large grains: blue, red. | Homogeneous grain size for red, larger differences for blue but similar averages |
| No homogeneity for yellow and black, large size black for MI. No black at MO | No homogeneity for yellow and black. |
| | Strong homogeneity for green (TI, DB |

**Table 7.3** Two groups based on pigment grain size measurements.

This may indicate differential labour input for the pigment production between Groups 1 and 2, with more time spent on pigment grinding and refining in Group 2. It may further illustrate that the pigments may not have been produced by the painters themselves. I will come back to this point later. The fact that black and yellow show far less homogeneity in both groups may have been related to manufacturing difficulties, or to the fact that their fineness was unimportant. However, green, a mixture of blue and yellow, forms a striking example of homogeneity at both Tiryns and Tell el-Dab'a (Table 6.4). Direct contact between both sites based on this factor alone cannot be deduced, but it helps to demonstrate the connection of the Dab'a paintings to the later Aegean mainland paintings. This connection, rather than a connection with the LM IA paintings, is more in line with Bietak's redating of these paintings (see Cline 1998) and now fits with the radiocarbon dating of the Santorini eruption at 1628–1600 BC (Friedrich *et al.* 2006). Thus, there seems to be more labour input into the pigments on the mainland, Tell el-Dab'a, Qatna and Phylakopi, in comparison to the other sites. This is consistent with the points made above (1, 2, 4). The odd one in the groupings is the later material from Hattusha which would fit better with the later Greek mainland and Qatna material chronologically. The differences between the two groups may be chronologically or geographically based or both, as appears to be the case for points 1 and 2.

6a. *Table 7.4 presents the XRD results of the blue pigments* and where relevant, these are complemented by SEM, EPMA, TXRF, and MRS. The earlier phases and sites appear to be characterized by amphiboles and a limited use of Egyptian blue. By contrast, no amphiboles were used at Tell el-Dab'a, Gla, Orchomenos, Tel Kabri, Qatna and Hattusha.

Materials transfer with early use of amphiboles at Miletus is clear and its later use at the mainland paintings is suggested (Chapter Six and Brysbaert and Perdikatsis 2008). The use of Egyptian blue as a chronological indicator has been put forward in the past

| *Pure Egyptian blue (EB)* | *Amphiboles (R)* | *EB x amphiboles* | *Other* |
|---|---|---|---|
| Gla ** | Knossos? * | Knossos | Thebes-indigo |
| Orchomenos * | Palaikastro | (Akrotiri) | Hattusha-azurite |
| Tell el-Dab'a *, ** | Phylakopi * | Phylakopi | Gla-*lapis lazuli* |
| Hattusha | Miletus | | Alalakh-indigo |
| Tel Kabri | Mycenae ? *, ** | | |
| Qatna | Tiryns ? | | |
| Tell Sakka | Thebes ? * | | |
| Alalakh ? (Woolley 1955:233) | (Akrotiri) * | | |

Table 7.4 Summary of all analyzed blue pigments. Egyptian blue was used on all sites where blue was analyzed apart from Palaikastro and Miletus. * = presence of tin in Egyptian blue, ** = presence of arsenic in Egyptian blue.

(Philippakis *et al.* 1976), but has been rejected by Dandrau (1999: 17-18). The table above and the data presented in Chapter Six, however, show that chronological factors may have played a role in the use of amphiboles. This may need to be explained, however, in the framework of technological transfer in a regional or interregional fashion over time, possibly expressing different levels of access to pigment materials and maybe even expressions of taste. This then may possibly be seen in combination with economic factors, while previous authors saw the use of Egyptian blue purely related to economic factors as a result of the Thera eruption (Philippakis *et al.* 1976: 150-53). Again, a connection between the Cyclades and the mainland is formed, but also one between Crete and the Cyclades and also with Miletus, which is extremely important in the discussion on the transfer of techniques, materials and knowledge. Considering that we can attest to the earliest amphibole use being at Knossos, and possibly also at Palaikastro in MM IIIA, we can now see evidence of materials, native to the Aegean, moving to places outside that area. This may suggest that artisans were directly responsible for this, taking their pigments with them, i.e. to Miletus, but also to the Cyclades and possibly to the mainland. At the same time, the elite may indirectly be held responsible for what artists could use and take with them. As such, they were creating, each in a different local way, a form of inequality in access to materials, labour, knowledge and produce.

The presence of tin and arsenic in Egyptian blue deserves discussion for several reasons, even though the results were not obtained under the best conditions (see Chapter Six) (see Brysbaert 2003b). Firstly, tin was found in Egyptian blue as early as MM II at Knossos (Cameron *et al.* 1977: 141, 176), and also at Mallia (Poursat and Loubet 2005). Basically, this means that copper alloys with tin may have been already recycled for pigment production during MM IB-II on Crete, a practice that would span a period from then until the LH IIIB period. Tin also occurs in the Tell el-Dab'a material, which, despite the redating of the material, may still be the earliest presence of tin in copper alloys in Egypt (see El Goresy 2000: 65). This important finding implies, first, that the early presence of tin in Egyptian blue may be a positive indicator that Egyptian blue was produced in the Aegean (Foster 1979: 11 for Cretan origin of Mycenaean faience workshops), independent of what was happening in the Near East and Egypt, since no tin has been found in Egyptian blue in Egypt before the 18th Dynasty. The earliest tin in Egyptian blue was reported from the Late Bronze Age (15th–14th centuries BC) at Ras Shamra, Syria (Matoïan and Bouquillon 2003: 334-38). Second, it may also indicate that the Egyptian blue used at Tell el-Dab'a was imported from the Aegean by travelling artisans (Brysbaert 2002b: 102). Why they did not bring amphiboles is not clear.

A second issue is the arsenic in the copper, noticed here in the material from three later sites. This is in contrast to the work in Egypt by El Goresy (2000: 64-67), who sees the alloy of copper with arsenic as an early chronological indicator which clearly disappears from the 19th Dynasty onwards.

Apart from the discussion above, tin in Egyptian blue is again an indicator of recycling, a trend so strongly evident within the Aegean Bronze Age that it demonstrates, in my view, far-reaching knowledge of materials and their properties by their manipulators (Brysbaert 2003b: 176). How other pigments fit into transfer discussions may be summarized as follows.

6b. *The XRD analysis of the* ***ochre-based colours*** *red, orange, yellow and* brown shed no light on whether these pigments were carried by travelling painters or found locally, although they were most probably locally collected in view of their omnipresence. Trace element analysis may reveal more.

6c. *The* ***black*** *mineral pigments* appear to be both haematite and pyrolusite. Both were found at Phylakopi, the former also at Thebes and Qatna, and the latter at Orchomenos. This may again show contacts between the Cyclades, the mainland and the east through the art of painted plaster.

6d. ***Purple***, ranging from deep red to a bluish-purple and magenta hue, was found at all sites apart from Orchomenos, Tell el-Dab'a and Monastiraki (see Table 6.7b, Chapter Six). It seems that red mixed with black organic particles was the most common way of painting purplish hues, practised on Crete, Miletus, Tel Kabri, Qatna and Hattusha, and on the mainland (Thebes), next to the use of dark red haematite and some occurrence of blue and red particles (Hattusha, Alalakh, Tel Kabri).

Transfer of knowledge may be seen in this evidence, especially between Miletus and Crete (Knossos sample 39: Cameron *et al.* 1977: 142), where both haematite and riebeckite are combined for purple. The main surprises are the possible *murex* purple and *lapis lazuli* found at Gla (Brysbaert and Perdikatsis 2008; Brysbaert 2006: 252). The scenes painted with these supposedly costly materials were found in storage areas of the apartments rather than, as one might expect, in a megaron, the centre of public and/or religious activities within a palace or, in the case of *murex*, even decorating the dress of an important female figure, as at Xesti 3, at Akrotiri (Doumas 1999: 136-40; Chryssikopoulou and Sotiropoulou 2003). How can this finding be interpreted? Do the paintings in these less important areas of a building complex belong to a trial area for a painter? Was that room decorated upon demand? Did the room have a different function? Was *murex* purple not that exclusive by that time, but then why does it not appear more often? For the moment, no easy answer can be given for this odd find. The use of *lapis lazuli* has been previously reported for a grey surface on the sarcophagus of Ayia Triada, which may have represented a discoloured *lapis lazuli* pigment, but no analytical data are available for this analysis. Interestingly, this pigment, compatible with *al fresco* painting, was not used more frequently, even though we know of the presence of *lapis lazuli* in the Aegean at Thebes (Porada 1981–82). Can we assume a link between the recycling of the seal stones found at Thebes and the pigment use at Gla? In any case, the most exciting aspect of finding *lapis lazuli* used positively on painted plaster in the *al fresco* technique is the fact that its earliest use can now be dated to minimum 1800 years earlier than the next early use of this pigment known on painted plaster (see bibliography in Brysbaert 2006).

In summary, the information above strongly suggests that technological transfer of both knowledge and materials did take place. It has also indicated directional information, but it cannot simply be summarized as one single straight route. Taking together all the evidence of my own research, the following may be stated:

- The direction of technological transfer seems to go from west to east.
- Most innovations in terms of techniques, surface manipulation, plaster mixtures, pigment mixing and refinement seem to start on Crete (see also Cameron 1975; Cameron *et al.* 1977; Perdikatsis *et al.* 2000; Dandrau 1999; 2000; 2001).
- There is no uniformity on Crete in technological sophistication during one given period: for instance, we see differences in the range of colour palette and surface manipulation at Knossos and Monastiraki.
- There is no one straightforward transfer of materials and techniques from Crete to the mainland. Certain aspects on the mainland are seen in the Cyclades first before they 'arrive' on the mainland.
- Miletus and Alalakh seem to fall most often in the same category as the early Minoan sites, both for material and techniques.
- Tell el-Dab'a, Qatna, Hattusha and Tel Kabri seem to fall most often in the same category as the later mainland sites despite the earlier date of Tel Kabri.
- Tell Sakka did not show any indicators for the *al fresco* painting technique in the plaster manipulation nor in the use of plaster since several of its paintings were not done on (pure) lime plaster (*contra* Feldman 2007: 50). This technological divergence, combined with additional large differences in style and iconographic themes, indicates a different painted plaster tradition.
- Monastiraki did not show indicators of *al fresco* painting so the earliest *al fresco* paintings seem to come after MM II, while the first string impressions are affirmed at Knossos in MM IIA (Cameron 1975: 16). So, by the start of MM III-LM IA, or in its transitional phase, the indications from sites

such as Knossos, Phaestos, Palaikastro, Phylakopi, Akrotiri, Miletus, Tel Kabri and Alalakh are that *al fresco* painting was in use as a technique in the larger eastern Mediterranean, and it originated on Crete.

- Despite local and regional variety, which is to be expected on this material because of its immovable nature (and partial reliance on local supplies of materials and maybe labour too), the MM III-LM IA phase sees a whole package of technological features appearing on a range of sites that had painted plaster. These features vary in intensity between sites, but one can still speak of a package of features strongly indicating the presence of *al fresco* painting more or less exclusively. Moreover, materials moved clearly between places and non-visible knowledge was expressed materially over long distances. These are indicators for a common level of knowledge and a set of materials that travelled, despite the local varieties that one finds. In the light of these findings, it is very important to acknowledge both similarities and differences in the material culture remains from all sites involved. Only this way can we, subsequently, try to understand the technology as fully as possible and to grasp the implications that these phenomena may have upon our interpretation of the level of contact between groups of people within their chronological and geographical context.

Two more points need to be made. First, the discussion about technological transfer in this context is closely connected to the debate about the absolute date of the Thera volcanic eruption. The material in itself, however, is a crucial indicator in its own right. If one follows the high chronology in the footsteps of Manning (1999: 135-36) and others, now being strengthened by the recent radiocarbon date of an olive tree 'time capsule' (Friedrich *et al.* 2006), and Manning *et al.*'s (2006) recent short-lived samples dates, Alalakh, Tell Sakka, Tel Kabri and Miletus fall within the same chronological band as the LM IA sites on Crete. Tell el-Dab'a, Qatna and Hattusha paintings are of a later date. This means that the former group of sites in the east was later than the MM III paintings in the Aegean, which clearly showed enough signs to determine the *al fresco* technique combined with figurative iconography and its spread from then onwards. In any case, several technological features that led to the perfection of the *al fresco* technique were already tested at Knossos during the MM II phase. If the low chronology is followed, these sites in the east could be considered older than the LM IA ones (M. Novak 2003: pers. comm.), so they then *may* have influenced the Minoan MM III material. The material itself, however, answers this debate in a quite simple way. One part of this answer lies in the evidence outlined above and in Chapter Five, the other lies in the fact that over 80 sites with painted plaster depicting this iconography exist in the Aegean, with only six outside the Aegean (Tell Sakka does not count). *If* the technological transfer went from east to west, two problems arise. First, on which early site did this tradition start? The choice can only fall between Alalakh, Tel Kabri and Miletus, because Tell Sakka has a different iconography and technology, and Qatna, Hattusha and Tell el-Dab'a are later. None of the earlier ones has enough (preserved) material to explain the transfer of iconography and the sudden 'boom' of figurative painting in the Aegean at the start of MM III/LM IA. Second, why would this painting tradition become so well spread within the Aegean once it is introduced there, but not spread in the place of origin itself?

Since painted plaster is considered an elite art, it is difficult to imagine that eastern artisans would carry out so many paintings all over the Aegean within a relatively short time span, without leaving any trace of their activities in the otherwise tight recording system in the eastern Mediterranean palaces that covers the artisans' whereabouts (Zaccagnini 1983: 249). Sasson (1990: 23) states that no mention has been made in the Mari tablets about any aspect of its wall paintings, and Winter (2004: pers. comm.) was not aware of any tablets in any of the eastern or Egyptian archives, whether Ugaritic, Hittite, or Akkadian, mentioning painters as artisans. Linear B tablets do not mention painters either, but they deal mainly with metal working, and textile, oil and wine production, as prestige goods, all of which fall under palatial control, even though some evidence indicates activities within these industries that are not controlled by the palace administration. It has been pointed out that hardly any textual evidence refers to the deployment of prestige goods outside the Mycenaean polities (but see e.g. Bech Nosch 2000; and Landenius Enegren 2000 for Knossos), and suggestions have been made that such transactions may have been kept on perishable materials such as parchment (Shelmerdine 2001: 361-62). Dakouri-Hild (2005: 210) also stated that some crafts taking place at or near the palace, and thus under its strict control, may have been so closely tied up with the palace administration that they were not recorded in the same way, i.e. on clay tablets or, indeed, were not recorded at all. If these suggestions are true, they would stand in contrast to the recording systems in the eastern Mediterranean palaces, where clay tablets seem to have been *the* material onto which all activities, records, transactions and letters were written. The small wooden diptych with ivory hinges, and possibly covered with wax to receive the text, found at the Uluburun shipwreck, is another indication that keeping records was not just done on clay tablets. In addition, if we assume that the painting transfer came from the east and then boomed in the Aegean, one may argue that the technology was handed down to Aegean painters who then spread it all over the Aegean in the short time span mentioned. But being an elite art form, I cannot support the suggestion that such a complex craft, which involved quite an intricate set of motor and other skills in order to be able to carry out a decent painting, was handed down that quickly especially when, most likely, it was under the control of the elites.

The second point is the relationship between the Aegean and eastern Mediterranean paintings and the paintings from Mari (Amiet 1980; Parrot 1958). These, whether considered contemporary with Alalakh level VII paintings or not, have often been associated with the latter as being painted in the same technique, illustrating a similar iconography, or at least influencing the idea or fashion of painting on plaster (Knapp 1998; Winter 2000 and many more; most recently Bergoffen 2005). In any case, these paintings do not form part of a well-established Near Eastern wall painting tradition based on presently known evidence (as stated by Manning and Hulin 2005: 276), since there exist hardly five incidences of painted plaster of the relevant period known so far (see Moortgat

1969; Nunn 1988). Despite the fact that there are some iconographic similarities (the dado painting with spirals mainly), this is the absolute end point of comparison (*contra* most recently Manning 2005). Many paintings at Mari have been executed on mud plaster (see Moortgat 1969: 72 for the paintings from the Audience Chamber), covered with a thin wash of white material, possibly lime (we have no analytical evidence of this) and can, hence, never have been executed in the *al fresco* technique (confirmed by Parrot 1958: 109). At most, one can state they were done *al secco*, not in tempera (Feldman 2007: 58), since no analyses are known to confirm the presence of egg yolk as a binder necessary for tempera painting. As a result, they cannot be compared with Aegean and eastern Mediterranean paintings where the point of comparison is precisely the *al fresco* painting technique, combined with a certain iconography and style. It is also quite clear that the human figures depicted on the paintings at Mari, as is also the case with the figures found on the Tell Sakka paintings (whether animal or human), have nothing in common with those on the paintings under study in this book. In fact, Moortgat (1969: 83) mentions the presence of goats against a tree as a scene found at Mari, which is known from both Tell Sakka and now also from Tell Burak (Lebanon, Kamlach 2007: pers. comm.), but not elsewhere in the Aegean or eastern Mediterranean. Therefore, I believe that the Mari paintings (and the Tell Sakka paintings for that matter) need to be seen within their own context, from technological, iconographic and stylistic viewpoints.

Thus, based on my own research and several earlier technological studies, and considering the *chaîne opératoire* on production processes, the technological transfer of the production of painted plaster seems to have moved from west to east, but several impulses may have been going in multiple directions locally. This was the case for instance between the Cyclades and the Greek mainland, from Minoan Crete to the Greek mainland, and from Crete to the Cyclades. Such back-and-forth interactions between different regions have also been confirmed in other crafts such as pottery (Knappett 2003). Even the connection between painted plaster from Boeotia and sites in the Argolid (eastern Peloponnese) with painted plaster is visible in a whole range of features. This is supported by one of the trade routes described by Graziado (1998: 33). Consequently, it is not far-fetched to assume that palatial plan similarities between sites in the Argolid and Boeotia may support common traits in their decoration too (Younger 2005).

The concept of the *chaîne opératoire* can essentially be seen in two other areas apart from production: distribution or spread of objects and their consumption. Applying this to painted plaster, we see that the production and distribution of the entire end product actually fall together, because the paintings are made *in situ*, thus determining their distribution while they are produced. Their consumption then follows the production and distribution patterns. An elite's consumption of luxury items, in order to confirm and reaffirm their social status versus their society and other elites, was already a well-established behavioural trait in the Near East. These consumption behaviours were evident from as early as the Uruk period via the use of monumental architecture and large-scale reliefs (Ross 2005), but were not known through painted plaster. So, one could say that the *chaîne opératoire* for production, distribution *and* consumption of the painted plaster went from west to east, even though the consumption of luxury goods by the elites in general was already a known strategy in the east.

In this context I would like to state that there is, in fact, not one single *chaîne opératoire* system operating on its own where composite objects are concerned, such as painted plaster (Brysbaert 2007a: 328-35). As soon as there is a level of cross-craft interaction involved, there is a crossing of *chaînes opératoires* as well. This is best illustrated by the fact that pigments and dyes may be made for more than one craft, for example, painted plaster, pottery painting, textile dyeing, etc. This crossing of *chaînes opératoires* thus takes place at the material and social level (Ingold 1993).

The *chaîne opératoire*, furthermore, needs to be seen as (1) the processes involved in the transformation from raw materials into end products, and (2) the human involvement in each of these transformations, thus forming the social aspect of technologies. However, since objects and humans are united in creative processes (Kohring 2006: 101), it seems unnecessary to separate both aspects of the *chaîne opératoire*, just as it seems essential to make clear that both aspects are implied, not just the first. One can illustrate this point by looking at how closely the plasterers and painters would have had to work together in terms of materials and timing. Very smooth communication would have been required between them in order to work speedily and accurately to get the *al fresco* work done. Another example relates to the potential interaction, both professionally and socially, between artisans of different crafts in areas of palatial workshops, or between artisans of different age groups, and how they might exchange professional knowledge whilst also forging social relations. These examples can be extended, for instance, into consumption, relating to how elites and commoners potentially interacted over restrictions as to who could see the paintings and also the eventual viewing of them, and even between the elites and the (absent) artisans when viewing the work in progress and upon its completion.

Looking at all three aspects of the *chaîne opératoire* (production, distribution and consumption) gives more opportunity to understand the entire picture and its social implications. For example, the fact that the consumption of luxury items as a strategy existed in the Near East before it did in the Aegean may have influenced similar Aegean strategies and, moreover, it may have allowed for an easier acceptance of the use of painted plaster as a luxury item, even though its style, character and technology came from the Aegean. A certain level of familiarity of the local population with something new seems to have been one of the keys in accepting new concepts or objects (Ross 2005).

In conclusion, the movement of technological transfer from west to east is supported by the evidence of a combination of several factors together, namely that the *al fresco* painting technique and the movement of certain pigments, both strictly relating to the medium of relatively pure lime plaster, eventually travel from west to east. With this last statement I stress that I *only* refer here to *technological* sharing and transfer (see Figure 7.1 for Greek mainland). Feldman (2006: 4) did not think that 'attributing' the first use of this technique to the Aegean was possible by employing instrumental analysis, because, in her study, she did not take into account objects or art forms that

| *Site* | *Pigment grains size* | *Primary calcite grain size* | *EB, ochres, carbon black* | *Plaster mix* | *Painting technique* | *Amphibole blue* |
|---|---|---|---|---|---|---|
| MY | • | • | • | • | • | • ? |
| TI | • | • | • | • | • | • ? |
| OR | • | • | • | • | • | |
| TH | • | • | • | • | • | • ? |
| GL | • | • | • | • | • | |

**Figure 7.1** Technology chart of shared and transferred elements for the Greek Mainland.

| *Site* | *Dado* | *Friezes* | *Architecture* | *♀procession* | *Battle* | *Bull leaping* | *Boar hunt* | *Fig/8 shield* | *Sphinx* | *Other* |
|---|---|---|---|---|---|---|---|---|---|---|
| MY | • | • | • | • | • | • | | • | | Ikria |
| TI | • | • | • | • | | • | • | • | • | Deer |
| OR | • | • | • | | • | | • | | | |
| TH | • | • | • | • | • | | | | ? | |
| GL | • | • | ? | | | | | | | Dolphin |

**Figure 7.2** Iconography chart of shared and transferred themes and motifs for the Greek Mainland.

were immovable. The figurative iconographic transfer, as stated above, is far less stable and moves much more easily, even though many themes and motifs are shared (see Figure 7.2 for Greek mainland), and may have already moved independently, to some extent, via other media.

## Relationship between Technology, Iconography and Style

Late Bronze Age Aegean iconography consists of male and female figures, animals and plants, landscapes, seascapes, etc. Despite several iconographic parallels, however, its style may be distinguished from Egyptian and Near Eastern paintings. Ross (2005: 327) has convincingly argued that imagery such as representations on walls and stelae in the Near East was not meant to be art in the aesthetic sense, but had, as its main 'raison d'être', to convey information and ideas representing order and organization in the society. I believe we can extend this to the Aegean Bronze Age world in the light of the discussion developing here. Moreover, we should also consider how Aegean representations present actual reality at face value, or whether these form ambiguous, selective (Palyvou 2004) or hybrid representations (Chapin 1997: 17; 2004c: 60-61) or even misrepresentations of events and conditions, in order to manifest specific elite ideologies of status expressions (Ross 2005: 328). Such thoughts, however, only make sense if the imagery is accessible to some extent, since the message within was meant to affect the public in the first place.

I argue here for a very strong interconnection between technology, iconography and style with respect to the paintings we have so far studied in some detail. In fact, I see this interconnection as so strong that I do not believe that one would have taken place without the other. This interconnection, moreover, did not take place in a vacuum either, as we see from the issues below:

(1) By the start of MM IIIA the plaster composition had reached its optimum mixture of almost pure lime plaster, conveying a very white, shiny and fine-textured plaster when worked properly and when dry. By contrast, wet lime plaster looks dull and grey while it is being painted (personal observation).

(2) By the start of MM III, the first and completely developed figurative paintings appeared at Knossos (Immerwahr 1990: 21). This has been seen as the final stage of an evolution in cultural development, a linear formation process insofar as its stylistic development is concerned (Blakolmer 1999: 41-42 but cf. p.49). Despite this view, no trials exist depicting humans in this medium, a phenomenon difficult to explain from an iconographic point of view (but see Blakolmer 1999: 49). Figurative art in a specific style had already existed for much longer on other media such as seals (Immerwahr 1990: 26), textiles and figurines (Fitton 1984; Getz-Preziosi 1985). Human figures were also found on contemporary pottery from the Cyclades and the mainland but far less from Crete (Immerwahr 1990: 18). One can, furthermore, suggest that the artistic changes such as those that took place at this stage (i.e. choosing a new medium, large-scale representations, etc.), act as an indicator for changes in political, social, economic and religious circumstances (Ross 2005: 330, see also Chapter Five). This is supported by other visual forms and expressions of control and leadership, namely monumental architecture, relief stone vessels, writing, seal use, etc. In fact, the imagery thus supported and supplemented the physical presence of the leader, or what represented this leadership (of worldly and/or divine nature) in a community that became socially more complex (see also Bennet 2001: 34, on image analysis at Pylos), i.e. going from a more heterarchical socio-political structure in the First Palace Period towards a more centralized hierarchical structure in the Second Palace Period (see summary on this by Chapman 2005). It is obvious from its style and iconography that figurative art in the Aegean was influenced by Egyptian art (Blakolmer 1999: 49; Warren 1995: 2), but continued to have its own distinctive character. That Egyptian art and influence cannot be the total answer to Aegean figurative painting is clear from the fact that the one palace, Mallia, that had intensive contact with Egypt from MM IB-II onwards (Watrous 2001: 201-202), did not produce much in terms of paintings later on.

(3) Minoan Crete, especially in MM IIIA-LM IB, is part of the international trade network (but see Manning and Hulin 2005: 278-79, 290) in its constant need for metals and other goods, while exchanging its own less recognizable products (wine, [perfumed] oil, shoes and possibly textiles) in return. The same period sees several interesting innovations and changes. Even though several of these started during MM IB-II, they became greatly expanded in the form of extensive monumental architecture incorporating proper planning of open spaces and drainage (Dickinson 1994: 60). There

is evidence for luxurious decoration of the palaces and extensive use of wheel-made pottery (Knappett 1999: 101), textile dyeing, etc. These can be seen as technological adaptations to new secure circumstances whereby the existing elites had groups of artisans working in their service, providing for their personal and professional needs (see Michaelidou and Voutsa 2005). The elites who realized the value of their highly skilled artisans could control these workers closely, or tie them to the palatial administration by providing them with agricultural produce or land as remuneration for their work (Shelmerdine 2001: 356, 359). Such a form of dependency on the economically strong elite may have provided artisans with the necessary stability they needed, while they worked on improving their final product by employing the best possible materials and techniques for optimal performance. This may have led to forms of standardization (see Berg 2004), but also of conservative use of materials and techniques, once perfected. Technological changes may also have taken place as a result of the growing contacts with the eastern Mediterranean regions and their possible cross-craft interaction (Brysbaert 2007a). Such growing international contacts are seen in changes in the iconographic vocabulary (see the adapted sphinx depictions in 18th Dynasty Egypt) and, in my view, should not be limited to hybridization only in iconography. For instance, the string impressions on Minoan and Mycenaean painted plaster may have been an adaptation of the red string lines in paint on Egyptian paintings, which may have been seen by specific people and passed on to others.

(4) We also see a growing dependency of the elites upon the consumption of ever changing luxury goods and/or services, in order to achieve the constant affirmation of their social status in the eyes of other elite groups with whom they compete for power (on agent-based competition: Axelrod 1997) and in the eyes of their own society. This phenomenon manifested itself in monumental architecture with large-scale painted decoration (see Cameron 1987 on Knossian palatial thematic *system*, my emphasis; Hägg 1985 on pictorial palatial programmes), in the potential ritual character of these paintings in relation to these elites (Kritseli-Providi 1982; Gesell 2000; Ayia Triada sarcophagus: Burke 2005; Kontorli-Papadopoulou 1996 on paintings with religious meaning; most recently Wright 2006: 56), and in the possibility that artisans may have been forced to be present during ritual feasting in elite contexts (after Dakouri-Hild 2005).

I believe that the coappearance of figurative iconography, expressed in its peculiar Aegean hybrid style with its specific sets of ideological messages and the perfection of the plaster mixture combined with a broader colour palette, is not coincidental, given the concurrent changing socio-political context during the formation of the second palaces on Crete. I believe that this iconographic expansion on painted plaster could not take place earlier than the moment this plaster mixture was perfected. First, the plaster wall had to stay damp long enough for such intricate (large) scenes to be executed *al fresco*. This requires the necessary lime plaster mixture and ambient environment. Second, even pure wet plaster mixtures are grey and dull in appearance. The more filler is added (except marble dust or white calcium carbonate), the more its dry and overall appearance will be removed from pure white.

Even though prestige items, in general, tend to be socially exclusive and politically loaded, in this specific case one could argue that the actual economic value of the materials used did not indicate any status that matched the significance or meaning of the iconographic representations ordered by the elites. However, its pure scale, its labour input, artisans that came from afar (see now also Manning and Hulin 2005: 270-71), the likely elite control over the import of raw materials, knowledge of artisans and the control over where such paintings could be executed were all added values to the thematic programmes. As such, I argue that technology, style and iconography had messages to convey to the people who were to view and experience these paintings. Even the control of the elite over who was allowed to see these paintings added extra value and meaning, since the medium of painted plaster was not transportable (in contrast to seals, textiles, pottery) and could thus only be seen *in situ*.

The plasterers may have been experimenting with the plaster mixture and the lime-filler ratio until they reached the almost pure calcium carbonate lime plaster. The actual difference between primary and secondary calcite in a matrix, however, is visible at a microscopic level and to a lesser degree at a macroscopic level where calcium carbonate filler has been added to the slaked lime *intentionally* before it was applied to the wall. This filler (primary calcite) of crushed limestone and crushed (*Murex)* shells was wholly compatible with the slaked lime (secondary calcite), and formed a keying and strengthening factor, one that fully integrated visibly with the matrix. Another point is the improved mastery of surface manipulation during this period by careful surface floating and flattening (see first experiment of Chapter Seven in Brysbaert 2004b; 2004a).

Three distinctive styles of Aegean figurative art have been recognized: miniature, life-size and three-dimensional rendering of human figures. The last is not discussed here, since no such samples were taken (but see Kaiser 1976). The other two styles both occurred from the earliest phase of figurative painting onwards and both coexisted at Knossos, Ayia Arini, Pylos and Akrotiri. They also existed widely in the Aegean and beyond (Tell el-Dab'a and Tel Kabri). Found in elite contexts, all these paintings had the active power of bringing together people from a specific level of society by generating a common social identity through such an expressive form of vocabulary.

Unfortunately, no technological features (see sections 1 and 2 in Chapter Six) of these two types of paintings can be compared. On the one hand, samples may have been from both types, but they could not be distinguished since their original position in the painting was no longer preserved, and areas which were clearly miniature or life-size were not sampled since they often had potential for further iconographic reconstruction. On the other hand, I argue that painting life-size and painting miniature may not have differed that much technologically. The main parameter of importance is the time the artist needs to carry out the painting *al fresco* and this ties in with his expertise, dexterity, motor skills, and co-ordination with others (both plasterers and other painters) to execute this work. Therefore, painting the level of detail of miniature paintings may have taken the same amount of time as painting life-size. Some life-size paintings were also very detailed, such as the ladies' skirts at Knossos or the turtles at Qatna, but surface preparation with grids to aid the repetitive painting of specific motifs will have speeded up this process (Shaw 2000a; 2003 for Knossos, Qatna: personal observations).

From the several hundred plaster analyses carried out, there was no chemical evidence for two different groups of plaster mixtures. As regards pigments, one could suggest that a larger grain size may have been employed for life-size paintings since less detailed and fine work is carried out. One should remember, however, that the grain size stands in inverse relation to its staining power for most pigments apart from the blue. Thus, the finer the pigment grain size of a given amount of pigment, the more surface may be covered. As Chapter Six has made clear, the size of the pigment grains of the same colour at a given site does vary, but we cannot go as far as correlating this grain size and the painting type until we can clearly prove that the specific samples come from different types of paintings.

Apart from the styles discussed above, I believe it is also possible to fix Aegean paintings as a *technological style* in the same way as it is defined by Lechtman and Steinberg (1979) and Lechtman (1993). The material discussed has enough traits to be recognized as culture-specific (Lechtman and Steinberg 1979: 154). The technique of painting *al fresco* does not exist in Egyptian painting until much later than the Bronze Age and no examples were found in the Near East, apart from the ones discussed within this study. It seems clear, thus, that the *al fresco* technique developed locally in the Aegean first before it spread. Moreover, I want to argue that decoration painted *al fresco*, in contrast to *al secco* painting, is so intricately interwoven with its technological backing support component (*in casu* the lime plaster coat) that it cannot be divorced from it without losing its structural form. As has been seen in the *al fresco* painting experiments, the application of the paint occurs on damp plaster, which, at that stage, is grey and dull rendering, in turn, the appearance of the applied paint greyish and dull. Hence, one can hardly speak of an attractive wall decoration at that stage. As the wall dries (the surface 'crust' may take a few hours, while the entire wall may take months: J. Coopman 2004: pers. comm.), however, it turns bright white and the colours take on their anticipated hues. In fact, the more the wall dries out over time, the more radiant the colours become. Thus, the painted decoration is only complete when the actual wall, its structural support, is dry. This means that not just the plasterers and painters were actively engaged in making the painting, but the structure itself as well. The actual painting is thus actively engaged in the power of coming to exist and of inviting admiration. From the moment that the wall is dry, the painting cannot be taken off without structurally damaging the wall (scratching the paint off the plaster will also remove the top surface of the plaster or the plaster entirely. *Al secco* paintings can be removed from their plaster backing by dissolving the medium and so removing the pigments). As such, decoration and structural support in *al fresco* paintings are completely interwoven. Some people are determined to find *al secco* elements on these paintings, but this does not at all contradict the fact that many *al fresco* elements have been detected (see Chapter Six), up to the final details (see the Tell el-Dab'a and Thebes examples for instance). I suggest that such *al secco* additions, *if* really there (but see Chapter Six with references to Serpico and White 2000a; 2000b; Newman and Serpico 2000, on the difficulties in analyzing aged organic compounds), were potentially carried out later on, maybe at a stage where the owner wanted a 'refreshment' of the visual scene. In any case, none of the *al secco* additions interferes with the technological style of these paintings, nor with any of the practical or social implications of the *chaînes opératoires* involved in executing these paintings. *If*, in a specific case, there are *al secco* additions, they do not contradict the fact that most of these paintings are carried out or started *al fresco*; instead, they are just that: additions.

## Social Agency and Technology: Social Context

Human actors and their relationships with both people and their material world form the key to understanding the social dynamics behind any action. Moreover, all actions are culture-specific and, therefore, have to be contextualized to make sense of them. While the previous part dealt with the specific material in its archaeological context, this part shows how and why things were done by placing the human actors in the scene.

Technology, in the first place, is a social process, because it involves people's daily activities, their relationships, their beliefs. Via technology, they negotiate their identities, confirm or change them. In fact, I see such negotiation as a technology in its own right. Moreover, the connection between technology and social activities demonstrates that technological change may evolve into or affect social change and dynamics and vice versa (see Wedde 2005 on Minoan and Mycenaean shipbuilding), since technology is never autonomous or external to society and is thus shaped by its social context. Technology, as an ever present expression of material and social actions, is thus both a physical/material (matter) and social/mental (mind) phenomenon (Gravina 2006: 4), while people's bodies form the interconnector or facilitator between the two.

As outlined in Chapter Two, the questions related to people's actions are discussed and answered within the framework of six relationships. Much help in answering these questions came from the experimental work, literature on traditional crafting and the interview conducted with a professional traditional plasterer. From this, it seems that many traditional crafting aspects of plastering do not differ that much in relation to the locality where the work was performed.

### Artisans in Relation to the Elite

Emerging from the archaeological material and my analyses, I have argued that the painted plaster tradition developed on Crete and initially moved from there to the Cycladic islands (Phylakopi, Akrotiri, later Kea, etc.), and to the east (Alalakh, Miletus, Tel Kabri). Later on, it seems to have moved to the mainland and to the east and south (Qatna, Tell el-Dab'a, Hattusha), possibly via the Cyclades again. Even though there is no textual evidence (see Jasink 2005: 61-62), I do believe that the material evidence is strong enough to suggest that painters and plasterers from Crete (and later on possibly from the mainland) went, upon demand, to other places to execute the paintings they were known for.

Locally, these artisans did not seem to leave anything behind, which is logical if we assume that they went for a specific period (weeks or months rather than years) to another locale to execute these paintings. They would not leave behind their own tools and equipment, apart perhaps from the occasional little lump of pigment. If they were invited to carry out work for short periods, one would expect them to use the daily utensils made available to them locally. The lack of evidence of any possible presence of foreigners in elite courts outside the Aegean does not mean that there were no foreigners there for specific periods. Likewise, the lack of evidence of foreigners does not necessarily mean that local artists painted the walls there after they were trained at Knossos or elsewhere in the Aegean. In fact, I reject the idea of a training school in Knossos for two reasons. First, no trial pieces have been found at Knossos in amounts that may support such theories, and no painting 'atelier' can be clearly delineated in any of the evidence from the site. Second, when artisans are specialized in a specific craft that is much in demand, they may not want to share their monopoly in knowledge so easily by setting up training schools because they would lose their economic and social position and hence their power. Chapin (1997: 10-11) also expresses the difficulties with trying to identify regional 'schools' based on stylistic grounds. Moreover, they may be under the strict control of their elite who see in them a gift and exchange value that may disappear once their knowledge is spread. A parallel argument may be illustrative: no Egyptian blue recipes were found to have been written down until Classical times, and when they were passed on, the recipe was not complete (Forbes 1966a: 224-25), possibly to avoid easy reproduction.

Whether the artisans worked for one specific ruler or elite group is unclear, but they did work only for the elite. This fully developed art form started at the rise of the second palaces on Crete (MM IIIA), possibly at Knossos, and disappeared completely with the collapse of the Mycenaean palaces, apart from the Lady of the Lilies painting from Mycenae dating to LH IIIC (Maran 2006c: 127-28; but E. French 2007: pers.comm. places it in LH IIIB). We do not know if there was one specific personality as ruler at Knossos, as seen later at Mycenaean courts (*wanax, lawagetas, basileus*, etc.). There are iconographic depictions of important male figures, such as on the Master Impression from Chania (Hallager 1985), but we do not know their status. Hamilakis (1998: 233-34) has suggested, instead, that competition between different elite groups may have occurred, which is supported by specific evidence pointing towards several elite class groups in most settlements on Crete and, from LM IA onwards, more or less all over the Aegean. I argue that the painters (and possibly some of the plasterers) were under the control of this elite class because we do not find such painted plaster outside elite contexts. These elites most likely controlled production, distribution and consumption of the paintings as luxury items, and the best way to do so would be by controlling the artisans themselves. We do not know these elites well enough yet though to understand the extent of their power over their society. Presumably, political, economic and military power was in their hands. Furthermore, the fact that by LM IA both Phaestos and Knossos were major religious centres too (Dickinson 1994: 275-76), suggests that their rulers may have held religious power as well (Rehak and Younger 2001: 394, 402). Whether the artisans actually lived within the palaces or elite complexes is harder to answer, but considering their likely restricted mobility, they may not have, at least not on a permanent basis. When these artisans were dispatched for a task away from the homeland, they must have worked full-time as the *al fresco* work dictated (see below). When they were working at home, one may assume that they were employed full-time when on an assignment.

The Minoan elites from LM IA onwards, and the Mycenaean elites later on, needed to express, confirm and reconfirm their political power towards their own local society and towards other and foreign elites. As they were integrated into international trade and contacts during large parts of the Late Bronze Age, with several places overseas (but see Manning and Hulin 2005: 283), they had to have exchange value in their hands for the procurement of much needed metals and several other valuable materials to come their way. Minoan and Mycenaean elites' economic power may have consisted of subsistence control, expressed in the form of palatial distribution centres, of wealth and prestige control in the form of strictly controlled prestigious goods' acquirement or production (Voutsaki 2001: 197) and, as far as Mycenaean palaces are concerned, of control of trade routes via which many trade goods passed (Sherratt 2001: 230). Some of this subsistence produce may have been redirected to artisans working for the palaces (Michaelidou and Voutsa 2005: 17).

The elite needed both primary and secondary commodities in order to express its political position, to legitimize social order and its ideology and to (re)confirm its social status within the society. Artisans with their highly valued abilities fitted perfectly into such power-expressing strategies (Peregrine 1991). Textiles, oil, wine and perfume were the goods that may have travelled eastwards, in return for more tangible materials. None of these eastward-travelling materials have proved to be very visible in the archaeological record; and services are another example of invisible goods. In Egypt, trainable people were considered useful commodities with good future profit value (Michaelidou and Voutsa 2005: 25-26). By examining textual evidence from other contemporary cultures whom the Aegean societies had multiple contacts with, we may assume that the Aegean elites took part in such service exchange and service gifts, especially if they had something extraordinary to present. There are further indications regarding the possibility that the elites used their artisans in religious feasts and ceremonies in the locations where they were supposed to be performing their craft. Some crafts were even more closely linked with the religious sphere, since they took place in cultic buildings (Michaelidou and Voutsa 2005: 22; Dakouri-Hild 2005: 211). This connection between religious affairs, elite contexts, and artisans (see Maran and Stavrianopoulou 2007: 292, n. 65) must have given the artisans a specific status, at least in the eyes of other artisans and their society (see below).

Thus far, the Aegean and foreign elites look omnipotent in their relationship to artisans, but there may be more to it. As already explained, the appearance of the actual plastered surface changed during its painting. The patrons must have felt in awe as they saw the paintings taking on their final dazzling appearance. Even while the artisans were at work, much of what they did was virtually invisible, the fillers in their plaster (uncalcined lime, crushed shells), the divisions of the surface to be painted (string lines versus painted ones such as in Egypt), and the final painted surface (turning brighter as time went on). It is possible that such paintings may have been imbued with aspects of magic (see Helms 1993: 22), together with their producers, while it may also be possible that elites did not allow this magical status of artisans, simply to protect their own social status. While we have no further evidence on how this painting activity was perceived in the past, watching *fresco* painters at work even now is a tremendous experience. We have no idea who may have watched their performance, but at least some inspection by their patron may be expected. Painting is performing, and performing is entertaining; it claims the attention and admiration of others. In this sense, Lemonnier (1993: 19) states that, sometimes not only does the artifact mark a particular social status or identity but the whole set of processes leading to the formation and consumption of the artifact does as well.

Much of the Aegean religion (and most likely the majority of Bronze Age religions) may have been part of the performance (Rehak and Younger 2001: 439). In this lay the power of the artisans. This power extended towards foreign and home elites for whom the artists held some form of magical capabilities and knowledge that no one else possessed.

## Artisans in Relation to their Society

The question as to who these artisans were may be answered in many ways, reflecting upon their multiple identities. They may have been highly skilled and sufficiently initiated in the religious and symbolic meaning of the elite's iconographic expressions and visions of power and cosmology to be able to depict such specific scenes for them. In other crafts, they seem to have been specifically called upon to perform during religious ceremonies (see above).

The artisans, more precisely the painters and plasterers, were probably men of an age that enabled them to travel to other lands on reasonably long boat journeys and to carry out the work locally, some of which was physically hard. Although the work itself may have taken no more than a few weeks, such journeys may have taken the artisans away from home for much longer, hardly an easy situation for pregnant women or women with small children in their care, who were, most likely, left behind (on gender in the Aegean: e.g. Kopaka 1997; also Talalay 2005: 142 on Rehak's gender work).

In any such assembly, we might expect men of different age groups, including the most experienced ones, specifically the chief plasterer and his helper, and the most experienced (few) painters and their helpers. These helpers will have been younger adolescents in training to become artisans themselves. Since this craft appears to have been reserved for the elite dwellings only, we may assume that passing on the knowledge was a very delicate matter and may have been under very strict control by the elite. Whether the elite had a say in who could become a trainee is impossible to state. The artisans themselves may at least have been able to suggest people based both on whom they could trust and whom they considered fit for the work (talent, kin relations, age considerations).

Some of the labour divisions must have been task-specific. Pigment preparation into cakes could have been done in advance and away from the painting scene (see travelling pigments at Miletus and the Greek mainland, possibly even at Tell el-Dab'a and many other places). Such work may, therefore, have been carried out by practically anyone within the society (but see Cameron 1975: 40-43), whether related to the artisans or not, possibly serving more than one craft group simultaneously. While carrying out experimental replication work in Italy, I received some very patient help from a five-year-old boy who, without being asked, ground raw stone into pigment for hours (Figure 7.3). His hands clearly suffered less than my own, and he continued this work for several days. Moreover, his work was excellent and he hardly needed any explanation because he had been carefully watching me and so knew how to continue, perfectly imitating my own actions. His situated way of learning, by looking and imitation, has been discussed in an ethnographic context by Baxter (2005: 51). In a society where wealth was the lifestyle for a limited few, as in prestige economy systems such as the Bronze Age eastern Mediterranean, children's labour was probably not an anomaly (see Michaelidou and Voutsa 2005: 25; Baxter 2005: 66; papers in Sofaer Derevenski 2000).

Cameron's idea (1975: 41-42) of plaster painters originally being the best pot painters and furnishing decorators, who had moved into this new craft, is appealing but poses specific problems. These can be summarized as: (1) the scale of the surface to be painted, (2) the difference in background colour of the media, (3) the level of painted detail, (4) the use of a differing colour palette (blue, green, etc.), (5) the problems of painting the wall or ceiling (vertical or from above) and (6) transmitting imagery from three-dimensional (pottery, seals) to two-dimensional surfaces, etc., all of which influenced the end result.

Minoans may have put more emphasis on the collective group than upon people as individuals (that is until LM II-IIIB: on warrior graves: Popham 1994: 90). This is evidenced in depictions on any media, burials in collective tombs and possibly by being ruled by a class of elite, rather than by one specific person, and in this Minoan palatial structures acted as codes for social integration (Wright 2006: 55). Minoans' individual actions, however, were crucial to society. As an overall societal characteristic, the fact that individual artisans (Cherry 1992; 1999; Morris 1993) or even groups appear to have been unimportant stands in apparent contrast to the value given to the fact that they came from far away (the power of distance and the 'other', see Helms 1993; Knapp 1998; Manning and Hulin 2005: 274-75). The most important aspect was most probably the artwork they produced, possibly combined with the distance (and vagueness) of their origin (the more unknown, the more powerful), which may have added value to their product.

Painters and plasterers were not mentioned in the limited Linear B tablet archives. This is probably because such work and the transactions involved were not recorded on these tablets; it is even possible that it was too closely controlled by the palaces (see Dakouri-Hild's analogous argument for textiles and furniture working: 2005: 210). Alternatively, it is possible that they were not important in the economic system, and perhaps the painters and plasterers were not employed full-time all year round. Near Eastern documents did not mention them either, but this suggests to me that the artisans did not come from there. In societies whose operation was under such strict bureaucratic control, the artisans would have been mentioned if they had belonged to one of the royal courts. Their absence from the Near Eastern tablets so far may speak in silent terms about their social status and, in my view, hints at their geographical origins being elsewhere.

The relationship of the artisans with their society may be most clearly depicted when we consider how they would have been received upon return from a long, distant journey. In non-literary societies where storytelling, singing and playing music may have been major forms of entertainment and information exchange, possibly at different levels, we can understand that the travellers, having worked at royal courts in foreign lands, had much to share with their own people. Their social status within their community will have been based on aspects of knowledge about distant places, about elite life styles and possibly also upon their association with religious and magic practices and performances.

The same artisans may have been viewed very differently by the societies of those distant lands. As an analogy to this, we may refer to Cline's (2005) discussion on imported and exported objects, where status changed in relation to the new context of use, function, meaning, significance and uniqueness. This changing status implies a multivalent nature of the artisans from at least an etic point of view. When on a mission, these artisans may have lived within the palatial sphere for the time it took to carry out the work. They may have been fed and housed by their employer, and hence it is difficult to know how much they would have interacted with the local society during their visit. Zaccagnini (1983: 253-54) informs us about artisans that do not return to their country of origin for a range of reasons. If this were the case, may we expect local changes within the elite architecture in the country being visited, whereby the local elite imitate the ruler's way of living? Hence, may we assume that, over time, more people would get access to the paintings as a result of artisans who stayed behind (Wiener's 'Versailles effect': 1984: 17)? The archaeological evidence cannot support such assumptions as far as the Near East is concerned, and the remains of Amarna and Malkata in Egypt are insufficiently strong cases, technologically, to be considered in such a light.

In relation to the organization of the plasterer's craft, plastering an area of 50 sq m containing four rooms, five doorways, five windows and walls of three metres in height, with a maximum of two coats, would have taken two experienced plasterers a minimum of eight full working days, each of eight hours (J. Coopman 2004: pers. comm.; see Rehak and Younger 2001: 439 for sizes). Painters would have had to start applying the paint soon after the plasterers had finished in order to make good use of the damp wall. Procuring limestone, calcining and slaking could have taken much longer but can be done in advance. Calcining and loading a kiln will have been carried out by more than one person because the limestone needed refining into smaller fragments before loading could take place. Slaking itself needs at least two experienced people, one to mix the quicklime well with water, while the other stirs constantly to avoid lumps of unslaked material. In fact, letting slaked lime putty mature for a longer period of time improves the material: four years is said to be ideal (P. Gibbons 2002: pers. comm.). Young adolescents may have joined these plasterers throughout in order to learn the craft.

For practical reasons, I would argue that none of the plaster production and preparation would have been done by the painters, and this has been confirmed by information about later *fresco* work performed until early this century (S. Hocombe 2004: pers. comm.). First, people who quarry the limestone, calcine and slake it carry out heavy physical labour, and as a result their hands become very rough and unsteady. Such conditions were not conducive to painting. Pigment grinding must have been done on a much smaller, but equally intensive scale. The painters themselves may have done this work, some time in advance, in order to have steady hands during the painting work or else apprentices, children, etc. carried it out. Pigment grinding may have served more than one craft at the same time (potters, weavers, stone sculptors, etc.), and hence may have been organized within the community, perhaps while receiving the necessary raw materials and tools from the palace/elites. Second, as explained above, the plasterers themselves cannot be the painters because their work only just precedes the painting work and steady hands were needed for the painting. I, therefore, suggest that every team of painters may have had at least two plasterers together with them who would prepare the walls to the wishes of the painters. As such, the travelling painters did not rely on local plasterers for this work because of the uniformity required for the plaster mixture and its application (*contra* Von Rüden's opinion for Qatna: 2006: 236). My observations suggest a team of painters *and* plasterers, who knew to what extent they could rely on each other to perform optimally. Since only two plasterers were needed for slaking and plastering (but possibly a few more for quarrying and calcining), their mobility would not have been problematic.

Tools, instruments and equipment (Figure 7.4) have not been discussed in detail in this research simply because they have been so well covered by Evely (1993; 2000). That no actual lime kilns have been found (but see now at Qantir, near Tell el-Dab'a, Egypt, Rehren 2004: pers. comm.) may be due to the fact that they were either set up at the quarrying sites (see also Kassianidou and Knapp 2005: 234 for metals) or because such large installations were used for a variety of purposes, such as, for instance, pottery firing, even though we have no evidence for this. Wingate (1985: 70-73) shows a whole range of possible kiln types, some of which were of a perishable nature. If such types of lime burning had taken place in the Middle and Late Bronze Age, it may be virtually impossible ever to find remains of them (in contrast to Neolithic lime burning: Perlès 2001: 191-96).

Finally, such a mixed team of people must have been very well organized and disciplined in order to be able to carry out the painting within the given time provided by the dampness of the plaster. This length of time was to some extent unpredictable being dependent on weather changes, but was in part controllable by producing a certain thickness of plaster layer. The plaster composition became a fixed matter once the 'perfect recipe' was found. As regards painting on a very thin *intonaco*, I would suggest both that this was possibly carried out during a period of predominantly stable weather, and also that it is possible that the thick plaster layer had dried out before the painting was finished. The only time of the year when the Mediterranean weather is suitable to allow a minimum of eight hours to paint is late spring and early summer when the ambient environment is relatively stable (see experiment 1, Brysbaert 2004a; 2004b). Climatic changes are very variable during early spring and autumn, while winter in general may be too damp for the plaster to dry out sufficiently to allow the paint to stay in place. When too wet a paint is applied to a wall, it simply runs off in dribbles, or the plaster surface gets disrupted by the paint brush, a sign that the painter did not wait long enough before starting the paint job. Most paintings, however, were done without *intonaco* on a much thicker plaster layer (Figure 7.5). This, combined with a well-manipulated surface and plaster mixture, may have allowed for much longer painting periods, perhaps up to several days (Laurie 1988: 193-94). In that case, seasonal weather considerations may have hardly mattered.

## The Elite in Relation to their Society

Elites cannot exist without the labourers in society, whose task lies in the provision of subsistence, goods and products to the elite for their own survival. Such subsistence and other provisions, then, form ways of remunerating other dependants like the elites' skilled artisans. In that sense, Bronze Age eastern Mediterranean societies were no different and the political and economic control of the elite and palaces seemed closely interlinked. The Linear B tablets from Pylos (E-series), for instance, refer to the palace that gave out land to people in return for produce or services (Shelmerdine 2001: 360). Subsistence wealth, obtained through a redistributive system such as the Mycenaean palaces, was partly responsible for the fact that the elite were powerful and stayed that way. Sherratt (2001: 224-31, esp. 215) provides an alternative explanation for the role and thus the power of the Mycenaean palaces (or lack thereof: their insecurities hidden in static use of writing and painted plaster) by referring to their function as control posts over strategic routes that facilitated passage of goods and trade actions between several regions. Moreover, the potential of figurative iconography with its specific style and technology as statements of the elite's power on both a local/regional and international scale was tremendous (see below), but see Sherratt's (2001: 215) position in this debate.

So far, none of the above discussion has touched upon why the change from abstract to figurative painting, combined with an improvement and perfection of the technology, took place. The relationship between the elite and its people explains this question.

Skibo (1992: x) draws a direct connection between change in decorative style and technological attributes and the use of a specific material. By MM IIIA-LM IB, there seems to be more centralized and intensified religious ritual and display within the palaces (Wiener 1990: 144), most notably in the larger palaces of Phaestos and Knossos. After the earthquake destruction (see also Niemeier 1994: 74) at the end of MM IIB that marked the rebuilding and repair of the previous palace at Knossos, a range of new materials and techniques may have been introduced as a result of both experimentation and of realizing the failure of the materials previously used (see MacGillivray 1994: 55). The elite occupying these renovated palaces may have used the mural art form as a large-scale medium to communicate a new way of thinking regarding the importance of religion within the society (Gates 2004; Panagiotopoulos 2006: 39). The elite were most certainly in control of this changing and powerfully controlling aspect of society.

This strong interconnection between elite and religious power continued in the Mycenaean palaces (Maran 2001a; Maran 2006b: 78; Maran and Stavrianopoulou 2007: 286). Clearly, patrons and social change were the main impulse for the technological experimentation, change, and innovation that underlay their new wishes for iconographic expression of their beliefs and power symbols. Given the constant need for the reaffirmation of their position in an increasingly centralized political system, stronger control was necessary in order to maintain specific power positions that would eventually collapse as a result of their own structure. Religious control and its forms of expression may have created very strong tools indeed. In this context we can, therefore, see and understand when a combination of narrative paintings and oral expressions, possibly accompanied by music (see lyre player painted on plaster in the throne room at Pylos, Lang 1969: 194; Thaler 2006: 102), and often set in a ceremonial (religious, magical) performing context, will have left a very overpowering impression upon people. Jasink (2005: 64-67) suggests how oral traditions in Mycenaean elite contexts may have been the ideal way—instead of written sources—to transmit messages from the palatial court to the people. This is confirmed by Markus (2006: 322) who states that language (visual, oral, or written) communicates everything about the production and consumption of the built environment. Some of these message transmissions, with musical instruments, may have been linked to religious and public activities. The way ideological materialization through paintings combined with performances forms an expression of power fits in with Renfrew's (2001: 136-37) concept of a 'hypostatic union of idea and material' (but see Markus 2006: 332 on the ambiguity over who *really* holds the power in relation to the built environment). Furthermore, Maran's (2006a: 11) discussion of 'the duality of space', (based on Low's 2001 theory cited in Maran 2006a: 9) places both the building and its furniture (in this case: the paintings) in an active and passive role in relation to their owners: the building and paintings are both the backdrop for action (e.g. ritual feasting), *and* they are the products of social practice (based upon the relationships between elite and artisans, and artisans in relation to each other, see below).

Apart from the above mentioned architectural changes that went on at Knossos, the transition from MM IB-II to MM IIIA-LM IB was also marked by several changes and innovations in a range of cultural spheres (Knappett 1999; see also Chapter One). Furthermore, the gradual decline of the strong first palaces of Mallia and Phaestos in favour of the rise of Knossian supremacy architecturally, administratively, and possibly as a religious centre, shows the strength of Knossos as a centralizing power from MM IIIA-LM IB onwards (Cadogan 1994: 68). In contrast to that, a cultural interruption was noted between MM IIIA-LM IB and LM II-IIIB on Crete in burial practices, even though many art forms continued. The first use of tombs for individual burials on Crete was in stark contrast to previous group burials (for a brief summary, see Cadogan 1994: 62-63; Popham 1994: 90 on warrior graves).

As already mentioned, the status of the individual during the Minoan period on Crete does not appear to have assumed great importance until LM II-IIIB, and this is reflected in many aspects of its society. This lack of emphasis on the individual changed, however, when mainland influences on Crete became stronger and more emphasis fell on individuality. Pydyn (1998) discussed this aspect of growing individualization for central Europe at the transition from Bronze Age to Iron Age and the resistance that accompanied this social shift. The main change in the material culture, however, seems to have happened during the previous transition phase on Crete, which may have induced the widespread use of the technological and iconographic innovation of the LM painted plaster. I suspect a preceding developmental phase took place, possibly at Knossos or Phaestos, during MM IB-II when a clear tendency for sumptuous stone-built architecture was an obvious elite expression of their power at Phaestos (Watrous 2001: 204), and when string-line impressions and pure lime plaster appeared at Knossos (Cameron 1975: 16).

The optimization of the painting techniques, the plaster mixture and pigment grain size from LM IA onwards may indicate increasing strictness of control over the quality of the production, possibly exercised by a socially well-organized and centralized power. The elite, economically strong by the start of LM, expressed their power and renewed this expression by the frequent change of display in prestige and wealth items. The possession of rare items (such as painted plaster) *and* the command over the access to the production of it, both of which they controlled, emphasized and reiterated their dominance as rulers over their society. Economic wealth (palace and linked villa system forming controlled distribution networks of subsistence and other goods, see Rehak and Younger 2001: 396) combined with prestigious goods assisted in keeping up the high profile of the elite involved.

Apart from the palaces, several larger 'villas' and elite buildings were decorated with painted plaster, some in an equally elaborate iconographic way to some of the palace decorations themselves (explained by 'the Versailles effect': Wiener 1984: 17). Equally important is that these paintings were accepted as prestige goods as long as the elite, with whom they were completely associated, were in power. As mentioned earlier, these paintings finally ceased to exist at the end of the LH IIIB period. This can be seen both as the power of the elite being in control of the art form and its producers until that time, but equally as the commoners' acceptance of such an art form as long as these elites were around. The artisans *could* in theory have continued their work after the collapse of the palatial system, but no one appeared to have been sufficiently interested, ideologically, economically or symbolically, in imitating or carrying on this specific wealth symbolism. In my view, the acceptance during the palatial periods of the art form as the elite's expression of power symbolism and the apparent indifference towards it after the palace collapse potentially points to forms of long-term unexpressed popular resistance against the elite. Similar examples of possible resistance are known from Phylakopi and Myrtos-Pyrgos much earlier (Whitelaw 2005; Cadogan 1992: 203).

## The Elite in Relation to Other Elites

This relationship has a double aspect to it: first, an aspect of friendship and diplomacy whereby both parties tried to keep each other pleased so that important transactions such as the exchange of metals, textiles, oil and wine could continue without too much disruption from either side. The second aspect is one of an underlying power display towards each other. Diplomatic friendships could not really exist if one or the other party was not of similar power or importance in at least some of the domains they shared (e.g. dominance over water and land routes for the passage of trade caravans).

In order to express both aspects, their interrelationships were maintained by gift giving and gift exchange, whether material or human (artisans, royal marriages: see Feldman 2003), and by wealth item possession and its overt display. Both aspects were achievable within the craft of painted plaster. Artisans, to render a service, seemed to have been sent away to execute beautiful paintings at eastern royal courts in order to maintain friendly relationships between elite groups (artisans as 'political personnel': Peregrine 1991: 1). At the same time, once these paintings decorated the surfaces of monumental architecture, they served as perfect large-scale display items to 'show off' power positions between elite groups and to demonstrate a sharing of equal positions, both locally and for visiting elite groups who got to see them. Inasmuch as the non-elite population may have been allowed to circulate within the palaces and mansions, social hierarchical distinctions could have been made clear here (see Kenoyer 2000: 88). Because of 'showing off' between elites and by elites towards their society, things and actions had to look exotic, expensive, durable (or 'permanent': Helms 2004: 124-25), different and hard to get. Simultaneously, these things had to be recognizable as an import to have an impact. Sometimes, one may wonder whether it was the 'distant origin', even of raw materials, and consequent inherent added value that was reason enough not to use local raw materials of the same kind. A similar argument can be made throughout the Bronze Age, for elite objects in most materials, including painted plaster. For instance, Egypt had a long-standing tradition of painted plaster *al secco* from the Early Bronze Age, but one still finds *al fresco* paintings at Tell el-Dab'a. The concept as such is recognizable, but also sufficiently different, to be accepted and to be considered 'exotic'. Moreover, at the same time or from a slightly earlier period, *al secco* paintings were executed right next to those done *al fresco* (Bietak and Forstner-Müller 2003). This indicates that a special effort was made to get *some* paintings done 'differently'.

In many ways, this technological transfer from west to east goes against the more common transfer of many aspects from east to west. Large undertakings such as palaces, metal technology and production, the use of administration by means of seals, sealings, possibly even the use of script on clay tablets, such as Linear A and B, and many aspects of the iconographic vocabulary used in the Aegean, may have had their origin or major influence from the east and Egypt. Put in this light, the origin of the *al fresco* technique and its accompanying technological transfer certainly does not follow the usually known, established path. We should not forget, however, that these examples are those clearly visible in the archaeological record. The resistance against such 'opposite' transfer, i.e. from west to east, may be supported by scholars that do not believe that peripheral areas produced anything of importance (Manning and Hulin 2005: 283-84, 291). In fact, calling it 'periphery' may already be too strong a straitjacket because what may be considered 'periphery' in one region may be 'core' within another network of locales (see also Killebrew 2005: 24). It is sufficient to state here that not necessarily all changes, innovations and improvements came from one direction and moved to another, as this study has amply shown (see for instance the issue of Egyptian Blue as a pigment, see Chapter Six).

In any case, one cannot even clearly say, in this specific case, that paintings are 'exported' from the Aegean and 'imported' into the Near East and Egypt. Cline (2005: 46) states that, to call something an 'import', it has to fulfill certain criteria, namely: (1) it must have been made or come from outside the receiver's country, and (2) it must have been perceived by the receivers (i.e. consumers) as having come from outside their own country. He also questions whether objects are real imports or local imitations (Cline 2005: 48). Through this query he does, in fact, enter into the discussion of different levels of 'Minoanization' (as outlined by Broodbank 2004 and applied to Phylakopi by Whitelaw 2005) or 'Mycenaeanization'. I agree with Cline's statement that one needs to combine both scientific and stylistic analysis of objects in order to assign the place of origin of questionable imports. In the specific case of the painted plaster, however, the situation is far less straightforward. Let us, for instance, put the paintings of Miletus, Qatna, Hattusha, Alalakh, Tell el-Dab'a, and Tel Kabri under the 'magnifying glass'. First of all, we tend to look at objects as a whole, which is as it should be, but when we dissect them and look at them from different angles the picture may be more revealing. Laffineur (2005) discusses a specific methodology in order to investigate imports and exports and suggests certain parameters in doing so. These are applied here to an analysis of the paintings, and the main constituent elements of the paintings are discussed below (Table 7.5). Another important factor is who actually consumes them and with what experience (elites, commoners, artisans, merchants, visiting elites, etc.). Finally, we should not forget that a range of parameters, indicating technological transfer, is intrinsic to the technological make-up of the paintings, some of which are invisible but nevertheless of great importance (see Jones 2002: 66, and discussions in Chapter Six). All these parameters need to be taken into consideration before a simple west-east or east-west direction of technological transfer is proclaimed.

| *Constituent elements* | *Alalakh* | *Hattusha* | *Miletus* | *Qatna* | *Tel Kabri* | *Tell el-Dab'a* |
|---|---|---|---|---|---|---|
| Materials of object (except plaster) | pigments | pigments | *pigments* | pigments bitumen | pigments | *pigments* |
| Techniques of manufacture | *pigments* plaster mix *fresco* | *pigments* plaster mix *fresco* | *pigments* *plaster mix* *fresco* | pigments *plaster mix* *fresco* | pigments plaster mix *fresco* | *pigments* *plaster mix* *fresco* |
| Shape/form | wall | wall | wall | wall | wall, floor | wall, floor |
| Style | *hybrid* | *hybrid* | *Hybrid* | *hybrid* | *hybrid* | *hybrid* |
| Meaning/function | *palace* | *temple* | *Elite* | *palace* | *palace* | *palace* |

Table 7.5 Main constituent elements of painted plaster outlined per site outside the Aegean. Italic entries indicate possible imported materials/concepts, others are local (a local viewpoint of 'local' versus 'import' is taken, i.e. in the place where the paintings were executed).

*Materials from which the painting is made*

Even if materials are locally procured (we assume here that heavy and bulky **plaster** is not taken from the Aegean on a journey to the place of execution and is, therefore, not entered in the table), they may still be considered as having come from far away in local eyes, or at least, be strange, exotic and therefore expensive, because these materials were not known in local daily contexts. Even if many **pigments** were procured and produced locally, this was not the case for all of them (see the use of amphibole blue at Miletus, indigo at Thebes and Alalakh, *lapis lazuli* at Gla, Egyptian blue containing tin at Tell el-Dab'a, and azurite at Hattusha).

*Techniques of manufacture*

We have seen that the **plaster mixture** used in the Aegean seemed to have followed a specific recipe and was made of specific materials in certain ratios. This 'recipe' was used on several sites outside the Aegean as well, and it is an invisible aspect of this craft once the painting was completely finished. Because of this, it can only have been understood by the artisans themselves and may, as such, have been seen as exotic. The **pigment working** may have been done before the artisans arrived in a new place. There is enough indication to suggest that the pigments were refined and ground in very similar ways on the Aegean and eastern Mediterranean sites. The technique of combining pigments and lime plaster, ***al fresco*** **painting**, was difficult to achieve to a sufficiently high level, and this, for me, forms a very strong argument that it was done by foreigners, both the painting *and* plastering. The act of painting itself and its end result were magical and highly attractive (for similar argumentation see Ingold 1993: 161), and the painting improved over time, during which nothing more was done to it.

*Shape/form of object*

The paintings were executed on wall and floor surfaces, as was the case in the Aegean. Paintings on walls were known in both Egypt and the Near East.

*Style*

The iconographic elements used were often recognizable to Near Eastern and Egyptian communities and their elites. The way they were depicted, however, was different from their own traditions, but was close to what was known as stylistic expression in the Aegean at that time, thus implying artisans from elsewhere. Therefore, the fact that we have a hybrid mixture of iconographic elements, most likely adapted to personal taste, together with a foreign style, made the paintings exotic but acceptable enough to exert the powerful messages that they were supposed to convey.

*Meaning/function*

These wall and floor decorations were all found in elite or palace structures, so there was at least a primary architectural function in common with the Aegean specifically to impress; to convey messages from the elite to others, etc. We do not know, however, who had access to them and whether the access given included a conscious strategy of specific social stratification between people. Furthermore, we need to establish that the meaning of these paintings was different for each group or person who viewed them; in fact, these paintings even had a meaning for those who were not allowed to view them. So their meaning becomes more or less exotic to people depending on their access to them and also on their previous experiences with possibly similar paintings. These paintings must have exerted a very powerful and even frightening impression upon people who were supposed to get the messages from them. This is an instance where objects clearly have power intrinsic to themselves, since they may evoke strong emotions from people, and I would argue here that the meaning of objects stands solely in relation to the people with whom they 'communicate'.

With respect to Cline's questions about imports, whether things are made locally or elsewhere, and whether objects are perceived as being local or from elsewhere, I feel that the only relevant question is the latter one. In case of the production of objects, I believe that too many aspects may muddle the straightforward answer to local or foreign production. In the case of the latter question, we can only guess how people perceived the paintings. We have strong evidence and arguments, however, when we come to compare their appearance and production in their actual context, that these paintings were ordered *because* they looked foreign and because they fitted a certain 'international fashion' that one needed to take part in in order to belong to the social 'club' of elites at the time.

As I said before, this discussion also leads into the difficult subject of 'Minoanization' (see Broodbank 2004) and, as an extension of this, 'Mycenaeanization'. In this specific case, when investigating even just the main constituent elements of the paintings, I believe that there are too many unknown factors at the human execution level for us to state whether these paintings are Minoan or Minoanized. We can then look at the different levels of Minoanization which may be easier to define, but, again, how many factors, quantitatively and/or qualitatively, do we need to collect before something is determined to fit a certain label? Trying to 'quantify Minoanization' is further complicated by the different opinions amongst scholars who study these materials. Extending this discussion to the later period of the painting tradition, especially looking at the paintings from Qatna, Hattusha and possibly Tell el-Dab'a, how would we ever be able to state whether these paintings were executed by people from Crete or from the Greek mainland? Moreover, even if they are executed by plasterers and painters from the mainland, when they were influenced by Minoan technology and iconography, where do we draw the line between 'Minoanized' and 'Mycenaeanized'?

Rather than creating more confusion I just want to demonstrate the fundamental complexity of assigning a specific 'origin or provenance' to material culture remains, something we still like to do all too often. We should try, instead, as Broodbank (2004) suggested, to name things Cretan if they are clearly from Crete (not Minoan), Cycladic and mainland Greek (not Mycenaean), if we feel that we have enough evidence to do so. In this context it is useful to cite Knapp and Blake (2005: 11), who say 'identities are devices employed for strategic action'. The constant need for us to assign labels to people, things and phenomena needs to be investigated as a social (and political) phenomenon in itself. Moreover, if they really need to be used, they should be defined unambiguously by the user, in order to prepare the receiver for what is coming.

## Artisans in Relation to Other Artisans

Very little has been published about cross-craft interaction within the Aegean (Laffineur 1995), which does not reflect the nature and intensity of the interactions that took place (but see now Brysbaert 2007a). I have pointed out (Brysbaert 2003b: 176) that craft divisions may have been different in the past. I have also mentioned the possibility that certain tasks may have been shared by more than one craft (e.g. pigment preparation). With these two statements, I suggest that the interaction between people may have been quite intensive within the domain of certain crafts and, as a result, on a social level. To give another example, kiln use may have been shared between potters, quicklime producers and glassy frit makers and, in fact, the temperature they need for complete calcining, complete clay transformation into ceramic and the formation of frit is around the same (Brysbaert 2007a: 339). Jones (2004: pers. comm.), however, suggests that frit and faience may have been produced in small furnaces, and Evely (2004: pers. comm.) does not agree on such kiln sharing between crafts at all, because he believes that each craft needed a different kiln type or fuelling system. In consideration of the above, the fact that a similar temperature was required and was actively employed for the dissociation of several materials may have meant that there was still knowledge exchange at the level of pyrotechnology, even though it may have been employed via different equipment.

Within the craft of painted plaster, cross-craft interaction has been noted, for instance, in the use of pigments. The production of Egyptian blue implies knowledge of metal and glass making, with copper providing Egyptian blue's colour. Furthermore, Egyptian blue was also found, for example, on the Kouros from Palaikastro (Moak 2000: 75), while Jones (1987) reports on copper-based colourants for faience products from Knossos. In Egypt, it was also used as a raw material from which to carve beads and other small objects (Rehren 2004: pers. comm.), and several uses for Egyptian blue at Ugarit are now published (Matoïan and Bouquillon 2003). Other pigments that crossed craft boundaries were *murex* purple and indigo blue, traditionally known in textile dyeing (Barber 1991: 227, 237-38), but now confirmed on painted plaster at Gla (*murex*: Brysbaert and Perdikatsis 2008; Brysbaert 2006: 260-64; Chryssikopoulou and Sotiropoulou 2003), at Thebes (indigo: Brysbaert and Vandenabeele 2004; 2006), and at Alalakh (indigo). Barber (1991: 237-38) mentions slaked lime as a fixative for dye stuffs on textiles and alludes to storage jars found at Tell Beit Mirsim (8th–7th centuries BC), originally containing hydrated lime for that purpose. Although this dates to a later period it confirms for her the possibility that lime was also used as a mordant on Egyptian textiles from the 12th and 18th Dynasties. In this context, we can refer to the vessels found at several sites in the Aegean (Knossos, Zakros, Akrotiri just to mention a few, see Evely 2000: 474; Shaw 1973: 213) that were found full or partially filled with what is now set calcium carbonate but must originally have been hydrated lime. These vessels have traditionally only been seen in the context of plastering, building and repair activities, but this may not have been their only purpose, and it would be worthwile to check if other textile-related production materials were found in the same or nearby contexts on these sites. The painters must have known that *murex* purple would be compatible with damp plaster for a blue/purple hue, and this compatibility was possibly understood through lime use in textile dyeing production. Equally, indigo was known as a stable material on both textiles (Barber 1991: 227; Pastoureau 2000: 17) and on lime plaster (Balfour-Paul 1998).

We do not know where such contacts may have taken place, but specific artisan quarters (e.g. Quartier Mu at Mallia or at Mochlos) will no doubt have witnessed episodes of some sharing. Even though artisans would have protected their knowledge (and may have had to) some sharing may have been to their benefit as well. Contacts between west and east from the Middle Bronze Age onwards will only have enhanced the exchange of knowledge and materials. A beautiful example of workshop and artisans sharing existed at Amarna where a glass workshop is next to a precious metal production site (Shortland 2000: 72).

Interaction between artisans seems to have resulted in the sharing of knowledge about every aspect of technology: raw material procurement and manipulation (e.g. Hodges 1970: 47), transformation into usable products and their use and application on different media. This is why, in my opinion, this last relationship may have been a crucial one for any technological change, innovation and/or transfer. Therefore, it forms one of the most important, but understudied, relationships of the six represented and discussed here.

Having discussed all professional aspects of sharing (or the conscious lack thereof) and contacts between artisans, we can further examine the fact that these people may have shared the same building site (plasterers and painters) or working quarters (such as pigment and dye preparation places) and may have had workshops next door to each other, or met while they were travelling. This means that the actual space in which they found themselves formed in itself a medium for action and interaction, memory building, construction, creativity and sharing. In particular, the total interdependency of the plasterers and painters within one room, which was necessary for the speedy completion of *al fresco* painting, must have created good means of communication, verbal and otherwise, and mutual understanding that the job would be carried out successfully (see McFadyen's concept of quick architecture applied here: 2006: 127). Maran (2006a: 12, referring to Rapopport 1982: 67) also refers to the built environment as a teaching medium where apprentices may see the building as a mnemonic device for appropriate behaviour once they learned specific skills (see also Wright 2006: 53: craft seen as an act of memory via apprenticeship and in production of space). As such, Maran (2006a: 12) claims that all architecture exerts power as part of 'forming' social relationships within. These built places, moreover, could form the material framework for specific social bonds between people. These were, for example, older painters and younger ones in training (see above), the growing interest between one craft and the other; possibly even arrangements between two families for the marriage of their relatives, and, in addition, elites who saw what took place and verbally 'advertized' the paintings and artisans so that future projects might arise. These places may also have seen the light of internal resistance against social inequalities between people or conformation to social status and identity.

## Human Actors in Relation to Material Culture

The relationship(s) between human actors and objects and/or material culture can and should be viewed from a variety of angles in order to comprehend the multiplicity and complexity of these relationships (see McFadyen 2006: 132, on blurring limits between materials and human bodies). Generally speaking, each relationship between a human being and any single object or group of objects is unique. As soon as one part of the equation changes, so does the relationship, in terms of intensity, duration, meaning, etc. Most importantly, these and all relationships are dynamic and are connected to their specific context. Contextual investigations of time and space, therefore, are crucial to interpretations of data, phenomena and behaviours.

Looking specifically into the relationship between human actors and non-human actors in the elite context of painted plaster, it becomes clear, as pointed out earlier, that the paintings would have had different impacts upon different people. As Gosden (2004: 34) stated: 'our main reactions to each other and the material world are emotional' but 'our most profound experiences are at once emotional and intellectual and the appeal of such experiences to several parts of our being gives these experiences their power' (Gosden 2004: 36). The artisans, therefore, who produced these paintings reacted on a different, more emotional level to them (or maybe not at all), than did the elites who ordered them and saw them being made. The iconographic representation, the style or even its technology may have stunned them or may even have left them completely cold, so long as the paintings had the intended effect upon those for whom they were meant.

I believe that this fits in fact very well with Sherratt's view (2001: 215, 230) that the figurative paintings, for instance, are a rather superficial application of palatial symbols in the Mycenaean palaces. The elite themselves were less concerned with any ideological potential that these paintings may have had, as long as they encoded the elite's self-definition and self-presentation to everyone else. Clearly, we can only conjecture about *how* people felt when they saw or experienced the paintings (in action: during their production, as stage backdrop for ceremonies or religious and/or political activities…). That they left impressions or stirred the emotions of some is clear, since these paintings were in fashion and were distributed accordingly within the eastern Mediterranean. The fact that they impressed means that they constituted meaning for the impressed and this meaning-making, the intrinsic power of these paintings, was carefully manipulated by their owners. Anthropologists such as Gell (1998) are convinced that objects exist to have effects on others. In the context of this case study, this effect of objects on people, subsequently, may have had physical, mental and emotional consequences for how people viewed the makers/designers and the owners of these objects, and it may even have altered the actions and thoughts of the beholders. These changing views, thoughts and actions, in turn, may have affected the social identities that people might contribute to some of the parties involved, even after they were no longer present. More specifically, the concept of distributed personhood is perfectly applicable in this context, being dependent upon the plasterers and painters during and after their production of the paintings, in people's (irrespective of their social identities) shaping of the social world around them (see Gosden and Marshall 1999: 173; Gosden 2004: 36).

So clearly, the paintings were powerful for what they were, where they were (their context) and by whom they were owned and consumed, but added power and value were interlocked in several other aspects as well. The fact that these paintings were made by artisans from elsewhere, possibly from distant places, that they looked different in comparison to what people were used to (but still familiar enough) and that they were made differently (invisible technology used for *al fresco* painting) all added to the power of these paintings over the onlooker.

When we next look at the materials used to produce these paintings, there is plenty of space for potential symbolic and economic added value and meaning. From the Egyptian (e.g. Pinch 1994) and Near Eastern literature we know that raw materials had certain powerful symbolic and sometimes medicinal and magical meanings, properties and powers (see e.g. Karimali 2005 for obsidian; Reese 1979–80 for lime plaster). To pre-industrial societies living in marine and agricultural environments, relying for their survival and subsistence mostly on natural phenomena and products, certain materials were bound to have such extra values, either intrinsically or by association. Egyptian blue, for instance, was said to be a very expensive material, but it was still made as a cheaper imitation for *lapis lazuli*. What are we to think then, of the discovery of actual *lapis lazuli* as a pigment on the Gla paintings; or of *murex* purple? Was the use of these materials only the expression of high economic value, or was it also indirectly referring to far and exotic lands (in the case of *lapis lazuli*) and artisans' superb mastery of material knowledge and compatibility (*murex* dye with lime as a fixative on textiles, or *Murex* shells with lime plaster as a physical keying material and chemical component)? The red paint, ochre, was known—and still is—for representing blood, fertility, life cycles, regeneration and birth. Without any textual evidence, however, we can only guess why certain materials were used, but it is very tempting to look beyond their pure physical presence.

The act of painting has been seen as a magical act of object transformation from one stage to another (see also the concept of objects' biographies: e.g. Gosden and Marshall 1999: 169-70). In fact, an object transforms into something else through the interaction with the painter, so gaining a different meaning and value and tending subsequently to have a different impact (thus power) upon the onlooker, owner, etc. When an object 'becomes', it lives, acts and interacts, it 'speaks' via its imagery and brings messages or finds the message challenged (Ross 2005: 328). It is dynamic within its present context and present relationships. It is exactly this viewpoint that gives power and an active role to the object, and material culture in general, and also to the artisans who are the medium through which change and shift of meaning is achieved, even in a dominant elite context.

Furthermore, objects such as these paintings can themselves be the medium (or stage setting) for human action, processes and developments, especially when they form the material context in which other relationships are acted out. These paintings are in fact integral to these actions and processes. Alternatively, as Kohring (2006: 100) states, 'objects and techniques in themselves become modifiers of social trajectories and their outcomes'. A throning ceremony (as an ideological manifestation, see Ross 2005: 328) of a ruler taking place in a room impressively painted by foreign artisans, with rhythmic drumming or music being performed, alcohol being served and incense clouding the air and stimulating the senses of those present could have been perceived very differently if there were no such paintings present, because these images may have supplemented and emphasized the presence of the ruler (see also Bennet 2001: 34). In societies where intoxicating products were accessible to (certain groups of) people, such paintings made of brilliant colours, depicting real or unreal and recognizably impressive and powerful imagery, may have come alive to those who took part in intensive celebrations or religious activities (see also Wright 2006: 56, who suggests that the painting scenes referred to real-life events, possibly connected to a cyclic calendar unknown to us). As such, the paintings took an active part in meaning production and both people and objects shaped each other's identities and values. We know that intoxicating products were in use in Minoan and Mycenaean contexts, so even without textual evidence documenting such hypothesized activities, we could easily imagine them taking place in those ceremonial and painted rooms. Such multisensory experiences may have created a bonding feeling of belonging for some people or, potentially, exactly the opposite.

The elites may have felt overconfident being powerful both worldly and religiously. The artisans may have been aware of the intended meaning of the paintings they made, or they may have been present during such specific ceremonies (see Dakouri-Hild

2005 on the presence of artisans at Thebes during feasting activities, thus connecting crafts to religious spheres) and may have even felt that they were important because of their potential and unique knowledge of creating something so powerful, even if the elites may have taken all responsibility for the production of these paintings. We could take this line of thought one step further and imagine that such paintings, each time produced in elite contexts, often depicting similar themes, executed in a similar style, possibly even by a known group of artisans, may have evoked similar feelings in each place: an emotional *koine*, as it were, accompanying the artistic *koine* (the latter outlined by Knapp 1998). This, however, could only hold true if the function and meaning of the paintings were the same in each case and we have no means of knowing this.

In which stage then lay most of the power of these paintings? According to Ingold (1993: 161), western thinking gives priority to the *end* product over the performance of *making* the product, while non-western societies prioritize performing activities. McFadyen (2006: 131) emphasizes that people remember best 'making' activities over the 'made', while Wright (2006: 50) mentions that 'the first performance of architecture is its construction'. Surely some combination of these may be possible too. In any case, the performing power of these paintings is not over once they are completed. The production of *al fresco* paintings is an ongoing affair since, even after they are painted, the walls continue to become whiter and shinier, the colours more brilliant and the painting more finished. Therefore, the plasterers and painters may have continued to amaze the onlooker, even after they were gone, with the painting potentially continuing to impress over time. The end product of the painting may have been important too within a subsequent performing context as a material expression of liminality at intersections of zones with different meanings (see Maran 2006b: 78, 82). Through their colours, themes, location, and in combination with other architectural features placed strategically, the paintings drew invited people towards the hearth (in the Great Megaron) or the 'heart' of the Mycenaean palace, and, as such, they formed part of a performative space. In the end, it is clear that the specific context dictates when the object gained its greatest importance and one (performance) does not need to rule out another (end product). Moreover, this is supported by Ingold's idea of taskscapes in which every activity, whether performing or not, takes place in a wider context of many other activities and thus gains its importance via its interconnection with other activities and their end products.

Therefore, we can say that no performance or activity stands in isolation, and the meaning and power of such activities or performances lie exactly in their relationship to others. When this relationship changes, the power and meaning may change too. This is, furthermore, also affected by a change in context. Therefore, if material expressions change, this may affect the entire relationship between material and human actors. If material expressions are stable, their power may lie in the fact that they continue to confirm existing power relationships. Without a single word, material expressions may have their power in letting others know who is in power, how and why. Objects such as these paintings really can be seen as the expressions and messengers of the ideology of the rulers/elite. The power embedded in these paintings can, furthermore, be reinforced by their location, their scale and their (restricted) access and how this contextual setting is controlled and manipulated by the paintings' owners.

## Conclusions

This chapter has brought together all the analytical and experimental data collected whilst being informed by theoretical considerations, by employing an agency-oriented approach to look more closely at the social identities of the artisans and their social practices and ways of communication and relationship-building through technology. In this way, it has been possible to offer some tentative answers to questions about the technological transfer in the painting tradition under study and about the forces responsible for this technological transfer. In effect, it is a case study about the broader social dynamics that flourished during the late Middle and Late Bronze Ages in the Aegean and the eastern Mediterranean, and how people participated and fitted within these. Chapter Eight seeks to place this case study in its broader context, highlighting its importance within the political, social and economic context or situation of the period to which these paintings belong.

# 8 Technology and Social Agency of Painted Plaster

This final chapter demonstrates the importance of studying human agency in the context of painted plaster in order to illuminate the importance of technology and trade in the Late Bronze Age eastern Mediterranean. It also shows how socio-political and economic changes in pre-industrial societies have had a direct impact upon the changes in material culture as a way of expressing, confirming and manipulating social identities and relationships.

## Introduction

In this book I aimed to answer two questions relating to the technological study of painted plaster. First, can evidence be found for technological transfer within the Aegean painted plaster tradition, and between this tradition and that in the eastern Mediterranean? If so, is this technological transfer in support of the well-studied corresponding case of iconographic transfer? Second, what are the forces behind such technological transfer? Based on the data available, can we investigate human agency and social identities within the sphere of craft specialization in the Late Bronze Age eastern Mediterranean?

The analytical and experimental study of the material remains at 17 sites, combined with a social perspective, provided adequate and representative data, enabling us to offer some answers to these questions. I also aimed to take technological studies out of their isolation by carrying out a comparative study between areas in the Aegean and the eastern Mediterranean. Moreover, I integrated the technological approach with the more traditional iconographic approach, so demonstrating that they must be considered together if one wants to understand the painting tradition as a whole and to see where its meaning lies within the society that created and enjoyed it. The key factor in this study was to reestablish human actors centre stage, where they had once performed for each other, with each other and through their material world, and where they once acted out the multiple social identities that shaped their individual and group relationships.

As a positive result of such interdisciplinary research, several data sets were created, some of them overlapping; and each time that the different approaches converged on the same conclusion, the working hypothesis became greatly enhanced (Wylie's [2000: 232] idea of triangulation as a way of strengthening interpretation). Finally, the application of different investigative techniques to this material yielded a set of workable criteria for the study of the *al fresco* technique that was systematically applied, while some criteria used previously were shown to be no longer valid.

## Technological Transfer in Support of Iconographic Transfer?

As demonstrated in Chapters Six and Seven, technological transfer can be clearly traced through the data collected from the painted plaster samples, and, in my opinion, these data indicate a transfer from west to east. These data should help to amplify the parameters of the debate about the direction of such transfer, which has so far been based mainly on iconography and on a few isolated technological studies.

Technological transfer appears to go, to a limited extent, in tandem with iconographic transfer and did not follow just one linear route (see Chapters Six and Seven). Most of the features studied here indicated that the *al fresco* painting technique was used extensively throughout the entire painting process, on both background and detailed work, on most sites from the MM III onwards. The only material that deviated from this was from EM Knossos and Monastiraki (MM I-II), and Middle Bronze Age Tell Sakka.

Several parameters within the technology of this painting tradition, for example very pure lime plaster, specific pigment use and grain size of pigments, indicate a remarkable level of homogeneity across the entire area, but this is only half the story. Regional and even inter-site differences are present everywhere, showing a level of regional originality within the craft. For instance, people in several places seem to have been using specific pigments, either because these pigments suited their purposes, or perhaps because they were available locally: *murex* purple at Gla, indigo at Thebes and Alalakh, celadonite mixed with chlorite at Palaikastro for green, and azurite at Hattusha. One can only attribute these differences to the actual artisans, who must have therefore mastered their craft very well.

On the one hand, local differences in iconography may reflect differences in personal taste of the elite or in the socio-political context in which the paintings were functioning, thereby reflecting differences at the level of consumption. Local differences in technology, on the other hand, reflect choices and changes made by the artisans, thereby reflecting differences at the level of production. Although technological transfer may not be complete from one site to another (see Chapter Six), Sherratt and Sherratt (2001: 17) argued that looking at alternative views of technology and technological change provide a useful reminder that the latter tends to be a matter of local tradition, which fits perfectly with the data collected for this study. Changes or continuities in both technology and iconography cannot be separated from each other since they are all the result of human action.

## Agency in Painted Plaster Production in the Late Bronze Age Eastern Mediterranean

In the study of the social dynamics and the forces behind technological transfer with regard to craft specialization, six important relationships could be distinguished (see Chapters Two and Seven). Such an approach brings human actors to the centre of the interaction between people and their material world. After investigating these six relationships, I came to the conclusion that, within an elite context of performing a specific craft, artisans were far from passive producers conforming to the elite demands. At a practical level, they could, up to a point, select their materials, improve their techniques over time and adapt to new circumstances by briefly reclaiming older traditions after MM II (Cameron *et al.* 1977). In a wider sense, however, they also had their own powerful role to play in relation to the elites themselves and to their own society. Judging from the local differences in the materials used, and from their thorough knowledge of the compatibility of materials in reuse and recycling aspects, their skills must have been obvious to any observer of their performance. Furthermore, it also became apparent that the paintings themselves, whether completed or in progress, played their own role as agents in forging relationships with and between people. These paintings must have impressed the people who saw them and, as such, may have created meaning for these viewers. This could then be carefully, but strategically, used and manipulated by the owners of the paintings in their relationships with others. It became clear that objects interact and are dynamic as a result of their relationships with other agents, such as for instance artists, and can thus be themselves the medium for more relationship-building. The making of objects, their technical and aesthetic development, is part of such processes.

As outlined in Chapters Two and Six, aspects of technology, iconography and style cannot be separated from each other if we want to understand the painting tradition as a whole, one in which human actors are centrally placed. These different aspects of the entire technology pointed clearly to a distinctive technological style for these paintings. During the early phases of the painting tradition, we still see decoration as a style of painting which is independent from the technology behind it. By this I mean that the support could have been done in any way, by any material, and the paint could have been applied at any time on this dry medium. This is very much the way that Egyptian painting was created *al secco*, where style of decoration and technology stand apart from each other. Although the themes and motifs of Minoan iconography were vast by then, as we see from other media, the possibilities of applying such iconography to walls were not explored as early as EM II-MM II periods because the technology was not yet ready, nor the elite, who needed this medium for the expression of power. The situation had changed by the start of MM III, when the technology of plaster was optimized and figurative painting introduced.

## Technology and Trade in an Eastern Mediterranean Bronze Age Context

Next, I would like to place this case study of painted plaster within the larger picture of the eastern Mediterranean Middle and Late Bronze Age, in order to demonstrate the role that this technology, in all its material and human aspects, played during this period that was marked by widespread networks of exchange, operating throughout the eastern Mediterranean and the Aegean. In this section I outline how this case study may contribute to our knowledge of the possible 'international era' of the productive Late Bronze Age eastern Mediterranean (and its preceding late Middle Bronze Age phases), during which elites exchanged or passed royal gifts and services to each other. By arguing that artisans were sent overseas to execute their expertise craft, this study may throw light upon trade and interaction patterns during this period. More specifically, the study of this material in terms of long-distance trade and exchange sheds light on export and import materials, partnerships and alliances, power strategies, and changing social relations (Knapp and Cherry 1994: 46). Bronze Age trade should be seen as having multi-dimensional and complex strategies, involving specific cultural and individual human factors, including both state-dominated and entrepreneurial aspects (Portugali and Knapp 1985: 65-67; Knapp and Cherry 1994: 127).

During the formative phase of the figurative painted plaster craft, it is interesting to note that innovations and changes in materials and techniques occurred during times when the palatial economy was strong and when other crafts developed in parallel (see Chapter Five), both during MM I-II and later, during MM IIIA-LM IB. These palatial economies, with whom the emerging Minoan elite may have had some contact through the process of gift-giving, may have grown under the (in)direct influence of Near Eastern palatial elites before and during MM I-II (Cherry 1986: 38-39). Several indications in the material evidence suggest that the elite of the first palaces (MM I-II) on Crete may have adopted certain religious and political institutions from the Near East in order to state, confirm and consolidate their socio-political status within their communities. Features such as monumental architecture, writing, centralized administration, pottery made by the fast wheel, and an intensified metal trade did not occur randomly all over the island but manifested themselves in places like Knossos, Phaestos and Mallia, where elites may have been in direct or close contact with eastern powers (Watrous 1987: 66-67, 70), but the evidence for this is not abundant (Cherry 1986).

Given early contacts between Mari and people from *Kaptaru* at Ugarit (Dalley 1984: 51) one could suggest that the idea of painted walls may have been picked up there and then. The same may be suggested with regard to possible direct contact of Minoans in Egypt at places where they may have been exposed to Egyptian paintings (Immerwahr 1985: 48-49), although it is far from clear whether they actually saw these paintings (see Morgan's comment in Immerwahr 1985: 50). In any case, the presence of Kamares pottery in both elite and labourers' contexts in Egypt (Kemp and Merrillees 1980) and in the Near East (Merrillees 2003) suggests possible direct contact between

Minoan Crete, Egypt and the Near East during the Middle Bronze Age. In both cases, however, there is a temporal gap between the possible influence/inspiration in terms of the paintings and the actual execution of figurative painted plaster *al fresco* on Crete (see Immerwahr 1985: 49-50); possible reasons for this delay are briefly summarized below (see also Chapter Six).

Early Knossian examples of string impressions and dado paintings, the development of a broader colour palette, and a good quality plaster mixture during MM IB-II all support the existence of the necessary supportive and wealthy environment for the development of the craft in an elite context. That this craft, the artisans and the materials involved were under palatial control may be deduced from the fact that these improvements did not feature at contemporary Monastiraki (see Chapter Six).

Following a series of destructive events affecting the first palaces at Knossos, Phaestos and Mallia, the high-quality, pure white plaster mixture used in MM IB appears to have been replaced again by coarser mixtures after MM IIA, while pure plaster was in use again by MM IIIA (Cameron *et al.* 1977: 148-49). This may have been caused by several factors:

- The more practical and functional needs of the people who survived these series of destructive events at this time
- The reconfiguration of social relationships as a result of these destructive events, which must have had various physical and social impacts upon people's lives during these periods (Phillips 2001: 75)
- Possible population shifts which may have affected any of the above factors.

Despite the scant evidence of painted plaster during the Middle Bronze Age on Crete, its specific spread does show a form of relatively close elite interaction, although there seems to be a hierarchy among them (Knossos versus Monastiraki). The potential influences from Egypt and the Near East on the painted plaster craft were probably made possible by the fact that Minoan Crete already used wall coverings in painted plaster (see EM III Knossos, etc.). Therefore, the artisans' possible adoption of specific iconography and possible string use on plaster may have been relatively easy, since the concept of painting on plaster was not totally new. The idea of combining local aspects with influences from outside into an innovative result was also seen in the same period for the architecture, religious material expressions and pottery making (see Watrous 1987; Knappett 1999).

The transition from the Middle to Late Bronze Age period saw the sudden emergence of figurative paintings carried out *al fresco* on very pure lime plaster. Only within a centralized wealthy elite context, such as for instance MM IIIA-LM IB Knossos, would artisans find a 'niche market' within which to experiment and develop this craft until it reached perfection, because their patrons would provide them with the necessary material support and social stability and security, possibly via the remuneration of subsistence or other goods (Sherratt 1997: 102-103). Knossos must have controlled the local sources of raw materials for both plaster and pigments and had possibly already been producing its own Egyptian blue since MM IIA. It is also possible that better organization of the artisans among themselves was part of the innovation. The fact that plasterers and painters had to work really closely together in order to make the *fresco* painting work, in contrast to the older pure *al secco* painting, is an innovation in itself. Such innovations emerging from social changes (i.e. cross-craft interaction) demonstrate the complete integration of technology within the social and political realm of that period (e.g. the elites needed to express their status on a large scale and required specialists—lots of them—to produce luxury items).

From LM IA onwards, this painting tradition spread over elite contexts on Crete, the Cyclades and also at Miletus, Alalakh, Tel Kabri, Qatna, Tell el-Dab'a, and at Hattusha. No textual evidence appears to support the hypothesis that artisans were sent from Crete (Knossos?) to any of these places, specifically for the execution of these paintings. It could be argued, therefore, that technological transfer was not the result of travelling artisans, but of royal gifts or other exchange patterns (for which again there is no textual evidence). The level of transfer (explained in detail in Chapters Six and Seven), however, shows that not only some materials, but also techniques and knowledge were involved. Apart from the materials, the other types cannot be explained without reference to the people who transferred the craft, because, contrary to gold granulation on jewellery for instance, this craft is not mobile on its own (Politis 2001; for a parallel argument regarding the spread of the pottery wheel, see Knappett 1999: 125-26).

Miletus clearly had strong ties with Minoan Crete, as seen through its array of pottery (Knappett 2003), which may help to explain the existence of painted plaster in its main building (Building A) carried out *al fresco* with clear Minoan motifs such as the white lilies and the griffins with notched plumes. Miletus was linked to Crete via the 'eastern string' of islands (Rehak and Younger 2001: 428), through which pottery moved extensively between several areas on Crete and several islands of the Dodecanese (Knappett 2003; Knappett and Nikolakopoulou 2005). Tel Kabri and Alalakh Level VII formed palatial centres that grew during the later Middle Bronze Age and fitted into the trading patterns of the eastern Mediterranean, where pottery such as small containers (juglets) and other objects were exchanged and imitated (Manning 1999: 135). Diplomatic and other contacts between Ahhiyawa and Hatti, on the one hand (during the 14th–12th centuries BC) (Niemeier 1999: 143-44; Bryce 2002: 191-94), and between Knossos, the Hyksos and Hatti (16th century BC at the earliest) through gift-giving, on the other hand, have been demonstrated (Mellink 1995: 86-89; see also Manning 1999: 136). Some form of (in)direct contact between Qatna and the Aegean is also shown via Aegean-style motifs found on objects from the Qatna Royal Tomb and via the presence of Mycenaean pottery from the Late Bronze Age palace. Qatna, furthermore, was situated at a strategic position on the crossroads of two main trade routes.

This painting tradition also reached the Greek mainland during LM II/LH IIB-IIIB. Many technological strands indicate a transfer from Crete outwards (Kythera, Laconia, southwest Messenia), but the Cyclades may also have played a part (see Chapter Six; and for the 'western string': Cherry and Davis 1982). The above suggests that a close relationship between elite technology and trade/exchange must be investigated for the Middle and Late Bronze Age eastern Mediterranean.

Because painted plaster was an elite craft, it was, therefore, strictly under elite control. They did this most effectively by controlling the artisans who were the only ones able to do anything with the raw materials *because* of their knowledge and skills. Clearly, a specific group of people, the elite, wished to express their social allegiances and rank via luxury items showing a common artistic language (Sherratt and Sherratt 2001: 21; Feldman 2006: 17).

The location of these paintings outside Minoan, Mycenaean and Cycladic territories indicates where the influential elite, who felt the need to show that they belonged to a specific social entourage, lived. This in fact illustrates the point made by Sherratt and Sherratt (2001: 28-29) regarding the shifting dominance of centres in the eastern Mediterranean, namely that the well-known major powers of Babylonia, Egypt and the Hatti now coexisted with a range of maritime eastern Mediterranean polities that formed important middlemen along the long-distance supply routes, providing certain types of goods that marked the Late Bronze Age elite consumers' 'fashion'. Most of these elite sites showing painted plaster were either close to the coast (Miletus, Tel Kabri), or were involved in trading along specific well-known routes (Alalakh, Tel Kabri, Qatna). Several of these sites were also connected by a common ceramic presence of small, narrow-necked Levantine containers that also made their way to LM IA Thera (Manning 1999: 135).

The marked increase in wealth both in Minoan Crete (MM IIIA-LM IB) and subsequently in the Mycenaean mainland (LH III) indicates their role in the international trading economy in which they clearly took part. The fact that elite and royal objects reached their palaces and mansions meant that they had the means of offering something in return, albeit not always as visible as the imports. It also suggests that some technologies were politically controlled by the elite.

The location of these paintings in elite contexts, outside the Aegean, also suggests that several, possibly unrelated technologies go hand in hand because they complement each other in the existing economic system and share, as a common context, the concentration of wealth. This can be seen in:

- The spread of painted plaster which illustrated and confirmed elite status and power
- The improvement in seafaring techniques (Wedde 2005) that must have developed as more sea trade took place during the Late Bronze Age
- The constant increase in metal production, trade, recycling and exchange showing a rising demand and an improving grasp of the technology by the artisans involved (Knapp 2000)
- The well-organized subsistence provision shown in centralized storage systems, at least from MM I-II onwards (Knappett 1999: 101).

These capital-based economies and luxury items became more widespread due to more demand in the Late Bronze Age. As a consequence of this growth, the production went up and luxury items were no longer in the sole possession of the elite, by which time some had lost their exclusivity. The elites, who depended upon the presentation and demonstration of luxury items for the confirmation and consolidation of their power position in society, eventually collapsed, one could argue, as the result of this economic and social shift, though this shift was certainly not the only reason. The 'age of internationalism' (quote from Killebrew 2005: 233) gradually disintegrated and with it disappeared its elite art forms. In the case of the painted plaster and the copper-based metallurgical industry, these elite-controlled crafts disappeared more or less entirely with the elite at the end of the Late Bronze Age.

Several studies have shown that contacts, exchange and trade in the Late Bronze Age eastern Mediterranean went in all directions and can best be described as a complex web of contacts between many larger and smaller maritime centres, from which goods were taken inland (Kantor 1947; Palaima 1991; Cline 1994; Lurz 1994; Knapp and Cherry 1994; Phillips 1995; 2004; Lambrou-Phillipson 1990; Cline and Harris-Cline 1998; Laffineur 1998; Weingarten 2000; Niemeier 2004; Kalogeropoulos 2005). Changes in contacts between regions occurred over time, as illustrated by Knapp and Cherry (1994: 42-47) for Cyprus. This reflected both demand for newly accessible materials (amber from the Baltic Sea, copper from the western Mediterranean) and shifting emphases among politico-social powers (from Minoan to Mycenaean). It may also reflect technological changes (increased and improved seafaring) that influenced many of the above factors.

The vast range of materials traded and exchanged is best illustrated by the finds from the Cape Gelidonia and the Uluburun shipwrecks (Bass 1986; Pulak 1988; 1997; 2005). These shipwrecks may also demonstrate that many goods came on board or were unloaded at different locations, as the ship continued its journey throughout the eastern Mediterranean. Such shiploads show that there was not just one direction for traded goods, as has already been mentioned above. They also indicate that more than just one social group may have been involved as tradesmen or merchants (see also Knapp and Cherry 1994: 143, citing Merrillees 1974), although certain social groups have been assumed for the crew that was present on the Uluburun ship before it sank. In many ways, the painted plaster case study fits this picture. As mentioned in Chapter Seven, only the *technological* transfer seems to have been directed from west to east, while iconographic motifs went in several directions, criss-crossing a range of media as carriers.

In discussing the social dynamics of Late Bronze Age Cyprus, Knapp and Cherry (1994: 128) suggested four types of trade mechanisms, and I have demonstrated where my study fits in one or more of these four categories below. As these authors stated, royal gift-exchange and freelance trade must have coexisted during the international era of the Late Bronze Age. In fact, the case of the painted plaster, with its travelling artisans, clearly shows that neither can be seen separately from one another. If these artisans had to travel from location A, their origin, to location B, where they were 'invited', this may have happened on board a ship navigated by either freelance entrepreneurs or merchants of an elite class (see Ugaritic documents, Knapp and Cherry 1994: 145), travelling with lots of other commodities, luxury goods or bulk goods (Muhly 1986: 56-59) and calling at different ports. Therefore, even if these artisans were part of a gift-giving system, in the form of offering their services, they were linked into, and limited by, other kinds of trade as well. This shows the complexity of discussing trade systems and mobility of artisans for the Late Bronze Age. Helms (1993: 37, 53) also refers to the fact that artisans may have served a dual purpose: as 'ambassadors for elite patrons...as well as skilled craftsmen'. We have no evidence at all to support or deny this possibility for the painted plaster artisans, but the idea is certainly attractive.

Knapp and Cherry (1994: 142) furthermore argue that freelance merchants may have been the producers of the commodity or raw material in demand. This certainly may have been possible for more widespread materials such as pottery and where the motivation of the merchant was profit-making. On the basis of the research presented here, it is unlikely that profit-making could explain the presence of the painted plaster solely in an elite context. Moreover, painted plaster was not the result of the work of one person, but of many, involving a long time to master sufficiently all aspects of it for completion of a painting.

One could argue that the painted plaster tradition may not be part of the royal gift-giving social practice because no written documents refer to it (none of the Mari tablets mention any aspect of its own wall paintings—Sasson 1990: 23). Textual evidence from the Linear B tablets may indicate aspects of foreign trade (see hypotheses suggested by Killen 1995: 215-16, esp. 220 [referring to *ku-pi-ri-ja* as 'Cypriot' or 'the Cypriote', in relation to the organization of external trade]; *contra* Muhly 1986: 58-59), but such evidence is limited by many factors, not least by its preservation and its recovery. On the one hand, everything about the painted plaster craft points towards royal and elite contexts and its circulation within them, whether by means of commercial trade or ceremonial exchange, or both. On the other hand, textual and other evidence demonstrates the presence of Minoans and Mycenaeans within the eastern Mediterranean trade networks. So, despite the fact that no textual evidence appears to prove royal or elite correspondence about artisans' services in regard to the execution of painted plaster, this study has shown that these artisans most probably travelled from west to east to decorate elite buildings upon elite demand.

To sum up, the present study has shown that elite gift-giving was crucial within the Middle to Late Bronze Age societies in the eastern Mediterranean and the Aegean. Not only was such a trade mechanism crucial for the emergence and growth of the elite on Crete and the Greek mainland, it also allowed them to maintain their position in the burgeoning trade networks of that period. Economic and political power were very much connected, so much so that gift-giving may be seen both as commercial and ceremonial, while both aimed at the (re)confirmation of power, social status and social identities.

Simultaneously, freelance merchants took part in the existing trade networks and may have been responsible for facilitating or limiting the mobility of artisans of several crafts (but not solely, cf. the case of royal merchants/messengers). In this specific case, it seems most unlikely that merchants could also have been the artisans or producers of the craft because of the non-mobile and complex technical nature of the craft itself. The overall craft and its artisans may have been under palace control, but there was space for local diversity as we can see in the subtle varieties of materials and techniques employed (see Chapters Six and Seven). Local differences could have been the result of an artisan's creativity or the local elite's demands (e.g. use of specific pigments made available to artisans), or a combination of both.

The increased social organization underlying this technology (whereby painters and plasterers worked more closely together) by MM III on Crete was mainly dictated by its complexity and, perhaps equally importantly, by the elite's religious ideology which was designed to employ specific large-scale imagery (*contra* seal imagery), in order to establish and confirm their power. Both aspects are accounted for by suggesting that this craft was elite-controlled. These paintings, combined with other finds and evidence of diplomatic contacts between several of the sites outside the Aegean, show a growing and intensifying interaction between people of the higher socio-political echelons of the Aegean and eastern Mediterranean during the Late Bronze Age. Knapp and Cherry (1994: 165-66) state that Mediterranean exchange systems were flexible and overlapping and in a state of continual dynamic transformation. At the same time, copper ox-hide ingots of the same weight spread all over the eastern Mediterranean. Standardized values for weights and mediated exchange rates demonstrate a certain degree of homogeneity, which may have functioned as a stabilizing influence in the constantly changing politico-economic relationships of the time.

Although the paintings, based on the evidence we have today, are not extremely widespread outside the Aegean, they do illustrate the increasing political-economic contacts of the region within the broader eastern Mediterranean. As an elite-controlled craft, the painted plaster craft disappears entirely when the elite economy of palaces and the interdependent regional trading systems collapsed at the end of the Late Bronze Age. This was the result of increased elite demands that also independently encouraged the growth of freelance trade alongside elite-controlled trade (most recently Sherratt 2001: 320; 2003: 37, 41-42 on decentralized bronze scrap circulation). This socio-political change that was more widespread than just the Aegean may also have been the reason why, in the 1st millennium BC, an increasing number of independent artisans operated outside both the palace and the temple in the Near East (Zaccagnini 1983: 264). As already suggested in Chapter Seven, it is quite possible that the painters and plasterers, once their elite market vanished at the end of the Late Bronze Age, went on to do other things, being freed from the tight control under which they had worked and lived.

## Where to Go from Here?

This book is not the final word on the subject of painted plaster. On the contrary, through its interdisciplinary approach it has opened up several paths for future research. Some of these future possibilities are briefly outlined below.

1. The sampling bias in this book, and in many previous studies, has been on palatial material from both the Aegean and eastern Mediterranean. Since this is an unbalanced approach to Bronze Age Aegean studies, it is crucial in the future, in order to understand all levels of society, to include material from non-palatial sites as well. New excavations, often focusing on non-palatial settlements, offer such new opportunities. The study of painted plaster from non-palatial sites should be approached in its own right from both an iconographic and a technological point of view. As a consequence, such material could be compared at a regional and an interregional level, as well as diachronically. A further step would be to compare data from these sites with data from palatial sites, to discover whether social stratification and/or different groups of artisans can be detected within such a craft. Knossos and Mycenae, where both palaces and elite houses are present, constitute opportune sites for such a study, while Pylos still presents a gap in our knowledge (under study by S. Stocker and team since 2000). Work on Mycenae and Tiryns in its entirety is on its way by the author. However, the sites at Galatas, Archanes and Zakros also deserve full attention. Some are currently under study (by L. Chryssikopoulou) and their publication is eagerly awaited.
2. It may be possible with newly excavated material, even if it is fragile, to distinguish more accurately between samples from miniature scenes and from life-size scenes. It would be worth investigating whether (and why) there are differences in the plaster composition, the manipulation of materials and the surface between the two different painting styles. If differences do exist, are they related to size or geographical location?
3. My research has been profoundly influenced by conservation considerations in terms of both sampling and analysis. I have tried to reduce sampling to a minimum by conducting a thorough macroscopic study, by applying surface XRD for the pigment analysis, by not touching iconographically rich scenes, and by introducing two non-destructive techniques. The aim of these two pilot studies was also to determine whether results comparable to the traditionally applied techniques could be achieved, and thereby partly replace such destructive techniques. The future for non-destructive techniques is promising and represents the way forward, especially as most countries now severely restrict their sampling permissions on archaeological materials. The application of these techniques must be refined to cover specific archaeological needs, since they were not initially designed for such work. This will happen successfully when archaeologists and scientists collaborate, sharing their expertise with each other (Brysbaert and Vandenabeele 2006). Furthermore, the application of such non-destructive techniques may enable us to study iconographically richer material in greater detail, simply because they are non-destructive.
4. As suggested in Chapter Four, more experimental work needs to be carried out. Experimental replication has proven its value in the past and has done so

again within this book. Apart from the list indicating future work in Chapter Four, the most important addition to this sort of work may be the life-size plastering and painting of the entire wall: it is exceedingly difficult to replicate on a vertical surface, but it very much reflects reality. Painting on small-scale panels, such as I carried out, needs to be taken out of the controlled laboratory environment in order to see what such work may tell us about the practical and social aspects of the painting tradition. Equally important for determining painting techniques is the ongoing study of experimental cross-sections in comparison to archaeological ones.

5. Geological reconnaissance of the materials involved in the painting tradition would be a very helpful tool to investigate, at a micro- and macro-level, where materials came from, how they travelled, how and where they were procured and what was eventually required in terms of labour input, transport and logistic planning to provide the surfaces with painted plaster.
6. As has been hinted at in Chapters Six and Seven, artisans did not live and work in isolation from their own material environment. Each time we investigate human actors in the field of craft specialization, we come across interactions that play a crucial role in the construction of the multiple social identities of the people involved. Moreover, these interactions are clearly demonstrated in the archaeological record and prompt cross-craft interactions that result in the exchange or transfer of materials and knowledge (Brysbaert 2007a). In any case, we should not forget that what we think of today as different crafts may not have been classified in the same way in the past. More work at an analytical level may reveal the degree of interaction between different crafts practised in a given time and space framework, and how such interactions were made feasible. Work of this nature is ongoing by the author. This topic and approach has recently brought together a group of archaeologists and computer scientists at Leicester, Exeter and Glasgow Universities who have started (with Leverhulme funding) an exciting new interdisciplinary project on cross-craft interaction studies in the Mediterranean and beyond (see www.le.ac.uk/museumstudies/staff and www.cs.le.ac.uk/TracingNetworks/.
7. It is now 18 years since Gunter (1990: 9-10) expressed her concern that too much art and craft was studied without consideration of the people who produced these works. Processes of technological change and innovation may be studied on purely technical bases (i.e. wet versus dry plaster to explain painting techniques), but only social factors (i.e. teamwork, elite support, artisans' mobility) enable us to understand better how and why these processes happen. This work has been influenced by a social approach in the development of a set of relationships, which has demonstrated its usefulness in offering views and answering questions about social issues relating to craft production and the relationships between all parties involved. In future technological studies, more attention should be paid to the social issues that are intricately tied up with the

actual archaeological materials. In this way, more meaningful discussions may be held about the very people who were involved in any actions of which we now only find the material remains.

8. The last, but not least, important point concerns the prompt publication of analytical results obtained in the field. Comparative work that is of crucial importance to archaeological research can only be conducted thoroughly and profitably when as many results as possible are available to those conducting similar work. Furthermore, the analytical results should provide the starting point for the necessary interpretative work that contextualizes these results and ultimately gives them meaning and value.

# Bibliography


(Web pages and articles are grouped at the end)

Akkermans, P.M.M.G., and G.M. Schwartz
2005 *The Archaeology of Syria: From Complex Hunter-Gatherers to Early Urban Societies (ca. 16,000-300 BC).* Cambridge: Cambridge University Press.

Amiet, P.
1980 *Glyptique Mésopotamienne Archaïque.* Paris: Centre Nationale de Recherche Scientifique.

Aslanidou, K.
2005 The Minoan wall paintings from Tell el-Dab'a/'Ezbet Helmi: the life-size male figures. In R. Laffineur and E. Greco (eds.), *EMPORIA. Aegean in the Central and Eastern Mediterranean. Proceedings of the 10th International Aegean Conference, Athens 14–18 April 2004.* Aegaeum 25: 463-69. Liège-Austin: Université de Liège-University of Texas.

Atkinson, T.D., R.C. Bosanquet, C.C. Edgar, A.J. Evans, D.G. Hogarth, D. Mackenzie, D. Smith and F.B. Welch
1904 *Excavations at Phylakopi in Melos.* Society for the Promotion of Hellenic Studies Supplement 4. London: Macmillan.

Axelrod, R.
1997 *The Complexity of Cooperation: Agent-Based Models of Competition and Collaboration.* Princeton: Princeton University Press.

Baines, J., and J. Málek
1988 *Atlas van het Oude Egypte.* Amsterdam: Agon.

Balfour-Paul, J.
1998 *Indigo.* London: British Museum Press.

Barber, E.J.W.
1991 *Prehistoric Textiles: The Development of Cloth in the Neolithic and Bronze Ages. With Special Reference to the Aegean.* Princeton: Princeton University Press.

Bar-Yosef, O.
1983 The Natufian in the southern Levant. In T. Cuyler-Young, P.E.L. Smith and P. Mortensen (eds.), *The Hilly Flanks and Beyond: Studies in Ancient Oriental Civilizations,* 11-42. Chicago: University of Chicago Oriental Institute.

Bass, G.
1986 A Bronze Age shipwreck at Uluburun (Kaş): 1984 campaign. *American Journal of Achaeology* 90: 269-96.

Baxter, J.E.
2005 *The Archaeology of Childhood: Children, Gender and Material Culture.* Walnut Creek, California: Alta Mira Press.

Bech Nosch, M.-L.
2000 Acquisition and distribution: ta-ra-si-ja in the Mycenaean textile industry. In C. Gillis, C. Risberg and B. Sjöberg (eds.), *Trade and Production in Premonetary Greece: Acquisition and Distribution of Raw Materials and Finished Products.* Studies in Mediterranean Archaeology Pocket Book 154: 43-62. Jonsered: P. Åström's Förlag.

Bell, M.R.
1983 Egyptian imitations of Aegean vases. Some additional notes. *Göttinger Miszellen* 63: 13-24.

Bender, B.
2001 Landscapes on-the-move. *Journal of Social Archaeology* 1: 75-89.

Bennet, J.
2001 Agency and bureaucracy: thoughts on the nature and extent of administration

in Bronze Age Pylos. In S. Voutsaki and T. Killen (eds.), *Economy and Politics in the Mycenaean Palace States. Proceedings of a Conference Held 1-3 July 1999 in the Faculty of Classics, Cambridge.* Cambridge Philological Society. Supplementary Volume 27: 25-37. Cambridge: Cambridge Philological Society.

2004 Iconographies of value: words, people and things in the Late Bronze Age Aegean. In J.C. Barrett and P. Halstead (eds.), *The Emergence of Civilisation Revisited.* Sheffield Studies in Aegean Archaeology 6: 90-106. Oxford: Oxbow Books.

Berg, I.

2004 The meanings of standardisation: conical cups in the Late Bronze Age Aegean. *Antiquity* 78: 74-85.

Bergoffen, C.J.

2003 The Cypriot pottery from Alalakh: chronological considerations. In M. Bietak (ed.), *The Synchronisation of Civilisations in the Eastern Mediterranean in the 2nd Millennium B.C. II. Proceedings of the SCIEM 2000-EuroConference, Haindorf, (2-7 May 2001)*, 395-410. Vienna: Verlag der Österreichischen Akademie der Wissenschaften.

2005 *The Cypriot Bronze Age Pottery from Sir Leonard Woolley's Excavations at Alalakh (Tell Atchana).* Contributions to the Chronology of the Eastern Mediterranean, 5. Vienna: Verlag der Österreichischen Akademie der Wissenschaften.

Betancourt, P.P., V. Karageorghis, R. Laffineur and W.-D. Niemeier (eds.)

1999 *MELETEMATA I-III. Studies in Aegean Archaeology Presented to Malcolm H. Wiener as He Enters His 65th Year.* Aegaeum 20. Liège-Austin: Université de Liège-University of Texas.

Bhattal, R.S.

2003 Ethnic identity: archaeological considerations. In A. Brysbaert, N. de Bruijn, E. Gibson, A. Michael and M. Monaghan (eds.), *SOMA 2002. Symposium on Mediterranean Archaeology.* British Archaeological Reports, International Series 1142: 1-6. Oxford: Archaeopress.

Bietak, M.

1992 Minoan wall-paintings unearthed at ancient Avaris. *Egyptian Archaeology* 2: 26-28.

1995 Connections between Egypt and the Minoan world: new results from Tell el-Dab'a/Avaris. In W.V. Davies and L. Schofield (eds.), *Egypt, the Aegean and the Levant: Interconnections in the Second Millennium BC*, 19-28. London: British Museum Press.

1996 *Avaris. The Capital of the Hyksos. Recent Excavations at Tell el-Dab'a.* London: British Museum Press.

2000a 'Rich beyond the dreams of Avaris: Tell el-Dab'a and the Aegean world—a guide for the perplexed': a response to Eric H. Cline. *Annual of the British School at Athens* 95: 185-205.

2004a The setting of the Minoan wall paintings at Avaris. In L. Morgan (ed.), *Aegean Wall Painting: A Tribute to Mark Cameron.* British School at Athens Studies 13: 83-90. London: British School at Athens.

2004b Seal impressions from the Middle till the New Kingdom. A problem for chronological research. In M. Bietak and E. Czerny (eds.), *Scarabs of the Second Millennium BC from Egypt, Nubia, Crete and the Levant: Chronological and Historical Implications. Papers of a Symposium, Vienna (10–13 January 2002)*, 43-55. Vienna: Verlag der Österreichischen Akademie der Wissenschaften.

Bietak, M., and N. Marinatos

2000 Avaris (Tell el-Dab'a) and the Minoan world. In A. Karetsou (ed.), *Kriti-Aigyptos. Polutismikoi Desmoi Trion Chilieton. Meletes,* 40-44. Athens: Ypourgeio Politismou-Archaiologiko Mouseio Herakleiou (in Greek).

Bietak, M., N. Marinatos and C. Palyvou

2000 The maze tableau from Tell el-Dab'a.

In S. Sherratt (ed.), *The Wall Paintings of Thera. Proceedings of the First International Symposium. Petros M. Nomikos Conference Centre, Thera, Hellas. 30 August–4 September 1997*, 77-90. Athens: Petros M. Nomikos and the Thera Foundation.

Bietak, M., and I. Forstner-Müller
2003 Ausgrabungen im Palastbezirk von Avaris. Vorbericht Tell el-Dab'a/'Ezbet Helmi Frühjahr 2003. *Egypt and Levant* 13: 39-50.

Binford, L.R.
1983 *In Pursuit of the Past: Decoding the Archaeological Record*. London: Thames and Hudson.

Bingöl, O.
1997 *Malerei und Mosaik der Antike in der Türkei*. Kulturgeschichte der Antiken Welt, 67. Mainz-am-Rhein: Verlag Philipp von Zabern.

Birtacha, K., and M. Zacharioudakis
2000 Stereotypes in Theran wall paintings: modules and patterns in the procedure of painting. In S. Sherratt (ed.), *The Wall Paintings of Thera. Proceedings of the First International Symposium. Petros M. Nomikos Conference Centre, Thera, Hellas, 30 August–4 September 1997*, 159-72. Athens: Petros M. Nomikos and the Thera Foundation.

Blake, E., and A.B. Knapp (eds.)
2005 *The Archaeology of Mediterranean Prehistory*. Blackwell Studies in Global Archaeology. Oxford: Blackwell.

Blakholmer, F.
1997 Minoan wall painting: the transformation of a craft into an art form. In R. Laffineur and P.P. Betancourt (eds.), *TEXNH. Crafsmen, Craftswomen and Craftsmanship in the Aegean Bronze Age. Proceedings of the 6th International Aegean Conference, Philadelphia, Temple University, 18–21 April 1996*. Aegaeum 16: 95-105. Liège-Austin: Université de Liège-University of Texas.
1999 The history of Middle Minoan wall painting: the 'Kamares' connection. In P.P. Betancourt, V. Karageorghis, R. Laffineur and W.-D. Niemeier (eds.), *MELETEMATA I-III. Studies in Aegean Archaeology Presented to Malcolm H. Wiener as He Enters His 65th Year*. Aegaeum 20: 41-51. Liège-Austin: Université de Liège-University of Texas.
2000 The functions of wall painting and other forms of architectural decoration in the Aegean Bronze Age. In S. Sherratt (ed.), *The Wall Paintings of Thera. Proceedings of the First International Symposium. Petros M. Nomikos Conference Centre, Thera, Hellas, 30 August–4 September 1997*, 393-412. Athens: Petros M. Nomikos and the Thera Foundation.

Bleed, P.
2001 Artifice constrained: what determined technological choice? In M.B. Schiffer (ed.), *Anthropological Perspectives on Technology*. Amerind Foundation New World Studies 5: 151-62. Albuquerque: University of New Mexico Press.

Blitzer, H.
1995 Minoan implements and industries. In J.W. Shaw and M.C. Shaw (eds.), *Kommos I. The Kommos Region and Houses of the Minoan Town I. The Kommos Region, Ecology, and Minoan Industries*, 403-535. Princeton: Princeton University Press.

Boas, F.
1955 *Primitive Art*. New York: Dover Publications.

Bosanquet, R.C., and R.M. Dawkins
1923 *The Unpublished Objects from the Palaikastro Excavations 1902–1906*. Part 1. British School at Athens Supplementary Paper 1, London: Mcmillan and Co.

Boulotis, C.
1997 Parastaseis poleon stin Aigaiaki techni tis 2ης chilietias p.X. pro-istorias agona istorimata. *Archaiologia* 62/Martio: 42-56 (in Greek).

Boulotis, C.
2000 Travelling fresco painters in the Aegean

Late Bronze Age: the diffusion patterns of a prestigious art. In S. Sherratt (ed.), *The Wall Paintings of Thera. Proceedings of the First International Symposium. Petros M. Nomikos Conference Centre, Thera, Hellas, 30 August-4 September 1997*, 844-58. Athens: Petros M. Nomikos and the Thera Foundation.

Bourke, S., and J.-P. Descoeudres (eds.)
1995 *Trade, Contact, and the Movement of Peoples in the Eastern Mediterranean. Studies in Honour of J. Basil Hennessy.* Mediterranean Archaeology, Supplement 3. Sydney: Meditarch.

Bourriau, J., and J. Phillips (eds.)
2004 *Invention and Innovation. The Social Context of Technological Change 2: Egypt, the Aegean and the Near East, 1650–1150 BC. Proceedings of a Conference Held at the McDonald Institute for Archaeological Research, Cambridge, (4-6 September 2002).* Oxford: Oxbow.

Bradley, S. (ed.)
1997 *The Interface between Science and Conservation.* British Museum Occasional Paper 116. London: British Museum Press.

Broodbank, C.
1993 Ulysses without sails: trade, distance, knowledge and power in the early Cyclades. *World Archaeology* 24: 315-31.
2000 *An Island Archaeology of the Early Cyclades.* Cambridge: Cambridge University Press.
2004 Minoanisation. *Proceedings of the Cambridge Philosophical Society* 50: 46-91.

Brothwell, D.R., and A.M. Pollard (eds.)
2001 *Handbook of Archaeological Sciences.* Chichester/New York: John Wiley and Sons.

Brown, A.
1994 *Arthur Evans and the Palace of Minos.* Oxford: The Ashmolean Museum.

Brumfiel, E.M.
1998 The multiple identities of Aztec craft specialists. In C.L. Costin and R.P. Wright (eds.), *Craft and Social Identity.* Archaeological Papers of the American Anthropological Association 8: 145-51. Arlington, Virginia: American Anthropological Association.

Brumfiel, E.M., and T.K. Earle
1987 Specialisation, exchange, and complex societies. In E.M. Brumfiel and T.K. Earle (eds.), *Specialisation, Exchange, and Complex Societies*, 1-9. Cambridge: Cambridge University Press.

Brundle, C.R., C.A. Jr. Evans and S. Wilson (eds.)
1992 *Encyclopedia of Materials Characterization. Surfaces, Interfaces, Thin Films.* Boston: Butterworth-Heinemann.

Bryce, T.R.
2002 The Trojan war: is there truth behind the legend? *Near Eastern Archaeology* 65 (3): 182-95.

Brysbaert, A.
1996 A Labyrinth of Problems in the Conservation of Cretan Sites: Is There a Leading Thread? Unpublished BSc Thesis, University College London, Institute of Archaeology.
2000 Bronze Age plaster finds from Palaikastro-Crete: options and limitations of a field laboratory. *Papers of the Institute of Archaeology* 11: 47-60.
2002a Colours in antiquity: colour and hue from Egypt to Byzantium, University of Edinburgh (10th–13th September 2001). *Papers of the Institute of Archaeology* 13: 135-39.
2002b Common craftmanship in the Aegean and east Mediterranean Bronze Age: preliminary technological evidence with emphasis on the painted plaster from Tell el-Dab'a, Egypt. *Egypt and the Levant* 12: 95-107.
2003a Does DIY work? Experimentation and the archaeology of technology in an Aegean Bronze Age context. In A. Brysbaert, N. de Bruijn, E. Gibson, A. Michael and M. Monaghan (eds.), *SOMA 2002. Symposium on Mediterranean Archaeology*. British Archaeological Reports, International Series, 1142: 7–16. Oxford: Archaeopress.

2003b Rotating angles in measuring the Aegean Bronze Age. The technology of Bronze Age painted plaster from the Aegean and eastern Mediterranean. In K.P. Foster and R. Laffineur (eds.), *METRON. Measuring the Aegean Bronze Age. Proceedings of the 9th International Aegean Conference, New Haven, Yale University, 18–21 April 2002*. Aegaeum 24: 167-78. Liège-Austin: Université de Liège-University of Texas.

2004a Take it or leave it? Implications and results of destructive versus non-destructive analysis of Bronze Age painted plasters in the Eastern Aegean. In L. Cleland, K. Stears with G. Davies (eds.), *Colours in the Ancient Mediterranean World*. British Archaeological Reports, International Series, 1267: 9-15. Oxford: Archaeopress.

2004b Technology and Social Agency in Bronze Age Aegean and eastern Mediterranean Painted Plaster. Unpublished PhD Dissertation, Dept. of Archaeology, University of Glasgow, Scotland.

2006 Lapis Lazuli in an enigmatic 'purple' pigment from a thirteenth-century BC Greek wall painting. *Studies in Conservation* 51.4: 252-66.

2007a Cross-craft and cross-cultural interactions during the Aegean and eastern Mediterranean Late Bronze Age. In S. Antoniadou and A. Pace (eds.), *Mediterranean Crossroads*, 325-59. Athens: Pierides Foundation.

2007b Murex uses in plaster features in the Aegean and eastern Mediterranean Bronze Age. *Journal of Mediterranean Archaeology and Archaeometry* 7.2: 29-51.

2007c A technological approach to the painted plaster of Tell el-Dab'a, Egypt: microscopy and scientific analysis. In M. Bietak, N. Marinatos, C. Palivou (eds.), *Taureador Scenes in Tell el-Dab'a (Avaris) and Knossos*, 151-62. Vienna: Verlag der Österreichischen Akademie der Wissenschaften.

2008 The Bronze Age painted plaster from Thebes: a technological study. *Journal of Archaeological Science* (Forthcoming)

Brysbaert, A., and A. Lindenlauf

2000 Colour in Ancient Greece. International conference on the role of colour in Ancient Greek Art and Architecture, 700-31 BC, Thessaloniki, (12th–16th April 2000). *Papers of the Institute of Archaeology* 11: 110-15.

Brysbaert, A., and P. Vandenabeele

2004 Bronze Age painted plaster in Mycenaean Greece: the testing and application of micro-raman spectroscopy in a technological study. *Journal of Raman Spectroscopy* 35 (8/9): 686-93.

2006 Bronze Age painted plaster in Mycenaean Greece: an application of micro-raman spectroscopy within the context of collaborative research. In A. Denker, A. Adriaens, M. Dowsett and A. Giumlia-Mair (eds.), *COST Action G8: Non-Destructive Testing and Analysis of Museum Objects. COST. European Cooperation in the Field of Scientific and Technical Research*, 102-108. Stuttgart: Fraunhofer IRB Verlag.

Brysbaert, A., K. Melessanaki and D. Anglos

2006 Pigment analysis in Bronze Age Aegean and eastern Mediterranean painted plaster by Laser Induced Breakdown Spectroscopy (LIBS). *Journal of Archaeological Science* 33.8: 1095-1104.

Brysbaert, A., and V. Perdikatsis

2008 Bronze Age painted plaster from the Greek mainland: a comparative study of its technology by means of XRD analysis and optical microscopy techniques. In Y. Facorellis, N. Zacharias, K. Polikreti, T. Vakoulis, Y. Bassiakos, V. Kiriatzi, E. Aloupi (eds.), *Archaeometry Studies in the Aegean: Reviews and Recent Developments. Proceedings of the 4th HSA Symposium on Archaeometry*. British Archaeological Reports, International Series 1746: 421-29. Oxford: Archaeopress.

Bulle, H.
1907 *Orchomenos I. Die Älteren Ansiedlungsschichten*, München: Verlag der K.B. Akademie der Wissenschaften.

Burke, B.
1999 Purple and Aegean textile trade in the early second millennium BC. In P.P. Betancourt, V. Karageorghis, R. Laffineur and W.-D. Niemeier (eds.), *MELETEMATA I-III. Studies in Aegean Archaeology Presented to Malcolm H. Wiener as He Enters His 65th Year*. Aegaeum 20: 75-82. Liège-Austin: Université de Liège-University of Texas.
2005 Materialisation of Mycenaean ideology and the Ayia Triada sarcophagus. *American Journal of Archaeology* 109: 403-22.

Burke, H.
1999 *Meaning and Ideology in Historical Archaeology. Style, Social Identity, and Capitalism in an Australian Town*. New York: Kluwer Academic/Penum Publishers.

Buxeda I Garrigós, J., R.E. Jones, V. Kilikoglou, S.T. Levi, Y. Maniatis, J. Mitchell, L. Vagnetti, K.A. Wardle and S. Andreou
2003 Technology transfer at the periphery of the Mycenaean world: the cases of Mycenaean pottery found in Central Macedonia (Greece) and the plain of Sybaris (Italy). *Archaeometry* 45 (2): 263-84.

Cadogan, G.
1992 Myrtos-Pyrgos. In J.W. and E.E. Myers and G. Cadogan (eds.), *The Aerial Atlas of Crete*, 202-208. Berkeley LA: University of California Press.
1994 An Old Palace period Knossos state? In D. Evely, H. Hughes-Brock and N. Momigliano (eds.), *Knossos: A Labyrinth of History. Papers Presented in Honour of Sinclair Hood*, 57-68. Oxford: Oxbow.

Cameron, M.A.S.
1964 Some Unpublished Fresco Fragments from the Minoan Palace at Knossos, Crete. Unpublished MA Thesis, University of Liverpool, UK.
1965 Four fragments of wall paintings with Linear A inscriptions. *Kadmos* 4: 7-15.
1967 Notes on some new joins and additions to well known frescoes from Knossos. In W.C. Brice (ed.), *Europa: Studien zur Geschichte und Epigraphik der Frühen Ägäis. Festschrift für Ernst Grumbach*, 45-74. Berlin: W. de Gruyter.
1968 The painted signs on fresco fragments from the 'House of the Frescoes'. *Kadmos* 7: 45-64.
1971 The Lady in Red: a complementary figure to the Ladies in Blue. *Archaeology* 24: 35-43.
1972 Appendix IV. The plasters. In P. Warren (ed.), *Myrtos: An Early Bronze Age Settlement in Crete*. British School at Athens Supplementary Volume 7: 305-14. Oxford: British School at Athens, Thames and Hudson.
1975 A General Study of Minoan Frescoes, with Particular Reference to Unpublished Wall Paintings from Knossos. Unpublished PhD Dissertation, Newcastle-Upon-Tyne, UK.
1976 On theoretical principles in Aegean Bronze Age mural restoration. In P.P. Betancourt (ed.), *Aegean Art and Archaeology in the Late Bronze Age. Temple University Aegean Symposium. A Symposium Sponsored by the Dept. of Art History, Temple University, Philadelphia, Pennsylvania, (Feb. 27, 1976)*, 20-41. Philadelphia: Temple University.
1978 Theoretical interrelations among Theran, Cretan and Mainland frescoes. In C. Doumas (ed.), *Thera and the Aegean World I. Papers Presented at the Second International Scientific Congress, Santorini, Greece, August* 1978, 579-92. London: Thera and the Aegean World.
1987 The 'Palatial' thematic system in the Knossos murals. Last notes on Knossos frescoes. In R. Hägg and N. Marinatos (eds.), *The Function of the Minoan Palaces. Proceedings of the 5th International Symposium at the Swedish Institute in Athens (10–16 June, 1984)*, 320-28.

Stockholm: Swedish Institute in Athens.

Cameron, M.A.S., and R.E. Jones

1976 A note on the identification of fresco material from the British campaign at Palaikastro, 1902–1906. *Annual of the British School at Athens* 71: 15-20.

Cameron, M.A.S., R.E. Jones and S.E. Philippakis

1977 Scientific analysis of Minoan fresco samples from Knossos. *Annual of the British School at Athens* 72: 121-84.

Cameron, P.

1993 *Crete. Blue Guide*. London: A and C Black.

Caneva, G., M.P. Nugari and O. Salvadori

1991 *Biology in the Conservation of Works of Art*. Rome: ICCROM.

Caple, C.

2000 *Conservation Skills: Judgement, Method and Decision Making*. London: Routledge.

2001 Overview—degradation, investigation and preservation of archaeological evidence. In D.R. Brothwell and A.M. Pollard (eds.), *Handbook of Archaeological Sciences*, 587-93. Chichester-New York: John Wiley and Sons.

Carr, C., and J.E. Neitzel

1995 Future directions for material style studies. In C. Carr and J.E. Neitzel (eds.), *Style, Society, and Person. Archaeological and Ethnological Perspectives*, 437-59. New York: Plenum Press.

Carr, C., and J.E. Neitzel (eds.)

1995 *Style, Society, and Person. Archaeological and Ethnological Perspectives*. New York: Plenum Press.

Cather, S. (ed.)

1991 *The Conservation of Wall Paintings. Proceedings of a Symposium Organized by the Courtauld Institute of Art and the Getty Conservation Institute, London, (July 13–16, 1987)*. Los Angeles: The Getty Conservation Institute.

Caubet, A. (ed.)

1999 *L'Acrobate au Taureau. Les Découvertes de Tell el-Dab'a (Egypte) et l'Archéologie de la Méditerrannée Orientale (1800–1400 av. J.-C.). Actes du Colloque Organisé au Musée du Louvre par le Service Culturel (le 3 Déc. 1994)*. Paris: Louvre.

Caubet, A., and A. Kaczmarczyk

1987 Bronze Age faience from Ras Shamra (Ugarit). In M. Bimson and I. Freestone (eds.), *Early Vitreous Materials*. British Museum Occasional Paper 56: 47-56. London: British Museum Press.

Chapin, A.P.

1997 A re-examination of the Floral Fresco from the Unexplored Mansion at Knossos. *Annual of the British School at Athens* 92: 1-24.

2004a The fresco with multi-coloured rocks and olive branches from the Northwest Slope plaster dump at Pylos reexamined. In L. Morgan (ed.), *Aegean Wall Painting: A Tribute to Mark Cameron*. British School at Athens Studies 13: 123-30. London: British School at Athens.

2004b Power, privilege, and landscape in Minoan Art. In A.P. Chapin (ed.), *XARIS: Essays in Honour of Sara A. Immerwahr*. Hesperia Supplement 33: 47-64. Athens: American School of Classical Studies at Athens.

2004c *XARIS: Essays in Honour of Sara A. Immerwahr*. Hesperia Supplement 33. Athens: American School of Classical Studies at Athens.

Chapman, R.

2005 Changing social relations in the Mediterranean Copper and Bronze Ages. In E. Blake and A.B. Knapp (eds.), *The Archaeology of Mediterranean Prehistory*. Blackwell Studies in Global Archaeology, 77-101. Oxford: Blackwell.

Cherry, J.F.

1986 Polities and palaces: some problems in Minoan state formation. In C. Renfrew and J.F. Cherry (eds.), *Peer Polity Interaction and Socio-Political Change*, 19-45. Cambridge: Cambridge University Press.

1992 Beazely in the Bronze Age? Reflections on attribution studies in Aegean prehistory. In R. Laffineur and J.L.

Crowley (eds.), *EIKΩN: Aegean Bronze Age Iconography: Shaping a Methodology. Proceedings of the 4th International Aegean Conference, University of Tasmania, Hobart, Australia, 6-9 April 1992.* Aegaeum 8: 123-44. Liège: Université de Liège.

1999 After Aidonia: further reflections on attribution in the Aegean Bronze Age. In P.P. Betancourt, V. Karageorghis, R. Laffineur and W.-D. Niemeier (eds.), *MELETEMATA I-III: Studies in Aegean Archaeology Presented to Malcolm H. Wiener as He Enters His 65th Year.* Aegaeum 20: 103-10. Liège-Austin: Université de Liège-University of Texas.

Cherry, J.F., and J.L. Davis

1982 The Cyclades and the Greek Mainland in LC I: the evidence of the pottery. *American Journal of Archaeology* 86: 333-41.

Childs, S.T.

1998 Social identity and craft specialisation among Toro iron workers in western Uganda. In C.L. Costin and R.P. Wright (eds.), *Craft and Social Identity.* Archaeological Papers of the American Anthropological Association 8: 109-21. Arlington, Virginia: American Anthropological Association.

1999 'After all, a hoe bought a wife': the social dimensions of iron working among the Toro of East Africa. In M.-A. Dobres and C.R. Hoffman (eds.), *The Social Dynamics of Technology: Practice, Politics, and World Views,* 23-45. Washington and London: Smithsonian Institution Press.

Chilton, E.S.

1999 Material meanings and meaningful materials. An introduction. In E.S. Chilton (ed.), *Material Meanings: Critical Approaches to the Interpretation of Material Culture,* 1-6. Salt Lake City: University of Utah Press.

Chryssikopoulou, E.

2004 Technologia ton chromaton sto proistoriko Aigaio, oi toichografies tou Akrotiriou Thiras. In Verein zur Förderung der Aufarbeitung der Hellenischen Geschichte (ed.), *Althellenische Technologie und Technik: Tagungsband der Tagung in Ohlstadt (21.–23. März 2003),* 487-99. Weilheim: Verein zur Förderung der Aufarbeitung der Hellenischen Geschichte.

Chryssikopoulou, E., V. Kilikoglou, V. Perdikatsis, S. Sotiropoulou, K. Birtacha and M. Zacharioudakis

2000 Making wall paintings: an attempt to reproduce the painting techniques of Bronze Age Thera. In S. Sherratt (ed.), *The Wall Paintings of Thera: Proceedings of the First International Symposium.* Petros M. Nomikos Conference Centre, Thera, Hellas, 30 August-4 September 1997, 119–27. Athens: Petros M. Nomikos and the Thera Foundation.

Chryssikopoulou, E., and S. Sotiropoulou

2003 To iodes stin paleta Thuraiou zografou. In A. Vlachopoulos and K. Birtacha (eds.), *ARGONAUTIS. Timitikos Tomos jia ton Kathigiti Christo Douma. Apo tous Mathites tou sto Panepistimio Athinon (1980–2000),* 490-504. Athens: Kathemerini (in Greek).

Ciliberto, E.

2000 Analytical methods in art and archaeology. In E. Ciliberto and G. Spoto (eds.), *Modern Analytical Methods in Art and Archaeology,* 1-10. New York: John Wiley and Sons.

Clark, J.E., and S.D. Houston

1998 Craft specialisation, gender, and personhood among the Post-Conquest Maya of Yucatan, Mexico. In C.L. Costin and R.P. Wright (eds.), *Craft and Social Identity.* Archaeological Papers of the American Anthropological Association 8: 31-46. Arlington, Virginia: American Anthropological Association.

Cline, E.H.

1994 *Sailing the Wine-Dark Sea: International Trade and the Late Bronze Age Aegean.* British Archaeological Reports, International Series 591. Oxford: Tempus Reparatum.

1998 Rich beyond the dreams of Avaris: Tell el-Dab'a and the Aegean—A guide for the perplexed. *Annual of the British School at Athens* 93: 199-219.

2005 The multivalent nature of imported objects in the ancient Mediterranean world. In R. Laffineur and E. Greco (eds.), *EMPORIA: Aegean in the Central and Eastern Mediterranean. Proceedings of the 10th International Aegean Conference, Athens 14–18 April 2004*. Aegaeum 25: 45-51. Liège-Austin: Université de Liège-University of Texas.

Cline, E.H., and D. Harris-Cline (eds.)

1998 *The Aegean and the Orient in the Second Millennium. Proceedings of the 50th Anniversary Symposium, Cinncinnati, 18–20 April 1997*. Aegaeum 18. Liège-Austin: Université de Liège-University of Texas.

Coburn, A., E. Dudley and R. Spence

1989 *Gypsum Plaster: Its Manufacture and Use*. London: Intermediate Technology Publications.

Coles, J.

1979 *Experimental Archaeology*. London: Academic Press.

Collon, D.

1994 Bull-leaping in Syria. *Egypt and Levant* 4: 81-88.

2000 Syrian glyptic and the Thera wall paintings. In S. Sherratt (ed.), *The Wall Paintings of Thera: Proceedings of the First International Symposium. Petros M. Nomikos Conference Centre, Thera, Hellas, 30 August–4 September 1997*, 283-94. Athens: Petros M. Nomikos and the Thera Foundation.

Conkey, M.W.

1993 Experimenting with style in archaeology: some historical and theoretical issues. In M.W. Conkey and C.A. Hastorf (eds.), *The Use of Style in Archaeology*, 5-17. Cambridge: Cambridge University Press.

Conkey, M.W., and C.A. Hastorf

1993 Introduction. In M.W. Conkey and C.A. Hastorf (eds.), *The Use of Style in Archaeology*, 1-4. Cambridge: Cambridge University Press.

Conneler, C.

2006 The space and time of the *chaîne opératoire*: technological approaches to past landscapes. In B. Gravina (ed.), *Technologies: Changing Matter, Changing Minds*. Archaeological Review from Cambridge 21 (1): 38-49. Cambridge: Department of Archaeology.

Cornale, P., Principalle, Rosanó P., and M. Tagliapietra

2002 Appendix II. Laboratory analysis. In N. Scheftelowitz and R. Oren (eds.), *Tel Kabri. The 1986–1993 Excavation Seasons. Aharon Kempinski*, 287-88. Tel Aviv: Emery and Claire Yass Publications in Archaeology.

Costin, C.L.

1991 Craft specialisation: issues in defining, documenting, and explaining the organisation of production. In M.B. Schiffer (ed.), *Archaeological Method and Theory* 3: 1-56. Tucson: University of Arizona Press.

Costin, C.L.

1996 Craft production and mobilisation strategies in the Inka Empire. In B. Wailes (ed.), *Craft Specialisation and Social Evolution: In Memory of V. Gordon Childe*. University Museum Monograph 93: 211-28. Philadelpia: University Museum of Archaeology and Anthropology, University of Pennsylvania.

Costin, C.L.

1998a Introduction: craft and social identity. In C.L. Costin and R.P. Wright (eds.), *Craft and Social Identity*. Archaeological Papers of the American Anthropological Association 8: 3-16. Arlington, Virginia: American Anthropological Association.

1998b Housewives, chosen women, skilled men: cloth production and social identity in the Late Prehispanic Andes. In C.L. Costin and R.P. Wright (eds.), *Craft and Social Identity*. Archaeological Papers of the American Anthropological Association 8: 123-41. Arlington,

Virginia: American Anthropological Association.
2005 Craft Production. In H.D.G. Maschner and C. Chippindale (eds.), *Handbook of Archaeological Methods*, 1034-1107. Lanham/New York: Altamira Press.

Costin, C.L., and R.P. Wright (eds.)
1998 *Craft and Social Identity*. Archaeological Papers of the American Anthropological Association, 8. Arlington, Virginia: American Anthropological Association.

Crawford, B.H.
1973 Just perceptible colour differences in relation to level of illumination. *Studies in Conservation* 18: 159-66.

Croft, S., H. McCurdy and C. Thompson
1994 *The Hutchinson Encyclopedic Dictionary*. Oxford: Helicon.

Cronyn, J.M.
1992 *The Elements of Archaeological Conservation*. London: Routledge.

Crowley, J.L.
1989 *The Aegean and the East: An Investigation into the Transference of Artistic Motifs between the Aegean, Egypt, and the Near East in the Bronze Age*. Jonsered: Paul Åström's Förlag.

Cullen, T.
2001a Introduction: voices and visions of Aegean prehistory. In T. Cullen (ed.), *Aegean Prehistory: A Review*. American Journal of Archaeology Supplement 1: 1-18. Boston: Archaeological Institute of America.

Cullen, T. (ed.)
2001b *Aegean Prehistory: A Review*. American Journal of Archaeology, Supplement 1. Boston: Archaeological Institute of America.

Dabney, M.K.
1997 Craft production consumption as an economic indicator of site status in regional studies. In R. Laffineur and P.P. Betancourt (eds.), *TEXNH: Crafsmen, Craftswomen and Craftsmanship in the Aegean Bronze Age. Proceedings of the 6th International Aegean Conference, Philadelphia, Temple University, 18–21 April 1996*. Aegaeum 16: 467-71. Liège-Austin: Université de Liège-University of Texas.

Dakouri-Hild, A.
2001 The House of Kadmos in Mycenaean Thebes reconsidered: architecture, chronology, and context. *Annual of the British School at Athens* 96: 81-122.
2005 Breaking the mould? Production and economy in the Theban state. In A. Dakouri-Hild and S. Sherratt (eds.), *AUTOCHTHON. Papers Presented to O.T.P.K. Dickinson on the Occasion of His Retirement. Institute of Classical Studies, University of London, 9 November 2005*. British Archaeological Reports, International Series 1432: 207-24. Oxford: Archaeopress.

Dakouri-Hild, A. and S. Sherratt (eds.)
2005 *AUTOCHTHON. Papers Presented to O.T.P.K. Dickinson on the Occasion of His Retirement. Institute of Classical Studies, University of London, 9 November 2005*. British Archaeological Reports, International Series 1432. Oxford: Archaeopress.

Dalley, S.
1984 *Mari and Karana. Two Old Babylonian Cities*. London: Longman.

Dandrau, A.
1999 La peinture murale minoenne, I. La palette du peintre Égéen et Égyptien à l'Âge du Bronze. Nouvelles données analytiques. *Bulletin de Correspondance Hellenique* 123 (1): 1-41.
2000 La peinture murale minoenne, II. La production des enduits: matériaux et typologie. *Bulletin de Correspondance Hellenique* 124 (1): 75-97.
2001 La peinture murale minoenne, III. Méthodes et techniques d'exécution. *Bulletin de Correspondance Hellenique* 125 (1): 41-66.

David, N., and C. Kramer
2001 *Ethnoarchaeology in Action*. Cambridge World Archaeology. Cambridge: Cambridge University Press.

Davies, W.V. (ed.)
2001 *Colour and Painting in Ancient Egypt*. London: British Museum Press.

Davies, W.V., and L. Schofield (eds.)
1995 *Egypt, the Aegean and the Levant: Interconnections in the Second Millennium BC.* London: British Museum Press.

Davis, E.N.
2000a The organisation of the Theran artists. In S. Sherratt (ed.), *The Wall Paintings of Thera: Proceedings of the First International Symposium. Petros M. Nomikos Conference Centre, Thera, Hellas. 30 August-4 September 1997*, 859-72. Athens: Petros M. Nomikos and the Thera Foundation.
2000b The Egyptian influence on Minoan figure painting. In A. Karetsou (ed.), *Kriti-Aigyptos: Polutismikoi Desmoi Trion Chilieton. Meletes,* 64-70. Athens: Ypourgeio Politismou-Archaiologiko Mouseio Herakleiou (in Greek).

Davis, J.L.
1982 The earliest Minoans in the south-east Aegean. *Anatolian Studies* 32: 33-41.
2001 Review of Aegean prehistory I: the islands of the Aegean. In T. Cullen (ed.), *Aegean Prehistory: A Review.* American Journal of Archaeology Supplement 1: 19-94. Boston: Archaeological Institute of America.
2008 Hall 64 at Pylos: The Context of the Wall-Paintings (paper presented at the Round Table, Sheffield 25–27 January 2008).

Davis, J.L., and J.F. Cherry
1984 Phylakopi in Late Cycladic I: a pottery seriation study. In J.A. McGillivray and R.L.N. Barber (eds.), *The Prehistoric Cyclades. Contributions to a Workshop on Cycladic Chronology*, 148-61. Edinburgh: Department of Classical Archaeology.

DeMarrais, E.
2004 The materialization of culture. In E. DeMarrais, C. Gosden and C. Renfrew (eds.), *Rethinking Materiality: The Engagement of Mind with the Material World.* McDonald Institute Monographs, 11–22. Oxford: Oxbow.

Denninger, E.
1974 What is 'Bianco di San Giovanni of Cennino Cennini'? *Studies in Conservation* 19: 185-87.

Dever, W.G.
1997 Settlement patterns and chronology of Palestine in the Middle Bronze Age. In E.D. Oren (ed.), *The Hyksos: New Historical and Archaeological Perspectives*, 285-301. Philadelphia: University Museum, University of Pennsylvania.

Dickinson, O.
1994 *The Aegean Bronze Age*. Cambridge: Cambridge University Press.

Dobres, M.-A.
1999 Of paradigms and ways of seeing. Artifact variability as if people mattered. In E.S. Chilton (ed.), *Material Meanings: Critical Approaches to the Interpretation of Material Culture*, 7-23. Salt Lake City: University of Utah Press.

Dobres, M.-A.
2000 *Technology and Social Agency. Outlining a Practice Framework for Archaeology*. Oxford: Blackwell.
2001 Meaning in the making: agency and the social embodiment of technology and art. In M.B. Schiffer (ed.), *Anthropological Perspectives on Technology,* 47-76. Albuquerque: University of New Mexico Press.

Doumas, C. (ed.)
1978 *Thera and the Aegean World I. Papers Presented at the Second International Scientific Congress, Santorini, Greece (August 1978).* London: Thera and the Aegean World.
(1980 *Thera and the Aegean World II. Papers Presented at the Second International Scientific Congress, Santorini, Greece (August 1978).* London: Thera and the Aegean World.

Doumas, C.
1999 *The Wall Paintings of Thera*. Athens: Thera Foundation/Petros M. Nomikos.

Drenkhahn, R.
1995 Artisans and artists in Pharaonic Egypt. In J.M. Sasson (ed.), *Civilisations of the Ancient Near East* I:

331-44. New York: Charles Scribner's Sons.

Driessen, J.

2005 On the use of upper floors in Minoan Neopalatial architecture. In I. Bradfer-Burdet, B. Detournay and R. Laffineur (eds.), *Kris Technitis. L'Artisan Crétois. Recueil d'Articles en l'Honneur de Jean-Claude Poursat, Publié à l'Occasion des 40 Ans de la Découverte du Quartier Mu.* Aegaeum 26: 83-88. Liège-Austin: Université de Liège-University of Texas.

Driessen, J., and C.F. Macdonald

1997 *The Troubled Island: Minoan Crete before and after the Santorini Eruption.* Aegaeum 17. Liège-Austin: Université de Liège-University of Texas.

Duell, P., and R.J. Gettens

1942 A review of the problem of Aegean wall painting. *Technical Studies* X-4: 179-223.

Du Mesnil du Buisson, R., Comte

1935 *Le Site Archéologique de Mishrifé-Qatna.* Collection de Textes et Documents d'Orient 1. Paris: Boccard.

Durand, J.-M.

1999 La façade occidentale du Proche-Orient d'après les textes de Mari. In A. Caubet (ed.), *L'Acrobate au Taureau. Les Découvertes de Tell el-Dab'a (Egypte) et l' Archéologie de la Méditerrannée Orientale (1800–1400 av. J.-C.). Actes du Colloque Organisé au Musée du Louvre par le Service Culturel (le 3 Déc. 1994)*, 149-64. Paris: Louvre.

Earle, T.

1993 Style and iconography as legitimation in complex chiefdoms. In M.W. Conkey and C.A. Hastorf (eds.), *The Use of Style in Archaeology*, 73-81. Cambridge: Cambridge University Press.

El Goresy, A.

2000 Polychromatic wall painting decorations in monuments of Pharaonic Egypt: compositions, chronology and painting techniques. In S. Sherratt (ed.), *The Wall Paintings of Thera: Proceedings of the First International Symposium. Petros M. Nomikos Conference Centre, Thera, Hellas, 30 August-4 September 1997*, 49-70. Athens: Petros M. Nomikos and the Thera Foundation.

Elsen-Novák, G.

2002 Die Altsyrische Glyptik aus Qatna. *Mitteilungen des Deutschen Orient-Gesellschaft zu Berlin* 134: 257-74.

Eriksson, K.O.

1995 Egyptian amphorae from Late Cypriot contexts in Cyprus. In S. Bourke and J.-P. Descoeudres (eds.), *Trade, Contact, and the Movement of Peoples in the Eastern Mediterranean. Studies in Honour of J. Basil Hennessy.* Mediterranean Archaeology, Supplement 3: 199-205. Sydney: Meditarch.

Evans, A.

1921 *The Palace of Minos at Knossos, Crete*, I. London: Macmillan.

1928 *The Palace of Minos at Knossos, Crete*, II. London: Macmillan.

1935 *The Palace of Minos at Knossos, Crete*, IV. London: Macmillan.

Evely, D. (ed.)

1999 *Fresco: A Passport into the Past. Minoan Crete Through the Eyes of Mark Cameron.* Athens: British School at Athens and N.P. Goulandris Foundation-Museum of Cycladic Art.

Evely, R.D.G.

1993 *Minoan Crafts: Tools and Techniques. An Introduction.* I. Studies in Mediterranean Archaeology 92 (1). Jonsered: P. Åström's Förlag.

2000 *Minoan Crafts: Tools and Techniques. An Introduction.* II. Studies in Mediterranean Archaeology 92 (2). Jonsered: P. Åström's Förlag.

Evely, D., H. Hughes-Brock and N. Momigliano (eds.)

1994 *Knossos: A Labyrinth of History. Papers Presented in Honour of Sinclair Hood.* Oxford: Oxbow.

Evershed, R.P., S. Charters and A. Quye

1995 Interpreting lipid residues in archaeological ceramics: preliminary results from laboratory simulations of vessel use and burial. *Material Research Society Symposium Proceedings* 352: 85-95.

Feldman, M.H.
2003 Ambiguous identities: the 'Marriage' vase of Niqmaddu II and the elusive Egyptian princess. *Journal of Mediterranean Archaeology* 15.1: 77-98.
2006 *Diplomacy by Design: Luxury Arts and an 'International Style' in the Ancient Near East, 1400–1200* BC. Chicago: University of Chicago Press.
2007 Frescoes, exotica, and the reinvention of the northern Levantine Kingdoms during the second millennium BCE. In M. Heinz and M.H. Feldman (eds.), *Representations of Political Power: Case Histories from Times of Change and Dissolving Order in the Ancient Near East*, 39-65. Winona Lake: Eisenbrauns.

Fitton, L. (ed.)
1984 *Cycladica. Studies in Memory of N.P. Goulandris*. London: British Museum) Press.

Forbes, R.J.
1966a *Studies in Ancient Technology*, III. Leiden: Brill.
1966b *Studies in Ancient Technology*, VII. Leiden: Brill.

Forrer, E.O.
1924 Vorhomerische Griechen in den Keilchrifttexten von Boghazköi. *Mitteilungen des Deutschen Orient-Gesellschaft zu Berlin* 63: 1-22.

Foster, K.P.
1979 *Aegean Faience of the Bronze Age*. New Haven and London: Yale University Press.

Foster, K.P., and R. Laffineur (eds.)
2003 *METRON. Measuring the Aegean Bronze Age. Proceedings of the 9th International Aegean Conference, New Haven, Yale University, (18–21 April 2002)*, Aegaeum 24. Liège-Austin: Université de Liège-University of Texas.

Frankfort, H.
1996 *The Art and Architecture of the Ancient Orient*. New Haven: Yale University Press.

French, E.B.
1965 Late Helladic IIIA2 pottery from Mycenae. *Annual of the British School at Athens* 60: 159-202.
2002 *Mycenae. Agamemnon's Capital. The Site and its Setting*. Strand: Tempus.
2005 The role of Mycenae. In R. Laffineur and E. Greco (eds.), *EMPORIA. Aegean in the Central and eastern Mediterranean. Proceedings of the 10th International Aegean Conference, Athens, 14–18 April 2004*. Aegaeum 25: 125-27. Liège-Austin: Université de Liège-University of Texas.

French, E., and K. Shelton
2005 Early palatial Mycenae. In A. Dakouri-Hild and S. Sherratt (eds.), *AUTOCHTHON. Papers Presented to O.T.P.K. Dickinson on the Occasion of His Retirement. Institute of Classical Studies, University of London, 9 November 2005*. British Archaeological Reports, International Series 1432: 175-84. Oxford: Archaeopress.

Friedrich, W.L., B. Kromer, M. Friedrich, J. Heinemeier, T. Pfeigger, S. Talamo
2006 Santorini eruption radiocarbon dated to 1627–1600 BC. *Science* 312: 548.

Gale, N.H.
1991 Copper oxhide ingots: their origin and their place in the Bronze Age metals trade in the Mediterranean. In N.H. Gale (ed.), *Bronze Age Trade in the Mediterranean. Papers Presented at the Conference Held at Rewley House, Oxford, in December 1989*. Studies in Mediterranean Archaeology 90: 197-239. Jonsered: P. Åström's Förlag.

Gates, C.
2004 The adoption of pictorial imagery in Minoan wall painting: a comparatist perspective. In A.P. Chapin (ed.), *XARIS. Essays in Honour of Sara A. Immerwahr*. Hesperia Supplement 33: 27-46. Athens: American School of Classical Studies at Athens.

Gell, A.
1998 *Art and Agency. An Anthropological Theory*. Oxford: Clarendon Press.

Gero, J.
1989 Assessing social information in material objects: how well do lithics measure

up? In R. Torrence (ed.), *Time, Energy and Stone Tools*, 92-105. Cambridge: Cambridge University Press.

Gesell, G.C.
2000 Blood on the horns of consecration? In S. Sherratt (ed.), *The Wall Paintings of Thera: Proceedings of the First International Symposium. Petros M. Nomikos Conference Centre, Thera, Hellas, 30 August-4 September 1997*, 947-57. Athens: Petros M. Nomikos and the Thera Foundation.

Gettens, R. J., and G.L. Stout
1966 *Painting Materials. A Short Encyclopaedia.* New York: Dover.

Getz-Preziosi, P.
1985 *Early Cycladic Sculpture: An Introduction.* Malibu: Getty Museum.

Gibbons P. (ed.)
1995 *Preparation and Use of Lime Mortars. An Introduction to the Principles of Using Lime Mortars.* Edinburgh: Scottish Lime Centre-Historic Scotland.

Gillard, R.D., S.M. Hardman, R.G. Thomas and D.E. Watkinson
1994 The detection of dyes by FTIR microscopy. *Studies in Conservation* 39: 187-92.

Gillis, C.
1994 Binding evidence. Tin foil and organic binders on Aegean Late Bronze Age pottery. *Opuscula Atheniensia* 20 (5): 57-61.

Gitlen, B.M.
1981 The cultural and chronological implications of the Cypro-Palestinian trade during the Late Bronze Age. *Bulletin of the American Schools of Oriental Research* 241: 49-59.

Given, M., and A.B. Knapp
2003 *The Sydney Cyprus Survey Project. Social Approaches to Regional Archaeological Survey.* Monumenta Archaeologica 21. Los Angeles: University of California, Cotsen Institute of Archaeology.

Goodenough, W.
2003 In pursuit of culture. *Annual Review of Anthropology* 32: 1-12.

Gooding, D.
1990 *Experiment and the Making of Meaning. Human Agency in Scientific Observation and Experiment.* Dordrecht: Kluwer Academic Publishers.

Goren, Y., I. Bunimovitz, I. Finkelstein and N. Na'aman
2003 The location of Alashiya: new evidence from petrographic investigation of Alashiyan tablets from El-Amarna and Ugarit. *American Journal of Archaeology* 107: 233-55.

Gosden, C.
1999 *Archaeology and Anthropology: A Changing Relationship.* London: Routledge.
2004 Aesthetics, intelligence and emotions: implications for archaeology. In E. DeMarrais, C. Gosden and C. Renfrew (eds.), *Rethinking Materiality: The Engagement of Mind with the Material World.* McDonald Institute Monographs, 33-40. Oxford: Oxbow.
2006 What Do Objects Want? Lecture presented at the School of Archaeology and Ancient History, University of Leicester, UK, February 2006.

Gosden, C., and Y. Marshall
1999 The cultural biography of objects. *World Archaeology* 31 (2): 169-78.

Gould, R.A.
2000 Ethnoarchaeology. In L. Ellis (ed.), *Archaeological Method and Theory: An Encyclopedia*, 181-87. New York/London: Garland.

Gourdin, W.H. and W.D. Kingery
1975 The beginnings of pyrotechnology: Neolithic and Egyptian lime plaster. *Journal of Field Archaeology* 2: 133-50.

Gowning, R., and A. Heritage (eds.)
2003 *Conserving the Painted Past. Developing Approaches to Wall Painting Conservation. Post-prints of a Conference Organized by English Heritage, London, 2-4 December, 1999.* London: James and James.

Gravina, B.
2006 Introduction. Technologies: changing matter, changing minds. In B. Gravina

(ed.), *Technologies: Changing Matter, Changing Minds*. Archaeological Review from Cambridge 21 (1): 1-5. Cambridge: Department of Archaeology.

Graziado, G.
1998 Trade circuits and trade-routes in the Shaft Grave period. *Studi Micenei ed Egeo-Anatolici* 40: 29-76.

Greenfield, P.
2000 Children, material culture and weaving: historical change and developmental change. In J. Sofaer Deverenski (ed.) *Children and Material Culture*, 72-86. London: Routledge.

Grimm, L.
2000 Apprentice flintknapping: relating material culture and social practice in the Upper Palaeolithic. In J. Sofaer Deverenski (ed.), *Children and Material Culture*, 53-71. London: Routledge

Grissom, C.A.
1986 Green earth. In R.L. Feller (ed.), *Artists' Pigments: A Handbook of Their History and Characteristics*, 1, 141-67. New York/Oxford: Oxford University Press.

Guichard, M.
1999 Les mentions de la Crète à Mari. In A. Caubet (ed.), *L'Acrobate au Taureau. Les Découvertes de Tell el-Dab'a (Egypte) et l'Archéologie de la Méditerranée Orientale (1800–1400 av. J.-C.). Actes du Colloque Organisé au Musée du Louvre par le Service Culturel (le 3 Déc. 1994)*, 165-77. Paris: Louvre.

Gunter, A.C.
1990 Artists in Near Eastern art. In A.C. Gunter (ed.), *Investigating Artistic Environments in the Ancient Near East*, 9-17. Washington: Arthur M. Sackler Gallery, Smithsonian Museum.

Güterbock, H.G.
1983 The Hittites and the Aegean world, part I. The Ahhiyawa problem reconsidered. *American Journal of Archaeology* 87: 133-38.

Güterbock, H.G.
1984 Hittites and Akhaeans: a new look. *Proceedings of the American Philosophical Society* 128: 114-22.

Hägg, R.
1985 Pictorial programmes in the Minoan palaces and villas. In P. Darcque and J.-C. Poursat (eds.), *L'Iconographie Minoenne. Actes de la Table Ronde d'Athènes (21–22 Avril 1983)*. Bulletin de Correspondance Hellenique Supplement 11: 209-17. Athens: Ecole Française d'Athène.

Hägg, R. (ed.)
1997 *The Function of the 'Minoan Villa'*. Stockholm: Svenska Institutet i Athen, Jonsered: P. Åström's Förlag.

Haider, P.W.
1996 Menschenhandel zwischen dem Ägyptischen Hof und der Minoisch-Mykenischen Welt? *Egypt and Levant* 6: 137-56.

Hall, J.M.
1997 *Ethnic Identity in Greek Antiquity*. Cambridge: Cambridge University Press.
2001 Social archaeology and the theatres of memory. *Journal of Social Archaeology* 1: 50-61.

Hallager, E.
1985 *The Master Impression. A Clay Sealing from the Greek-Swedish Excavations at Kastelli, Khania*. Studies in Mediterranean Archaeology 69. Göteborg: P. Åström's Förlag.

Hamilakis, Y.
1998 Consumption patterns, factional competition and political development in Bronze Age Crete. *Bulletin of the Institute of Classical Studies* 42: 233-34.

Hankey, V.
1993 A Theban battle axe. Queen Aahotep and the Minoans. *Minerva* 4 (3): 13-14.

Hankey, V.
1995 A Late Bronze Age temple at Amman airport: small finds and pottery. In S. Bourke and J.-P. Descoeudres (eds.), *Trade, Contact, and the Movement of Peoples in the eastern Mediterranean. Studies in Honour of J. Basil Hennessy.*

Mediterranean Archaeology Supplement 3: 169-85. Sydney: Meditarch.

Hankey, V., and A. Leonard Jr.

1998 Aegean LB I-II pottery in the east: 'who is the potter, pray, and who the pot?' In E.H. Cline and D. Harris-Cline (eds.), *The Aegean and the Orient in the Second Millennium. Proceedings of the 50th Anniversary Symposium, Cincinnati, 18–20 April 1997*. Aegaeum 18: 29-37. Liège-Austin: Université de Liège-University of Texas.

Hardy, D.A., C.G. Doumas, I.A. Sakellarakis, and P.M. Warren (eds.)

1990 *Thera and the Aegean World III-I: Archaeology. Proceedings of the 3rd International Congress, Santorini, Greece (3-9 September 1989)*. London: Thera Foundation.

Haskell, H., R.E. Jones, P.M. Day, H. Catling and J.T. Killen.

2005 *Transport Stirrup Jars of the Bronze Age Aegean and Eastern Mediterranean*. Fitch Laboratory Occasional Paper 5. London: British School at Athens.

Heaton, N.

1910 The mural paintings of Knossos: an investigation into the method of their production. *Journal of the Royal Society of Arts* LVIII: 206-12.

1912 On the nature and method of execution of specimens of painted plaster from the palace of Tiryns. In G. Rodenwaldt (ed.), *Tiryns II. Die Ergebnisse der Ausgrabungen des Instituts. Die Fresken des Palastes*. Athens: Eleutheroudakis and Barth.

Helms, M.W.

1993 *Craft and the Kingly Ideal: Art, Trade and Power*. Austin: Texas University Press.

2004 Tangible materiality and cosmological others in the development of sedentism. In E. DeMarrais, C. Gosden and C. Renfrew (eds.), *Rethinking Materiality: The Engagement of Mind with the Material World*, 117-27. Cambridge: McDonald Institute for Archaeological Research.

Hiller, S.

1990 The Miniature Frieze in the West House—evidence for Minoan poetry? In D.A. Hardy, C.G. Doumas, I.A. Sakellarakis and P.M. Warren (eds.), *Thera and the Aegean World III-I: Archaeology. Proceedings of the 3rd International Congress, Santorini, Greece, 3-9 September 1989*, 229-36. London: Thera Foundation.

2000 Thera ships, Egypt and Homer. In S. Sherratt (ed.), *The Wall Paintings of Thera: Proceedings of the First International Symposium. Petros M. Nomikos Conference Centre, Thera, Hellas, 30 August-4 September 1997*, 334-44. Athens: Petros M. Nomikos and the Thera Foundation.

2003 Beni Hassan, die Anfänge der Minoischen Freskenkunst und Fragen der Chronologie. In M. Bietak (ed.), *The Synchronisation of Civilisations in the Eastern Mediterranean in the 2nd Millennium B.C. II. Proceedings of the SCIEM 2000-EuroConference, Haindorf, (2-7 May 2001)*, 487-99. Vienna: Verlag der Österreichischen Akademie der Wissenschaften.

Hirsch, E.S.

1975 Infrared photography and archaeology: painted plaster floors at Gournia. *Archaeology* 28 (4): 260-66.

1977 *Painted Decoration on the Floors of Bronze Age Structures on Crete and the Greek Mainland*. Studies in Mediterranean Archaeology 53. Göteborg: P. Åström's Förlag.

Hitchcock, L.A.

2000 *Minoan Architecture: A Contextual Analysis*. Studies in Mediterranean Archaeology Pocket-Book 155. Jonsered: P. Åström's Förlag.

2005 'Who will personally invite a foreigner, unless he is a craftsman?': exploring interconnections in Aegean and Levantine architecture. In R. Laffineur and E. Greco (eds.), *EMPORIA. Aegean in the Central and eastern Mediterranean. Proceedings of the 10th International*

*Aegean Conference, Athens 14–18 April 2004.* Aegaeum 25: 691-99. Liège-Austin: Université de Liège-University of Texas.

Hocombe, S.
1999 *Fresco Painting for Home and Garden.* Devon: David and Charles.

Hodder, I.
1986 *Reading the Past. Current Approaches to Interpretation in Archaeology.* Cambridge: Cambridge University Press.
1993 Style as historical quality. In M.W. Conkey and C.A. Hastorf (eds.), *The Use of Style in Archaeology*, 44-51. Cambridge: Cambridge University Press.

Hodges, H.
1970 *Technology in the Ancient World.* London: Allen Lane, Penguin Press.

Hood, S.
1978 *The Arts in Prehistoric Greece.* Harmondsworth: Penguin Books.
2004 Dating the Knossos frescoes. In L. Morgan (ed.), *Aegean Wall Painting: A Tribute to Mark Cameron.* British School at Athens Studies 13: 45-81. London: British School at Athens.

Horden, P., and N. Purcell
2000 *The Corrupting Sea: A Study of Mediterranean History.* Oxford: Blackwell.

Housley, R.A., S.W. Manning, G. Cadogan, R.E. Jones and R.E.M. Hedges
1999 Radiocarbon, calibration, and the chronology of the Late Minoan IB phase. *Journal of Archaeological Science* 26 (2): 159-72.

Iakovidis, S.E.
1983 *Late Helladic Citadels on Mainland Greece.* Leiden: Brill.
1989 *Glas I. I Anaskafi 1955–1961.* Vivliothiki tis en Athinais Archaiologikis Etairias, 107. Athens: Archaiologiki Etairia. (in Greek).
1998 *Glas II. I Anaskafi 1981–1991.* Vivliothiki tis en Athinais Archaiologikis Etairias, 173. Athens: Archaiologiki Etairia. (in Greek).
2001 *Gla and the Kopais in the 13th Century BC.* Library of the Archaeological Society at Athens, 221. Athens: Archaeological Society.

Immerwahr, S.A.
1985 A possible influence of Egyptian art in the creation of Minoan wall painting. In P. Darcque and J.-C. Poursat (eds.), *L'Iconographie Minoenne. Actes de la Table Ronde d'Athènes (21–22 Avril 1983).* Bulletin de Correspondance Hellenique Supplement 11: 41-50. Athens: Ecole Française d'Athène.
1990 *Aegean Painting in the Bronze Age.* Philadelphia: Pennsylvania State University Press.

Ingold, T.
1993 The temporality of the landscape. *World Archaeology* 25 (2): 152-74.
1999 Foreword. In M.-A. Dobres and C.R. Hoffman (eds.), *The Social Dynamics of Technology: Practice, Politics, and World Views*, vii-xi. Washington and London: Smithsonian Institution Press.
2001 Beyond art and technology: the anthropology of skill. In M.B. Schiffer (ed.), *Anthropological Perspectives on Technology.* Amerind Foundation New World Studies 5: 17-32. Albuquerque: University of New Mexico Press.

Iversen, E.
1975 *Canon and Proportions in Egyptian Art.* Warminster: Aris and Phillips.

Jacobsen, I.
1994 *Aegyptiaca from Late Bronze Age Cyprus.* Studies in Mediterranean Archaeology 112. Jonsered: P. Åström's Förlag.

Jánosi, P.
2002 Bericht über die im Frühjahr 2001 Erfolgten Sondagen in Dorf 'Ezbet Helmi (Grabungsfläche H/I). *Egypt and Levant* 12: 195-210.

Jasink, A.M.
2005 Mycenaean means of communication and diplomatic relations with foreign royal courts. In R. Laffineur and E. Greco (eds.), *EMPORIA. Aegean in the Central and Eastern Mediterranean. Proceedings of the 10th International*

*Aegean Conference, Athens 14–18 April 2004*. Aegaeum 25: 59-67. Liège-Austin: Université de Liège-University of Texas.

Jones, A.
2002 *Archaeological Theory and Scientific Practice*. Cambridge: Cambridge University Press.

Jones, B.
1998 Minoan Women's Clothes: An Investigation of their Construction from the Depictions in Aegean Art. Unpublished PhD Dissertation, New York University (UMI, Ann Arbor).
2000 Revealing Minoan fashions. Cretan styles regain center stage. *Archaeology* May/June 2000: 36-41.
2003 Veils and mantles: an investigation of the construction and function of the costumes of the veiled dancer from Thera and the Camp Stool Banqueter of Knossos. In K.P. Foster and R. Laffineur (eds.), *METRON. Measuring the Aegean Bronze Age. Proceedings of the 9th International Aegean Conference, New Haven, Yale University, 18–21 April 2002*. Aegaeum 24: 441-50. Liège-Austin: Université de Liège-University of Texas.

Jones, B.
2005 The clothes-line: imports and exports of Aegean cloth(es) and iconography. In R. Laffineur and E. Greco (eds.), *EMPORIA. Aegean in the Central and eastern Mediterranean. Proceedings of the 10th International Aegean Conference, Athens 14–18 April 2004*. Aegaeum 25: 707-16. Liège-Austin: Université de Liège-University of Texas.

Jones, R.E.
1978 Analysis of Painted Plaster from Phylakopi. Unpublished report for the Phylakopi excavations.
1987 Minoan Faience from Knossos (Royal Road), Sellopoulo, Pyrgos and Kommos. Unpublished Interim Report.
1999 Analyses of frescoes, technical studies and replication experiments. In D. Evely (ed.), *Fresco: A Passport into the Past. Minoan Crete through the Eyes of Mark Cameron*, 141-73. Athens: British School at Athens and N.P.Goulandris Foundation-Museum of Cycladic Art.

Jones, R.E., and C. B. Mee
1986 Provenance studies of Aegean Late Bronze Age pottery. In R.E. Jones (ed.), *Greek and Cypriot Pottery: A Review of Scientific Studies*. The British School at Athens Fitch Laboratory Occasional Paper 1: 439-521. London: British School at Athens.

Jones, R.E., with E. Photos-Jones
2004 Technical studies of Aegean Bronze Age wall painting: methods, results and future prospects. In L. Morgan (ed.), *Aegean Wall Painting: A Tribute to Mark Cameron*. British School at Athens Studies 13: 199-228. London: British School at Athens.

Jones, S.
1997 *The Archaeology of Ethnicity: Constructing Identities in the Past and Present*. London and New York: Routledge.

Kaiser, B.
1976 *Untersuchungen zum Minoischen Relief*. Bonn: Rudolf Habelt.

Kaiser, I.
2005 Minoan Miletus: a view from the kitchen. In R. Laffineur and E. Greco (eds.), *EMPORIA. Aegean in the Central and eastern Mediterranean. Proceedings of the 10th International Aegean Conference, Athens 14–18 April 2004*. Aegaeum 25: 193-98. Liège-Austin: Université de Liège-University of Texas.

Kakoulli, I.
2002 Late Classical and Hellenistic painting techniques and materials: a review of the technical literature. *Reviews in Conservation* 3: 56-67.

Kakoulli, I., A. Kottaridou and N. Minos
2001 Materials and techniques of ancient monumental paintings: analysis of the painted throne from the 'Tomb of Eurydice', Vergina, Greece. In Y. Basiakos, E. Aloupi and Y. Facorellis (eds.),

*Archaeometry Issues in Greek Prehistory and Antiquity*, 261-74. Athens: Hellenic Society of Archaeometry-Society of Messenian Archaeological Studies.

Kalogeropoulos, K.
2005 Decorative schemes as an indication of artistic relations between Early Mycenaean Greece and the eastern Mediterranean. In R. Laffineur and E. Greco (eds.), *EMPORIA. Aegean in the Central and eastern Mediterranean. Proceedings of the 10th International Aegean Conference, Athens 14–18 April 2004*. Aegaeum 25: 393-403. Liège-Austin: Université de Liège-University of Texas.

Kanta, A.
1992 Monastiraki. In J.W. and E.E. Myers and G. Cadogan (eds.), *The Aerial Atlas of Crete*, 194-97. Berkeley, California: University of California Press.
1994–96 Monastiraki. *Kritiki Estia. Periodos D—Tomos* 5: 312-14.
2003 The Aegean world between East and West. Aspects of common cultural elements, 16th-6th c. BC. The Greek Mainland, Cyclades, Crete and the East in prehistory. A review of cultural relations and trade. In N.C. Stampolidis (ed.), *Sea Routes… From Sidon to Huelva. Interconnections in the Mediterranean*, 20-40. Athens: Museum of Cycladic Art.

Kantor, H.J.
1947 *The Aegean and the Orient in the Second Millennium* BC. Monographs on Archaeology and Fine Arts, IV. The Archaeological Institute of America. Monograph, 1. Boston: The Archaeological Institute of America (reprinted 1997).

Kaplan, M.
1980 *The Origin and Distribution of Tell el-Yahudiyeh Ware*. Studies in Mediterranean Archaeology 62. Göteborg: P. Åström's Förlag.

Kardulias, P.N.
2000 Adaptation on arid islands. *Near Eastern Archaeology* 63-1: 35-44.

Karetsou, A. (ed.)
2000 *Kriti-Aigyptos. Polutismikoi Desmoi Trion Chilieton. Meletes*. Athens: Ypourgeio Politismou-Archaiologiko Mouseio Herakleiou (in Greek).

Karimali, E.
2005 Lithic technology and use. In E. Blake and A.B. Knapp (eds.), *The Archaeology of Mediterranean Prehistory*, Blackwell Studies in Global Archaeology: 180-214. Oxford: Blackwell.

Kassianidou, V., and A.B. Knapp
2005 Archaeometallurgy in the Mediterranean: the social context of mining, technology, and trade. In E. Blake and A.B. Knapp (eds.), *The Archaeology of Mediterranean Prehistory*, Blackwell Studies in Global Archaeology: 215-51. Oxford: Blackwell.

Keller, C.
1994 Invention, thought, and process: strategies in iron tool production. In S.U. Wisseman and W.S. Williams (eds.), *Ancient Technologies and Archaeological Materials*, 59-70. Amsterdam: Gordon and Breach.

Kemp, B.J.
2000 The discovery of the painted plaster fragments at Malkata. In A. Karetsou (ed.), *Kriti-Aigyptos. Polutismikoi Desmoi Trion Chilieton. Meletes*, 45-46. Athens: Ypourgeio Politismou-Archaiologiko Mouseio Herakleiou (in Greek).

Kemp, B.J., and F. Weatherhead
2000 Palace decoration at Tell el-Amarna. In S. Sherratt (ed.), *The Wall Paintings of Thera: Proceedings of the First International Symposium. Petros M. Nomikos Conference Centre, Thera, Hellas, 30 August-4 September 1997*, 491-523. Athens: Petros M. Nomikos and the Thera Foundation.

Kemp, B.J., and R.S. Merrillees
1980 *Minoan Pottery in Second Millennium Egypt*. Mainz am Rhein: Verlag Philipp von Zabern.

Kempinski, A.
1983 *Syrian und Palästina (Kanaan) in der*

*Letzten Phase der Mittelbronze IIB Zeit (1650–1570 v. Chr.).* Wiesbaden: Harrassowitz.

Kempinski,, A. and W-D. Niemeier

1991 Kabri, 1989–1990. *Israel Exploration Journal* 41 (1-3): 188-94.

1992 Kabri, 1991. *Israel Exploration Journal* 42 (1-2): 260-65.

1993a Kabri, 1992. *Israel Exploration Journal* 43 (2-3): 181-84.

1993b Kabri, 1993. *Israel Exploration Journal* 43 (4): 256-59.

Kenoyer, J.M., M. Vidale and K.K. Bhan

1991 Contemporary stone beadmaking in Khambhat, India: patterns of craft specialisation and organisation of production as reflected in the archaeological record. *World Archaeology* 23 (1): 44-63.

Kenoyer, J.M.

2000 Wealth and socioeconomic hierarchies of the Indus Valley civilisation. In J. Richards and M. Van Buren (eds.), *Order, Legitimacy and Wealth in Ancient States.* New Directions in Archaeology, 88-110. Cambridge: Cambridge University Press.

Keramopoulos, A.

1909 I Oikia tou Kadmou. *Archaiologiki Efimeris* 1909: 57-122 (in Greek).

Keswani, P.S.

1996 Hierarchies, heterarchies, and urbanization processes: the view from Bronze Age Cyprus. *Journal of Mediterranean Archaeology* 9: 211-49.

Kilian, K.

1978 Ausgrabungen in Tiryns 1976. Bericht zu den Grabungen. *Archäologische Anzeiger* 1978: 449-98.

1979 Ausgrabungen in Tiryns 1977. Bericht zu den Grabungen. *Archäologische Anzeiger* 1979: 379-447.

1980 Zum Ende der Mykenischen Epoche in der Argolis. *Jahrbuch Römisches-Germanisches Zentralmuseum* 27: 166-95.

1981 Ausgrabungen in Tiryns 1980. Bericht zu den Grabungen. *Archäologische Anzeiger* 1981: 149-94.

1987 L'architecture des résidences Mycéniennes: origine et extension d'une structure pouvoir politique pendant l'Âge du Bronze Récent. In E. Levy (ed.), *Le Système Palatial en Orient, en Grèce et à Rome. Actes du Colloque de Strasbourg 19-22 juin 1985, Université des Sciences Humaines de Strasbourg.* Travaux du Centre de Recherche sur le Proche-Orient et la Grèce Antiques 9: 203-17. Leiden: Brill.

1988a Mycenaeans up to date. Trends and changes in recent research. In E.B. French and K.A. Wardle (eds.), *Problems in Greek Prehistory. Papers Presented at the Centenary Conference of the British School at Athens, Manchester (April 1986)*, 115-52. Bristol: Bristol Academic Press.

1988b Die 'Thronfolge' in Tiryns. *Archäologische Anzeiger* 103: 1-9.

1988c Ausgrabungen in Tiryns 1983/83. Bericht zu den Grabungen. *Archäologische Anzeiger* 1988: 105-51.

1996 Earthquakes and archaeological context at 13th century BC Tiryns. In S. Stiros and R.E. Jones (eds.), *Archaeoseismology*. Fitch Laboratory Occasional Paper, 7: 63-68. London: British School at Athens and Institute for Geological and Mineral Exploration.

Kilian, K., C. Podzuweit and W.-J. Weißhaar

1981 Ausgrabungen in Tiryns 1978, 1979. Mit 92 Abbildungen. *Archäologische Anzeiger* 1981: 149-256.

Killebrew, A.

2005 *Biblical Peoples and Ethnicity: An Archaeological Study of Egyptians, Canaanites, Philistine, and Early Israel 1300–1100 BCE.* Atlanta: Society of Biblical Literature.

Killen, J.T.

1985 The Linear B tablets and the Mycenaean economy. In A. Morpurgo Davies and Y. Duhoux (eds.), *Linear B: A 1984 Survey. Proceedings of the Mycenaean Colloquium of the International Federation of the Societies of Classical*

*Studies Dublin (27 Aug.–1 Sept. 1984)*. Bibliothèque des Cahiers de l'Institut de Linguistique de Louvain, 241-305. Louvain-La-Neuve: Cabay.

1995 Some further thoughts on 'collectors'. In R. Laffineur and W.-D. Niemeier (eds.), *POLITEIA. Society and State in the Aegean Bronze Age. Proceedings of the 5th International Conference. University of Heidelberg, Archäologisches Institut. 10-13 April 1994*. Aegaeum 12: 213-26. Liège-Austin: Université de Liège-University of Texas.

2001 Some thoughts on TA-RA-SI-JA. In S. Voutsaki and J. Killen (eds.), *Economy and Politics in the Mycenaean Palace States*. Cambridge Philological Society, Supplementary Volume 27: 161-80. Cambridge: Cambridge Philological Society.

Killick, D.

2004 Social constructionist approaches to the study of technology. *World Archaeology* 36 (4): 571-78.

Kim, J.

2001 Elite strategies and the spread of technological innovation: the spread of iron in the Bronze Age societies of Denmark and Southern Korea. *Journal of Anthropological Archaeology* 20: 442-78.

Kingery, W.D.

2001 The design process as a critical component of the anthropology of technology. In M.B. Schiffer (ed.), *Anthropological Perspectives on Technology*. Amerind Foundation New World Studies 5: 123-38. Albuquerque: University of New Mexico Press.

Kingery, W.D., P.B. Vandiver and M. Prickett

1988 The beginnings of pyrotechnology, part II: production and use of lime and gypsum plaster in the Pre-Pottery Neolithic Near East. *Journal of Field Archaeology* 15: 219-44.

Kiriatzi, E.

2003 Sherds, fabrics and clay sources: reconstructing the ceramic landscapes of prehistoric Kythera. In K.P. Foster and R. Laffineur (eds.), *METRON. Measuring the Aegean Bronze Age. Proceedings of the 9th International Aegean Conference, New Haven, Yale University, 18–21 April 2002*. Aegaeum 24: 123-30. Liège-Austin: Université de Liège-University of Texas.

Kitchen, K.A.

1982 *Pharaoh Triumphant: The Life and Times of Ramses II King of Egypt*. Warminster: Aris and Phillips.

Knapp, A.B.

1986 Production, exchange and socio-political complexity on Bronze Age Cyprus. *Oxford Journal of Archaeology* 5: 35-60.

1994 Emergence, development and decline on Bronze Age Cyprus. In C. Mathers and S. Stoddart (eds.), *Development and Decline in the Mediterranean Bronze Age*. Sheffield Archaeological Monograph 8: 271-304. Sheffield: John Collis Publications.

1996 Introduction. In A.B. Knapp (ed.), *Near Eastern and Aegean Texts from the Third to the First Millennia* BC. Sources for the History of Cyprus 2: 1-13. New York: Altamont.

1998 Mediterranean Bronze Age trade: distance, power and place. In E.H. Cline and D. Harris-Cline (eds.), *The Aegean and the Orient in the Second Millennium. Proceedings of the 50th Anniversary Symposium, Cincinnati, (18–20 April 1997)*. Aegaeum, 18: 193-207. Liège-Austin: Université de Liège-University of Texas.

2000 Archaeology, science-based archaeology and the Mediterranean Bronze Age metals trade. *European Journal of Archaeology* 3: 31-56.

Knapp, A.B., and J.F. Cherry

1994 *Provenience Studies and Bronze Age Cyprus. Production, Exchange and Politico-Economic Change*. Monographs in World Archaeology 21. Madison and Wisconsin: Prehistory Press.

Knapp, A.B., and E. Blake

2005 Prehistory in the Mediterranean: the connecting and corrupting sea. In

E. Blake and A.B. Knapp (eds.), *The Archaeology of Mediterranean Prehistory*, Blackwell Studies in Global Archaeology, 1-23. Oxford: Blackwell.

Knappett, C.

1999 Tradition and innovation in pottery forming technology: wheel-throwing at MM Knossos. *Annual of the British School at Athens* 94: 101-29.

2003 *Minoanisation in the South-East Aegean: Pottery Fabrics from Miletus and Iasos.* Paper presented at the Hellenic Society for Archaeometry International Conference, Athens, Greece.

Knappett, C., and I. Nikolakopoulou

2005 Exchange and affiliation networks in the Middle Bronze Age southern Aegean: Crete, Akrotiri and Miletus. In R. Laffineur and E. Greco (eds.), *EMPORIA. Aegean in the Central and eastern Mediterranean. Proceedings of the 10th International Aegean Conference, Athens 14–18 April 2004.* Aegaeum 25: 175-84. Liège-Austin: Université de Liège-University of Texas.

Knutsson, K.

1988 Patterns of tool use: Scanning Electron Microscopy of experimental quartz tools. *Societas Archaeologica Upsaliensis* 10: 9-20.

Koehl, R.B.

2000 Minoan rhyta in Egypt. In A. Karetsou (ed.), *Kriti-Aigyptos. Polutismikoi Desmoi Trion Chilieton. Meletes,* 94-100. Athens: Ypourgeio Politismou-Archaiologiko Mouseio Herakleiou (in Greek).

Kohl, P.

1987 The ancient economy, transferable technologies and the Bronze Age world system: a view from the northeastern frontier of the ancient Near East. In M. Rowlands, M. Larsen and K. Kristiansen (eds.), *Centre and Periphery in the Ancient World*, 13-24. Cambridge: Cambridge University Press.

Kohring, S.

2006 Let's NOT talk about technology? Bringing production into a discussion of technological knowledge. In B. Gravina (ed.), *Technologies. Changing Matter, Changing Minds.* Archaeological Review from Cambridge 21 (1): 98-116. Cambridge: Department of Archaeology.

Kontorli-Papadopoulou, L.

1996 *Aegean Frescoes of Religious Character.* Studies in Mediterranean Archaeology 117. Göteborg: P. Åström's Förlag.

Kopaka, K.

1997 'Women's arts—men's crafts?' Towards a framework for approaching gender skills in the Prehistoric Aegean. In R. Laffineur and P.P. Betancourt (eds.), *TEXNH. Crafsmen, Craftswomen and Craftsmanship in the Aegean Bronze Age. Proceedings of the 6th International Aegean Conference, Philadelphia, Temple University, 18–21 April 1996.* Aegaeum 16: 521-31. Liège-Austin: Université de Liège-University of Texas.

Kritseli-Providi, I.

1982 Toixografies tou Thriskeutikou Kentrou ton Mukinon. Unpublished PhD Dissertation, University of Athens, Greece (in Greek).

Kuhn, S.L.

2004 Evolutionary perspectives on technology and technological change. *World Archaeology* 36 (4): 561-70.

Kuhrt, A.

1995 *The Ancient Near East c. 3000-330 BC*, 2 vols. London: Routledge.

Laffineur, R.

1995 Craftsmen and craftsmenship in Mycenaean Greece: for a multimedia approach. In R. Laffineur and W.-D. Niemeier (eds.) *POLITEIA. Society and State in the Aegean Bronze Age. Proceedings of the 5th International Conference. University of Heidelberg, Archäologisches Institut, 10–13 April 1994.* Aegaeum 12: 189-99. Liège-Austin: Université de Liège-University of Texas.

1998 From west to east: the Aegean and Egypt in the early Late Bronze Age. In E.H. Cline and D. Harris-Cline (eds.), *The Aegean and the Orient in the Second Millennium. Proceedings of the 50th Anniversary Symposium, Cincinnati, 18–20 April 1997*. Aegaeum 18: 53-67. Liège-Austin: Université de Liège-University of Texas.

2005 Imports/exports in the eastern Mediterranean: for a specific methodology. In R. Laffineur and E. Greco (eds.), *EMPORIA. Aegean in the Central and Eastern Mediterranean. Proceedings of the 10th International Aegean Conference, Athens, 14–18 April 2004*. Aegaeum 25: 53-58. Liège-Austin: Université de Liège-University of Texas.

Laffineur, R., and P.P. Betancourt (eds.)

1997 *TEXNH. Crafsmen, Craftswomen and Craftsmanship in the Aegean Bronze Age. Proceedings of the 6th International Aegean Conference, Philadelphia, Temple University, 18–21 April 1996*. Aegaeum 16. Liège-Austin: Université de Liège-University of Texas.

Laffineur, R., and E. Greco (eds.)

2005 *EMPORIA. Aegean in the Central and Eastern Mediterranean. Proceedings of the 10th International Aegean Conference, Athens 14–18 April 2004*. Aegaeum 25. Liège-Austin: Université de Liège-University of Texas.

Lambrou-Phillipson, C.

1990 *Hellenorientalia: the Near Eastern Presence in the Bronze Age Aegean ca. 3000–1100 BC. Plus Orientalia: a Catalogue of Egyptian, Mesopotamian, Mitannian, Syro-Palestinian, Cypriot and Asia Minor Objects from the Bronze Age Aegean*. Göteborg: P. Åström's Förlag.

Landenius Enegren, H.

2000 Craft production at Knossos—raw materials and finished goods—the Linear B evidence. In C. Gillis, C. Risberg and B. Sjöberg (eds.), *Trade and Production in Premonetary Greece: Acquisition and Distribution of Raw Materials and Finished Products*. Studies in Mediterranean Archaeology Pocket Book 154: 29-42. Jonsered: P. Åström's Förlag.

Lang, M.L.

1969 *The Palace of Nestor at Pylos in Western Messenia*. Volume II. *The Frescoes*. Princeton: Princeton University Press.

Lapatin, K.D.S.

2001 *Chryselephantine Statuary in the Ancient Mediterranean World*. Oxford: Oxford University Press.

Laurie, A.P.

1988 *The Painter's Methods and Materials*. New York: Dover.

Leake, B.E.

1978 Nomenclature of amphiboles. *Mineralogical Magazine* 42: 533-63.

Lechtman, H.

1993 Technologies of power: the Andean case. In J.S. Henderson and P.J. Netherly (eds.), *Configurations of Power: Holistic Anthropology in Theory and Practice*, 244-80. Ithaca and London: Cornell University Press.

1996 Arsenic bronze: dirty copper or chosen alloy? A view from the Americas. *Journal of Field Archaeology* 23: 477-514.

1999 Afterword. In M.-A. Dobres and C.R. Hoffman (eds.), *The Social Dynamics of Technology: Practice, Politics, and World Views*, 223-32. Washington and London: Smithsonian Institution Press.

Lechtman, H., and A. Steinberg

1979 The history of technology: an anthropological point of view. In G. Bugliarello and D.B. Donner (eds.), *The History of Philosophy of Technology*, 135-62. Urbana: University of Illinois Press.

Lee, L., and S. Quirke

2000 Painting materials. In P.T. Nicholson and I. Shaw (eds.), *Ancient Egyptian Materials and Technology*, 104–20. Cambridge: Cambridge University Press.

Lemonnier, P.

1993 Introduction. In P. Lemonnier (ed.), *Technological Choices: Transformation in Material Culture Since the Neolithic*, 1-35. London: Routledge.

Leonard, A. Jr.

1994 *An Index to the Late Bronze Age Aegean*

*Pottery from Syro-Palestine*. Studies in Mediterranean Archaeology 114. Jonsered: P. Åström's Förlag.

Levi, D.
1956 The sarcophagus of Hagia Triada restored. *Archaeology* 9 (3): 192-99.

Lilyquist, C.
1997 A foreign vase representation from a new Theban tomb (the Chapel for MMA SA P2). In J. Phillips (ed.), *Ancient Egypt, the Aegean, and the Near East. Studies in Honour of Martha Rhoads Bell*, 307-43. San Antonio: Van Siclen Books.

Liverani, M.
1987 The collapse of the Near Eastern regional system at the end of the Bronze Age: the case of Syria. In M. Rowlands, M. Larsen and K. Kristiansen (eds.), *Centre and Periphery in the Ancient World*, 66-73. Cambridge: Cambridge University Press.

Liverani, M.
1990 *Prestige and Interest: International Relations in the Near East ca. 1600–1100 BC*. Padova: Sargon.

London, G.
2000 Ethnoarchaeology and interpretations of the past. *Near Eastern Archaeology* 63.1: 2-8.

Longacre, W.A., J.M. Skibo and M.T. Stark
1991 Ethnoarchaeology at the top of the world. New ceramic studies among the Kalinga of Luzon. *Expedition* 33-1: 4-15.

Lucas, A., and J.R. Harris
1962 *Ancient Egyptian Materials and Industries*. London: Edward Arnold.

Lurz, N.
1994 *Der Einfluß Ägyptens, Vorderasiens und Kretas auf die Mykenischen Fresken. Studien zum Ursprung der Frühgriechischen Wandmalerei*. Europäische Hochschulschriften, 38-48. Frankfurt am Main: Peter Lang.

MacGillivray, A.
1994 The early history of the palace at Knossos (MM I-II). In D. Evely, H. Hughes-Brock and N. Momigliano (eds.), *Knossos: A Labyrinth of History. Papers Presented in Honour of Sinclair Hood*, 45-55. Oxford: Oxbow Books.

MacGillivray, J.A., and L.H. Sackett
1984 An archaeological survey of the Roussolakos area at Palaikastro, *Annual of the British School at Athens* 79: 129-59.

MacGillivray, J.A., L.H. Sackett, J.M. Driessen and S. Hemingway
1992 Excavations at Palaikastro 1991. *Annual of the British School at Athens* 87: 121-52.

MacGillivray, J.A., L.H. Sackett and J.M. Driessen
1998 Excavations at Palaikastro 1994 and 1996. *Annual of the British School at Athens* 93: 221-68.

MacGillivray, J.A., J.M. Driessen and L.H. Sackett (eds.)
2000 *The Palaikastro Kouros: A Minoan Chryselephantine Statuette and its Aegean Bronze Age Context*. British School at Athens Studies 6. London: British School at Athens.

Majewski, L., and M. Reich
1973 Frescoes from Aghia Irini, Keos, part I, appendix. *Hesperia* 42: 297-300.

Malafouris, L.
2004 The cognitive basis of material engagement: where brain, body and culture conflate. In E. DeMarrais, C. Gosden and C. Renfrew (eds.), *Rethinking Materiality: The Engagement of Mind with the Material World*, 53-62. Cambridge: McDonald Institute for Archaeological Research.

Manning, S.W.
1995 *The Absolute Chronology of the Aegean Early Bronze Age: Archaeology, History and Radiocarbon*. Monographs in Mediterranean Archaeology 1. Sheffield: Sheffield Academic Press.
1999 *A Test of Time: The Volcano of Thera and the Chronology and History of the Aegean and east Mediterranean in the Mid Second Millennium BC*. Oxford-Oakville: Oxbow Books.
2005 Simulation and the Thera eruption: outlining what we do and do not know

from radiocarbon. In A. Dakouri-Hild and S. Sherratt (eds.), *AUTOCHTHON. Papers Presented to O.T.P.K. Dickinson on the Occasion of His Retirement. Institute of Classical Studies, University of London, 9 November 2005*. British Archaeological Reports, International Series 1432: 97-114. Oxford: Archaeopress.

Manning, S.W., S.J. Monks and G. Nakou

1994 The fatal shore, the long years and the geographical unconscious. Considerations of iconography, chronology, and trade in response to Negbi's 'The "Lybian landscape" from Thera: a review of Aegean enterprises overseas in the Late Minoan IA period'. *Journal of Mediterranean Archaeology* 7.2: 219-35.

Manning, S.W., and C. Bronk Ramsey

2004 Radiocarbon calibration from 0-26 cal kyr BP (dating the volcanic eruption at Thera). *Radiocarbon* 46.3: 1029-58.

Manning, S.W., and L. Hulin

2005 Maritime commerce and geographies of mobility in the Late Bronze Age of the eastern Mediterranean: problematisations. In E. Blake and A.B. Knapp (eds.), *The Archaeology of Mediterranean Prehistory*. Blackwell Studies in Global Archaeology: 270-302. Oxford: Blackwell.

Manning S.W., C. Bronk Ramsey, W. Kutschera, T. Ingham, B. Kromer, P. Steier and E.M. Wild

2006 Chronology for the Aegean Late Bronze Age 1700–1400 BC. *Science* 312: 565-59.

Maran, J.

2001a Political and religious aspects of architectural change on the upper citadel of Tiryns. The case of Building T. In R. Laffineur and Hägg (eds.), *POTNIA. Deities and Religion in the Aegean Bronze Age. Proceedings of the 8th International Aegean Conference, Göteburg, Göteburg University, 12–15 April 2000*, 113-22. Liège-Austin: Université de Liège-University of Texas.

2001b Zur Frage des Vorgängers des Ersten Doppelpalastes von Tiryns. In S. Böhm and K-V.von Eickstedt (eds.), *ΙΘΑΚΗ. Festschrift für Jörg Schäfer zum 75. Geburtstag am 25. April 2001*, 23-29. Würzburg: Ergon-Verlag.

2004a Architektonische Innovation im Spätmykenischen Tiryns: Lokale Bauprogramme und Fremde Kultureinflüsse. In Verein zur Förderung der Aufarbeitung der Hellenischen Geschichte (ed.), *Althellenische Technologie und Technik: Tagungsband der Tagung in Ohlstadt (21.–23. März 2003)*, 261-86. Weilheim: Verein zur Förderung der Aufarbeitung der Hellenischen Geschichte.

2004b The spreading of objects and ideas in the Late Bronze Age eastern Mediterranean: two case examples from the Argolid of the 13th and 12th centuries BC. *Bulletin of the American Schools of Research* 336: 11-30.

2006a Architecture, power and social practice—an introduction. In J. Maran, C. Juwig, H. Schwengel and U. Thaler (eds.), *Constructing Power: Architecture, Ideology and Social Practice*. Geschichte Forschung und Wissenschaft 19: 9-14. Hamburg: Lit Verlag.

2006b Mycenaean citadels as performative space. In J. Maran, C. Juwig, H. Schwengel and U. Thaler (eds.), *Constructing Power: Architecture, Ideology and Social Practice*. Geschichte Forschung und Wissenschaft 19: 75-91. Hamburg: Lit Verlag.

2006c Coming to terms with the past: ideology and power in Late Helladic IIIC. In S. Deger-Jalkotzy and I.S. Lemos (eds.), *Ancient Greece: From the Mycenaean Palaces to the Age of Homer*. Edinburgh Leventis Studies 3: 123-50. Edinburgh: Edinburgh University Press.

Maran, J., C. Juwig, H. Schwengel and U. Thaler (eds.)

2006 *Constructing Power: Architecture, Ideology and Social Practice*. Geschichte

Forschung und Wissenschaft 19. Hamburg: Lit Verlag.

Maran, J., and E. Stavrianopoulou

2007 Potnios Anir—reflections on the ideology of Mycenaean Kingship. In E. Alram-Stern and G. Nightingale (eds.), KEIMELION. *The Formation of Elites and Elitist Lifestyles from Mycenaean Palatial Times to the Homeric Period.* Akten des internationalen Kongresses vom 3. bis 5. Februar 2005 in Salzburg, 285-98. Vienna: Verlag der Österreichischen Akademie der Wissenschaften.

Markus, T.A.

2006 Piranesi's paradox: to build is to create assymetrics of power. In J. Maran, C. Juwig, H. Schwengel and U. Thaler (eds.), *Constructing Power: Architecture, Ideology and Social Practice.* Geschichte Forschung und Wissenschaft 19: 321-35. Hamburg: Lit Verlag.

Marinatos, N.

1994 The 'export' significance of Minoan bull-hunting and bull-leaping scenes. *Egypt and Levant* 4: 89-93.

Marthari, M.

2000 The attraction of the pictorial: observations on the relationship of Theran pottery and Theran fresco painting. In S. Sherratt (ed.), *The Wall Paintings of Thera: Proceedings of the First International Symposium. Petros M. Nomikos Conference Centre, Thera, Hellas, 30 August-4 September 1997*, 873-89. Athens: Petros M. Nomikos and the Thera Foundation.

Martin, G.T.

2001 Wall paintings in Memphite tombs of the New Kingdom. In W.V. Davies (ed.), *Colour and Painting in Ancient Egypt*, 101-105. London: British Museum Press.

Massari, G., and I. Massari

1993 *Damp Buildings Old and New.* Rome: ICCROM.

Mathieu, J.R.

2002 Introduction—experimental archaeology: replicating past objects, behaviours, and processes. In J.R. Mathieu (ed.), *Experimental Archaeology: Replicating Past Objects. Behaviours, and Processes.* British Archaeological Reports, International Series 1035: 1-11. Oxford: Archaeopress.

Matoïan, V., and A. Bouquillon

2003 Vitreous materials in Ugarit: new data. In T. Potts, M. Roaf and D. Stein (eds.), *Culture Through Objects: Ancient Near Eastern Studies in Honour of P.R.S. Moorey*, 333-46. Griffith Institute Oxford: Cambridge University Press.

Matson, F.R.

1989 Ceramics: the hub of ancient craft interplay. In P.E. McGovern, M.D. Notis and W.D. Kingery (eds.), *Cross-Craft and Cross-Cultural Interactions in Ceramics.* Ceramics and Civilization, IV: 13-30. Westerville-OH: American Ceramic Society.

Matthews, D.

1995 Artisans and artists in Ancient Mesopotamia. In J.M. Sasson (ed.), *Civilisations of the Ancient Near East I*, 455-68. New York: Charles Scribner's Sons.

McCarthy, B.

2001 Technical analysis of reds and yellows in the tomb of Suemniwet, Theban Tomb 92. In W.V. Davies (ed.), *Colour and Painting in Ancient Egypt*, 17-21. London: British Museum Press.

McFadyen, L.

2006 Building technologies, quick architecture and Early Neolithic longbarrow sites in southern Britain. In B. Gravina (ed.), *Technologies: Changing Matter, Changing Minds.* Archaeological Review from Cambridge 21 (1): 117-34. Cambridge: Department of Archaeology.

McGovern, P.E.

1989 Ceramics and craft interaction: a theoretical framework, with prefatory remarks. In P.E. McGovern, M.D. Notis and W.D. Kingery (eds.), *Cross-Craft and Cross-Cultural Interactions in Ceramics.* Ceramics and Civilization IV: 1-11. Westerville, Ohio: American Ceramic Society.

Meilunas, R.J., J.G. Bentsen and A. Steinberg
1990 Analysis of aged paint binders by FTIR spectroscopy. *Studies in Conservation* 35: 33-52.

Mellaart, J.
1967 *Çatal Hüyük: A Neolithic Town in Anatolia*. London: Thames and Hudson.
1970 *Excavations at Hacılar*. Occasional Publications of the British Institute of Archaeology at Ankara 9. Edinburgh: Edinburgh University Press.

Mellink, M.J.
1995 New perspectives and initiatives in the Hyksos period. *Egypt and Levant* 5: 85-89.

Merrillees, R.S.
2003 The first appearances of Kamares Ware in the Levant. *Egypt and Levant* 13: 127-42.

Mesnil du Buisson, R.M.E.L.
1935 *Le Site Archéologique de Mishrife-Qatna*. Paris: E. de Boccard.

Michaelidou, A., and K. Voutsa
2005 Merchants and merchandise: humans as a commodity in Aegean and oriental societies. In R. Laffineur and E. Greco (eds.), *EMPORIA. Aegean in the Central and Eastern Mediterranean. Proceedings of the 10th International Aegean Conference, Athens, 14–18 April 2004*. Aegaeum 25: 17-29. Liège-Austin: Université de Liège-University of Texas.

Militello, P.
1998 *Haghia Triada I. Gli Affreschi*. Monografie della Scuola Archeologia di Atene e delle Missione Italiane in Oriente, IX. Padova: Bottega d' Erasmo.
2001 *Gli Affreschi Minoici di Festos;* Palio O. 2001. *La Casa Tardo-Minoico I di Chalara a Festos*. Studi di Archeologia Cretese II. Padova, Bottega d' Erasmo.

Moak, M.
2000 The Palaikastro kouros. In J.A. MacGillivray, J.M. Driessen and L.H. Sackett (eds.), *The Palaikastro Kouros: A Minoan Chryselephantine Statuette and its Aegean Bronze Age Context*. British School at Athens Studies 6: 65-83. London: British School at Athens.

Momigliano, N., and D. Wilson
1996 Knossos 1993: excavations outside the south front of the palace. *Annual of the British School at Athens* 91: 1-57.

Mommsen, H., and J. Maran
2001 Production places of some Mycenaean pictorial vessels. The contribution of chemical pottery analysis. *Opuscula Atheniensia* 25-26: 95-106.

Moody, J.
2005 Unravelling the threads: climatic changes in the Late Bronze Age III Aegean. In A.L. D'Agata and J. Moody with E. Williams (eds.), *Ariadne's Threads: Connections between Crete and the Greek Mainland in Late Minoan III (LM IIIA2 to LM IIIC). Proceedings of the International Workshop Held at Athens Scuola Archeologica Italiana, 5-6 April 2003*. Tripodes 3: 443-70. Athens: Scuola Archeologica Italiana di Atene.

Moore, D.M., and R.C. Reynolds Jr.
1997 *X-Ray Diffraction and the Identification and Analysis of Clay Minerals*. Oxford/New York: Oxford University Press.

Moorey, P.R.S.
1985 *Materials and Manufacture in Ancient Mesopotamia: the Evidence of Archaeology and Art. Metals and Metalwork*. British Archaeological Reports, International Series 237. Oxford: British Archaeological Reports.

Moorey, P.R.S.
1994 *Ancient Mesopotamian Materials and Industries: The Archaeological Evidence*. Oxford: Clarendon Press.

Moorey, P.R.S.
2001 The mobility of artisans and opportunities for technology transfer between western Asia and Egypt in the Late Bronze Age. In A.J. Shortland (ed.), *The Social Context of Technological Change: Egypt and the Near East, 1650–1550 BC. Proceedings of a Conference Held at St. Edmund Hall, Oxford, 12–14 September 2000*, 1-14. Oxford: Oxbow.

Moortgat, A.
1959 *Alt-Vorderasiatische Malerei*. Berlin: Safari Verlag.
1969 *The Art of Ancient Mesopotamia. The Classical Art of the Near East*. London: Phaedon.

Mora, P., L. Mora and P. Philippot
1984 *Conservation of Wall Paintings*. London: Butterworths.

Moran, W.L.
1992 *The Amarna Letters*. Baltimore and London: John Hopkins University Press.

Morgan, L.
1985 Idea, idiom and iconography. In P. Darcque and J.-C. Poursat (eds.), *L'Iconographie Minoenne. Actes de la Table Ronde d'Athènes, 21–22 Avril 1983*. Bulletin de Correspondance Hellenique Supplement 11: 5-19. Athens: Ecole Française d'Athène.
1988 *The Miniature Wall Paintings from Thera: A Study in Aegean Culture and Iconography*. Cambridge: Cambridge University Press.
1990 Island iconography: Thera, Kea and Milos. In D.A. Hardy, C.G. Doumas, I.A. Sakellarakis and P.M. Warren (eds.), *Thera and the Aegean World III-I: Archaeology. Proceedings of the 3rd International Congress, Santorini, Greece, 3-9 September 1989*, 252-66. London: Thera Foundation.
1995 Minoan painting and Egypt: the case of Tell el-Dab'a. In W.V. Davies and L. Schofield (eds.), *Egypt, the Aegean and the Levant: Interconnections in the Second Millennium BC*, 29-53. London: British Museum Press.
2007 The painted plasters and their relation to the wall paintings of the Pillar Crypt (with contributions by Mark Cameron). In C. Renfrew, N. Brodie, C. Morris and C. Scarre (eds.), *Excavations at Phylakopi in Melos 1974-77*. British School of Athens Supplementary Volume 42, 371-97. London: British School at Athens.

Morgan, L. (ed.)
2004 *Aegean Wall Painting: A Tribute to Mark Cameron*. British School at Athens Studies 13. London: British School at Athens.

Morris, C.
1995 Symbols to power. Styles and media in the Inka state. In C. Carr and J.E. Neitzel (eds.), *Style, Society and Person: Archaeological and Ethnological Perspectives*, 419-33. New York: Plenum Press.
1993 Hands up for the individual! The role of attribution studies in Aegean prehistory. *Cambridge Journal of Archaeology* 3.1: 41-66.
2004 'Art makes visible': an archaeology of the senses in Minoan elite art. In N. Brodie and C. Hills (eds.), *Material Engagements: Studies in Honour of Colin Renfrew*, 31-44. Cambridge: McDonald Institute for Archaeological Research.

Morris, S.P.
1989 A tale of two cities: the miniature frescoes from Thera and the origins of Greek poetry. *American Journal of Archaeology* 93: 511-35.
1993 *Daidalos and the Origin of Greek Art*. Princeton: Princeton University Press.
2000 From Thera to Sheria: Aegean art and narrative. In S. Sherratt (ed.), *The Wall Paintings of Thera: Proceedings of the First International Symposium. Petros M. Nomikos Conference Centre, Thera, Hellas, 30 August-4 September 1997*, 317-33. Athens: Petros M. Nomikos and the Thera Foundation.
2003 Islands in the sea: Aegean polities as Levantine neighbours. In W.G. Dever and S. Gitin (eds.), *Symbiosis, Symbolisms, and the Power of the Past: Canaan, Ancient Israel, and their Neighbours from the Late Bronze Age through Roman Palestina. Proceedings of the Centennial Symposium W.F. Albright Institute of Archaeological Research and American Schools of Oriental Research, Jerusalem (29-31 May 2000)*, 3-16. Winona Lake: Eisenbrauns.

Mountjoy, P.A.
1983 *Orchomenos V. Mycenaean Pottery from Orchomenos, Eutresis and Other Boeotian Sites*. Bayerische Akademie der Wissenschaften. Philosophisch-Historische Klasse Abhandlungen. Neue Folge, Heft 89. Munich: Verlag der Bayerischen Akademie der Wissenschaften.
1993 *Mycenaean Pottery: An Introduction*. Oxford University Committee for Archaeology Monograph 36. Oxford: Oxford University Committee for Archaeology.

Muhly, J.D.
1986 The role of Cyprus in the economy of the eastern Mediterranean during the second millennium BC. In V. Karageorghis (ed.), *Acts of the International Symposium 'Cyprus Between the Orient and the Occident', Nicosia, 8–14 Sept. 1985*, 45-62. Nicosia: Department of Antiquities, Republic of Cyprus.
2003 Trade in metals in the Late Bronze Age and the Iron Age. In N.C. Stampolidis (ed.), *Sea Routes... From Sidon to Huelva. Interconnections in the Mediterranean, 16th-6th c. BC*, 141-50. Athens: Museum of Cycladic Art.
2005 Travelling craftsmen: Love 'em or leave 'em. In R. Laffineur and E. Greco (eds.), *EMPORIA. Aegean in the Central and Eastern Mediterranean. Proceedings of the 10th International Aegean Conference, Athens 14–18 April 2004*. Aegaeum 25: 685-90. Liège-Austin: Université de Liège-University of Texas.

Müller, K.
1930 *Die Architektur der Burg und des Palastes, Tiryns. Die Ergebnisse der Ausgrabungen des Instituts*, Vol. III. Augsburg: De. Benno Filser Verlag.

Müller-Karpe, A.
2003 Remarks on central Anatolian chronology of the Middle Hittite period. In M. Bietak (ed.), *The Synchronisation of Civilisations in the Eastern Mediterranean in the 2nd Millennium B.C. II. Proceedings of the SCIEM 2000-EuroConference, Haindorf, 2-7 May 2001*, 383-94. Vienna: Verlag der Österreichischen Akademie der Wissenschaften.

Munoz-Vinas, S.
2004 *Contemporary Theory of Conservation*. London: Butterworth-Heinemann, Elsevier.

Negbi, O.
1994 The 'Libyan landscape' from Thera. A review of Aegean enterprises overseas in the Late Minoan IA period. *Journal of Mediterranean Archaeology* 7: 73-111.

Neitzel, J.E.
1995 Elite styles in hierarchically organised societies. The Chacoan regional system. In C. Carr and J.E. Neitzel (eds.), *Style, Society, and Person: Archaeological and Ethnological Perspectives*, 393-417. New York: Plenum Press.

Neve, P.
1993 Die Ausgrabungen in Bogazköy-Hattusa 1992. *Archäologische Anzeiger* 1993: 621-52.
1996 *Hattuša. Stadt der Gőtter und Tempel. Neue Ausgrabungen in der Hauptstadt der Hettiter*. Zaberns Bilderbände zur Archäologie. Mainz-am-Rhein: Verlag Philipp von Zabern.
1999 *Die Oberstadt von Hattuša. Die Bauwerke I. Das Zentrale Tempelviertel*. Boğaszkoÿ-Hattuša XVI. Berlin: Gebr. Mann.
2001 *Die Oberstadt von Hattuša. Die Bauwerke II. Die Bastion des Sphinxtores und die Tempelviertel am Königs- und Löwentor*. Boğaszkoÿ-Hattuša XVII. Berlin: Gebr. Mann.

Newman, R., and M. Serpico
2000 Adhesives and binders. In P.T. Nicholson and I. Shaw (eds.), *Ancient Egyptian Materials and Technology*, 475-94. Cambridge: Cambridge University Press.

Newman, R., and S.M. Halpine
2001 The binding media of ancient Egyptian painting. In W.V. Davies (ed.), *Colour*

*and Painting in Ancient Egypt*, 22-32. London: British Museum Press.

Nicholson, P.T.

1997 The place of glass technology in 18th dynasty Egypt. In J. Phillips (ed.), *Ancient Egypt, the Aegean, and the Near East. Studies in Honour of Martha Rhoads Bell*, 377-87. San Antonio: Van Siclen Books.

Nicolakaki-Kentrou, M.

2000 Amenhotep III and the Aegean: new evidence on intimacy. The painted plaster fragments from Site K at Malkata. In A. Karetsou (ed.), *Kriti-Aigyptos. Polutismikoi Desmoi Trion Chilieton. Meletes,* 47-51. Athens: Ypourgeio Politismou-Archaiologiko Mouseio Herakleiou (in Greek).

Niemeier, B., and W-D. Niemeier

1997 Milet 1994–1995. Projekt: 'Minoisch-Mykenisches bis Protogeometrisches Milet': Zielsetzung und Grabungen auf dem Stadionhügel und am Athenatempel. Mit 82 Abbildungen. *Archäologische Anzeiger* 1997: 189-248.

Niemeier, B., and W-D. Niemeier

1999 The Minoans of Miletus. In P.P. Betancourt, V. Karageorghis, R. Laffineur and W.-D. Niemeier (eds.), *MELETEMATA I-III. Studies in Aegean Archaeology Presented to Malcolm H. Wiener as He Enters His 65th Year.* Aegaeum 20: 543-53. Liège-Austin: Université de Liège-University of Texas.

2000 Aegean frescoes in Syria-Palestine: Alalakh and Tel Kabri. In S. Sherratt (ed.), *The Wall Paintings of Thera: Proceedings of the First International Symposium. Petros M. Nomikos Conference Centre, Thera, Hellas, 30 August-4 September 1997*, 763-802. Athens: Petros M. Nomikos and the Thera Foundation.

2002 The frescoes in the Middle Bronze Age palace. In N. Scheftelowitz and R. Oren (eds.), *Tel Kabri. The 1986–1993 Excavation Seasons. Aharon Kempinski*, 254-98. Tel Aviv: Emery and Claire Yass Publications in Archaeology.

Niemeier, W.-D.

1987 Das Stuckrelief des 'Prinzen mit der Federkrone' aus Knossos und Minoische Götterdarstellungen. *Archäologische Mitteilungen* 102: 65-97.

1988 The 'Priest-King' fresco from Knossos. A new reconstruction and interpretation. In E.B. French and K.A. Wardle (eds.), *Problems in Greek Prehistory: Papers Presented at the Centenary Conference of the British School of Archaeology at Athens, Manchester, April 1986,* 235-44. Bristol: Classical Press.

1990 Mycenaean elements in the Miniature Fresco from Thera? In D.A. Hardy, C.G. Doumas, I.A. Sakellarakis and P.M. Warren (eds.), *Thera and the Aegean World III-I: Archaeology. Proceedings of the 3rd International Congress, Santorini, Greece, 3-9 September 1989*, 267-84. London: Thera Foundation.

1991 Minoan artisans travelling overseas: the Alalakh frescoes and the painted plaster floor at Tel Kabri (Western Galilee). In R. Laffineur and L. Basch (eds.), *Thalassa. L'Egée Préhistoire et la Mer. Actes de la Troisième Rencontre Egéenne Internationale de L'Université de Liège, Station de Recherches Sous-Marines et Océanographiques (StaReSO), Calvi, Corse, 23–25 avril 1990.* Aegaeum 7: 189–201. Liège: Université de Liège.

1994 Knossos in the New Palace period (MM III-LM IB). In D. Evely, H. Hughes-Brock and N. Momigliano (eds.), *Knossos: A Labyrinth of History. Papers Presented in Honour of Sinclair Hood*, 71-88. Oxford: Oxbow.

1999 Mycenaeans and Hittites in war in western Asia Minor. In R. Laffineur (ed.), *POLEMOS. Le Contexte Guerrier en Egée à l'Age du Bronze. Actes de la 7ième Rencontre Egéene Internationale. Université de Liège, 14–17 Avril 1998.* Aegaeum 19: 141-55. Liège-Austin: Université de Liège-University of Texas.

2004 When Minos ruled the waves: Knossian power overseas. In G. Cadogan,

E. Hatzaki and A. Vasilakis (eds.), *Knossos: Palace, City, State. Proceedings of the Conference in Heraklion organized by the British School at Athens and the 23rd Ephoria of Prehistoric and Classical Antiquities of Heraklion, in November 2003, for the Centenary of Sir Arthur Evans's Excavations at Knossos*. British School at Athens Studies 12: 393-98. London: British School at Athens.

2005 The Minoans and Mycenaeans in western Asia Minor: settlement, emporia or acculturation? In R. Laffineur and E. Greco (eds.), *EMPORIA. Aegean in the Central and Eastern Mediterranean. Proceedings of the 10th International Aegean Conference, Athens, 14–18 April 2004*. Aegaeum 25: 199-204. Liège-Austin: Université de Liège-University of Texas.

2006 Greece and Hatti. Relationship between the Mycenaeans and Hittites according to Recent Research. Paper presented at the German Archaeological Institute, Athens, Greece.

Nordqvist, G.

1997 Male craft and female industry. Two types of production in the Aegean Bronze Age. In R. Laffineur and P.P. Betancourt (eds.), *TEXNH. Craftsmen, Craftswomen and Craftsmanship in the Aegean Bronze Age. Proceedings of the 6th International Aegean Conference, Philadelphia, Temple University, 18–21 April 1996*. Aegaeum 16: 533-37. Liège-Austin: Université de Liège-University of Texas.

Novák, M.

2005 The chronology of the royal palace of Qatna. *Egypt and the Levant* 14: 299-317.

Novák, M., and P. Pfälzner

2001 Ausgrabungen in Tell Mišrife-Qatna 2000. Vorbericht der Deutschen Komponente des Internationalen Kooperationsprojektes. *Mitteilungen des Deutschen Orient-Gesellschaft zu Berlin* 133: 157-98.

2002 Ausgrabungen in Tell Mišrife-Qatna 2001. Vorbericht der Deutschen Komponente des Internationalen Kooperationsprojektes. *Mitteilungen des Deutschen Orient-Gesellschaft zu Berlin* 134: 207-46.

Nunn, A.

1988 *Die Wandmalerei und der Glasierte Wandschmuck im Alten Orient*. Handbuch der Orientalistik 7.1. Leiden: Brill.

O'Connor, D.

1997 The Hyksos period in Egypt. In E.D. Oren (ed.) *The Hyksos: New Historical and Archaeological Perspectives*. University Museum Monograph 96. University Museum Symposium Series 8: 45-67. Philadelphia: University Museum. University of Pennsylvania.

Olivier, J.P.

1987 Des extraits de contrats de vente d'esclaves dans les tablettes de Knossos. In J.T. Killen, J.L. Melena and J.-P. Olivier (eds.), *Studies in Mycenaean and Classical Greek Presented to John Chadwick*. Minos 20-22: 479-98. Salamanca: Universidad de Salamanca.

Oren, E.D.

1997a The kingdom of 'Sharuhen' and the Hyksos kingdom. In E.D. Oren (ed.), *The Hyksos: New Historical and Archaeological Perspectives*. Museum Monograph 96, University Museum Symposium Series 8: 253-83. Philadelphia: University Museum, University of Pennsylvania.

Oren, E. (ed.)

1997b *The Hyksos: New Historical and Archaeological Perspectives*. University Museum Monograph 96, University Museum Symposium Series 8. Philadelphia: University Museum, University of Pennsylvania.

2000 *The Sea Peoples and Their World: A Reassessment*. University Museum Monograph 108, University Museum Symposium Series 11. Philadelphia: University of Pennsylvania Press.

2002 Stratigraphy and architecture III: area D. In N. Scheftelowitz and R. Oren (eds.), *Tel Kabri. The 1986–1993 Excavation Seasons. Aharon Kempinski*, 55-72. Tel Aviv: Emery and Claire Yass Publications in Archaeology.

Özyar, A.
2006 A prospectus of Hittite art based on the state of our knowledge at the beginning of the 3rd millennium AD. In D.P. Mielke, U.-D. Schoop and J. Seeher (eds.) *Structuring and Dating in Hittite Archaeology. Requirements-Problems-New Approaches. International Workshop*, Istanbul, 26-27 November 2004. Deutsches Archäologisches Institut. Abteilung Istanbul. Byzas 4: 125-48. Istanbul: Yayınları.

Palaima, T.G.
1991 Maritime matters in the Linear B tablets. In R. Laffineur and L. Basch (eds.), *Thalassa. L'Egée Préhistoire et la Mer. Actes de la Troisième Rencontre Egéenne Internationale de L'Université de Liège, Station de Recherches Sous-Marines et Océanographiques (StaReSO), Calvi, Corse, 23–25 avril 1990.* Aegaeum 7: 273-310. Liège: Université de Liège.

Palyvou, C.
2004 Architecture in Aegean Bronze Age art: façades with no interiors. In L. Morgan (ed.), *Aegean Wall Painting: A Tribute to Mark Cameron*. British School at Athens Studies 13: 185-97. London: British School at Athens.

Panagiotopoulos, D.
2006 Der Minoische 'Hof' als Kulisse zeremonieller Handlung. In J. Maran, C. Juwig, H. Schwengel and U. Thaler (eds.), *Constructing Power: Architecture, Ideology and Social Practice*. Geschichte Forschung und Wissenschaft 19: 31-48. Hamburg: Lit Verlag.

Paoletti, D., and G. Schirripa Spagnolo
1995 The potential of portable TV holography for examining frescoes *in situ*. *Studies in Conservation* 40: 127-32.

Papadimitriou, A.
2001 *Tiryns: A Guide to Its History and Archaeology*. Athens: Hesperos.

Papaodysseus, C., T. Panagopoulos, M. Exarhos, D. Fragoulis, G. Roussopoulos, P. Rousopoulos, G. Galanopoulos, C. Triantafillou, A. Vlachopoulos, and C. Doumas
2006 Distinct, Late Bronze Age (c. 1650 BC) wall-paintings from Akrotiri, Thera, comprising advanced geometrical patterns. *Archaeometry* 48 (1): 97-114.

Parkinson, R., and L. Schofield
1995 Images of Mycenaeans: a recently acquired painted papyrus from El-Amarna. In W.V. Davies and L. Schofield (eds.), *Egypt, the Aegean and the Levant. Interconnections in the Second Millennium BC*, 125-26. London: British Museum Press.

Parrot, A.
1958 *Mission Archéologique de Mari II. Le Palais: Peintures Murales*. Paris: Librairie Orientaliste Paul Geuthner.

Pastoureau, M.
2000 *Blue: The History of a Colour*. Princeton and Oxford: Princeton University Press.

Patrianakou-Iliaki, A.
1978 The Excavation of wall-paintings at the site of Akrotiri. In C. Doumas (ed.), *Thera and the Aegean World I. Papers Presented at the Second International Scientific Congress, Santorini, Greece, August* 1978, 657-60. London: Thera and the Aegean World.

Patterson, T.C.
2005 Craft specialisation, the reorganization of production relations and state formation. *Journal of Social Archaeology* 5.3: 307-37.

Peltenburg, E.
1996 From isolation to state formation in Cyprus, c. 3500–1500 BC. In V. Karageorghis and D. Michaelidis (eds.), *The Development of the Cypriot Economy from the Prehistoric Period to the Present Day*, 17-43. Nicosia: Leventis Foundation.

Perdikatsis, V.
1998 Analysis of Greek Bronze Age wall painting pigments. In S. Colinart and M. Menu (eds.), *La Couleur dans la*

*Peinture et l' Emaillage de l' Egypte Ancienne. Actes de la Table Ronde, Ravello, 20–22 Mars 1997*, 103-108. Bari: Epuglia.

Perdikatsis, V., V. Kilikoglou, S. Sotiropoulou, and E. Chryssikopoulou

2000 Physiochemical characterization of pigments of Theran wall paintings. In S. Sherratt (ed.), *The Wall Paintings of Thera: Proceedings of the First International Symposium. Petros M. Nomikos Conference Centre, Thera, Hellas. 30 August-4 September 1997*, 103-18. Athens: Petros M. Nomikos and the Thera Foundation.

Peregrine, P.

1991 Some political aspects of craft specialisation. *World Archaeology* 23.1: 1-11.

Perlès, C.

2001 *The Early Neolithic in Greece: The First Farming Communities in Europe*. Cambridge: Cambridge University Press.

Petrie, W.M.F.

1903 *Abydos*. Memoir of the Egypt Exploration Fund, XXIV, part II. London: Egypt Exploration Fund.

Pfaffenberger, B.

1988 Fetishised objects and humanised culture: towards an anthropology of technology. *Man* n.s., 23: 236-52.

1992 Social anthropology of technology. *Annual Review of Anthropology* 21: 491-516.

Philippakis, S.E., V. Perdikatsis and T. Paradellis

1976 An analysis of blue pigment from the Greek Bronze Age. *Studies in Conservation* 21: 143-53.

Phillips, J.

1995 False analogies: the Minoan origin of some so-called 'Egyptianising' features. In *Pepragmena tou 2th Diethnous Kritologikou Sunedriou*, vol. A2: 757-65. Rethymnon: Dimos Rethymnis. Istoriki kai Laografiki Etaireia Rethymnis.

2001 Stone vessel production: new beginnings and new visions in New Palace Crete. In A.J. Shortland (ed.), *The Social Context of Technological Change. Egypt and the Near East, 1650–1550 BC. Proceedings of a Conference Held at St. Edmund Hall, Oxford, 12–14 September 2000*, 73-91. Oxford: Oxbow.

Phillips, J.

2004 The odd man out: Minoan scarabs and scaraboids. In M. Bietak and E. Czerny (eds.), *Scarabs of the Second Millennium BC from Egypt, Nubia, Crete and the Levant: Chronological and Historical Implications. Papers of a Symposium, Vienna, 10–13 January 2002*, 161-70. Vienna: Verlag der Österreichischen Akademie der Wissenschaften.

Photos-Jones E., R.E. Jones and A.J. Hall.

2003 Appendix 3. Technical report on painted plaster fragments from the Greek-Swedish excavations at Kastelli, Khania, Crete. In E. Hallager (ed.), *Greek-Swedish Excavations at Chania*, 2: 306-20. Stockholm: P. Åström's Förlag.

Pinch, G.

1994 *Magic in Ancient Egypt*. London: British Museum Press.

Politis, T.

2001 Gold and granulation: exploring the social implications of a prestige technology in the Bronze Age Mediterranean. In A.J. Shortland (ed.), *The Social Context of Technological Change. Egypt and the Near East, 1650–1550 BC. Proceedings of a Conference Held at St. Edmund Hall, Oxford, 12–14 September 2000*, 161-94. Oxford: Oxbow.

Polz, D.

1997 An Egyptian painter's utensils from Dra' Abu el-Naga. *Egyptian Archaeology* 10: 34-35.

Popham, M.R.

1994 Late Minoan II to the end of the Bronze Age. In D. Evely, H. Hughes-Brock and N. Momigliano (eds.), *Knossos: A Labyrinth of History. Papers Presented in Honour of Sinclair Hood*, 89-102. Oxford: Oxbow.

Porada, E.

1981–82 The cylinder seals found at Thebes in Boeotia. *Archiv für Orientforschung* 28: 1-70.

Portugali, Y., and A. B. Knapp
1985 Cyprus and the Aegean: a spatial analysis of interaction in the 17th–14th centuries BC. In A.B. Knapp and T. Stech (eds.), *Prehistoric Production and Exchange, the Aegean and the Eastern Mediterranean*. UCLA Institute of Archaeology Monograph 25: 44-78. Los Angeles: UCLA Institute of Archaeology.

Postrel, V.
2003 *The Substance of Style: How the Rise of Aesthetic Value is Remaking Commerce, Culture, and Consciousness*. New York: HarperCollins.

Poursat, J.-C., and M. Loubet
2005 Métallurgie et contacts extérieurs à Mallia (Crète) au Minoen Moyen II: remarques sur une série d'analyses isotopiques du plomb. In R. Laffineur and E. Greco (eds.), *EMPORIA. Aegean in the Central and eastern Mediterranean. Proceedings of the 10th International Aegean Conference, Athens, 14–18 April 2004*. Aegaeum 25: 117-22. Liège-Austin: Université de Liège-University of Texas.

Preucel, R.W., and I. Hodder
1996 Constructing identities. In R.W. Preucel and I. Hodder (eds.), *Contemporary Archaeology in Theory: A Reader*, 601-14. Oxford: Blackwell.

Preusser, F.
1991 Scientific and technical examination of the tomb of queen Nefertari at Thebes. In S. Cather (ed.), *The Conservation of Wall Paintings: Proceedings of a Symposium Organised by the Courtauld Institute of Art and the Getty Conservation Institute, London, July 13–16, 1987*, 1-12. Santa Monica: Getty Conservation Institute.
2000 Overview of Analytical Approaches to the Characterisation of Ancient Painted Surfaces. Paper presented at the Conference held in Thessaloniki, Aristotle University, Greece, 'Colour in Ancient Greece', 12–16 April, 2000.

Price, C.A.
1996 *Stone Conservation: An Overview of Current Research*. Los Angeles: Getty Conservation Institute.

Profi, S., L. Weier and S.E. Philippakis
1974 X-ray analysis of Greek Bronze Age pigments from Mycenae. *Studies in Conservation* 19: 105-12.
1976 X-ray analysis of Greek Bronze Age pigments from Knossos. *Studies in Conservation* 21: 34-39.

Pulak, C.
1988 The Bronze Age shipwreck at Uluburun, Turkey: 1985 campaign. *American Journal of Archaeology* 92: 1-37.
1997 The Uluburun Shipwreck. In S. Swiny, R.L. Hohlfelder and H. Wylde Swiny (eds.), *Res Maritimae: Cyprus and the Eastern Mediterranean from Prehistory to Late Antiquity*. Cyprus American Archaeological Research Institute Monograph Series 1: 233-62. Atlanta: Scholars Press.
2005 Who were the Mycenaeans aboard the Uluburun ship? In R. Laffineur and E. Greco (eds.), *EMPORIA. Aegean in the Central and Eastern Mediterranean. Proceedings of the 10th International Aegean Conference, Athens, 14–18 April 2004*. Aegaeum 25: 295-310. Liège-Austin: Université de Liège-University of Texas.

Pydyn, A.
1998 Universal or relative? Social, economic and symbolic values in Central Europe in the transition from the Bronze Age to the Iron Age. In D. Bailey (ed.), *The Archaeology of Value: Essays on Prestige and the Process of Valuation*. British Archaeological Reports, International Series 730: 97-105. Oxford: Archaeopress.

Rahmstorf, L.
2006 Zur Ausbreitung vorderasiatischer Innovationen in die frühbronzezeitliche Ägäis. *Prähistorische Zeitschrift* 81.1: 49-96.

Raiswell, R.
2001 Defining the burial environment. In

D.R. Brothwell and A.M. Pollard (eds.), *Handbook of Archaeological Sciences*, 595-603. Chichester/New York: John Wiley and Sons.

Ramsey, C.B., S.W. Manning and M. Galimberti

2004 Dating the volcanic eruption at Thera. *Radiocarbon* 46-1: 325-44.

Rassart-Debergh, M.

1997 Peinture Crétienne d'Egypte: techniques. In H. Béarat, M. Fuchs, M. Maggetti and D. Paunier (eds.), *Roman Wall Painting: Materials, Techniques, Analysis and Conservation. Proceedings of the International Workshop, Fribourg, 7-9 March 1996*, 193-97. Fribourg: Institute of Mineralogy and Petrography.

Raymond, A.

2005 Importing culture at Miletus: Minoans and Anatolians at Middle Bronze Age Miletus. In R. Laffineur and E. Greco (eds.), *EMPORIA. Aegean in the Central and eastern Mediterranean. Proceedings of the 10th International Aegean Conference, Athens 14–18 April 2004*. Aegaeum 25: 185-92. Liège-Austin: Université de Liège-University of Texas.

Read, H.H.

1971 *Rutley's Elements of Mineralogy*. London: Thomas Murby.

Redford, D.

2000 Egypt and western Asia in the Late New Kingdom: an overview. In E.D. Oren (ed.), *The Sea Peoples and Their World: A Reassessment*. University Museum Monograph 108, University Museum Symposium Series 11: 1-20. Philadelphia: University of Pennsylvania Press.

Reedy, T.J., and C.L. Reedy

1992 *Principles of Experimental Design for Art Conservation Research*. Getty Conservation Institute Scientific Program Report, January 1992. Newark/Marina del Rey: University of Delaware, Getty Conservation Institute.

Reents-Budet, D.

1998 Elite Maya pottery and artisans as social indicators. In C.L. Costin and R.P. Wright (eds.), *Craft and Social Identity*. Archaeological Papers of the American Anthropological Association 8: 71-89. Arlington, Virginia: American Anthropological Association.

Reese, D.

1979–80 Industrial exploitation of *Murex* shells: purple-dye and lime production at Sidi Khrebish, Benghazi (Berenice). *Lybian Studies* 11: 79-93.

Rehak, P., and J. Younger

2001 Review of Aegean prehistory VII: Neopalatial, Final Palatial, and Postpalatial Crete. In T. Cullen (ed.), *Aegean Prehistory: A Review*. American Journal of Archaeology Supplement 1: 383-473. Boston: Archaeological Institute of America.

Renfrew, A.C.

1978 Phylakopi and the Late Bronze Age I period in the Cyclades. In C. Doumas (ed.), *Thera and the Aegean World I. Papers Presented at the Second International Scientific Congress, Santorini, Greece, August* 1978, 403-21. London: Thera and the Aegean World.

1985 *The Archaeology of Cult: The Sanctuary at Phylakopi*. British School at Athens Supplementary Volume 18. London: BSA, Thames and Hudson.

2001 Symbol before concept: material engagement and the early development of society. In I. Hodder (eds.), *Archaeological Theory Today*, 122-40. Cambridge: Polity Press.

2003 *Figuring It Out. What Are We? Where Do We Come from? The Parallel Visions of Artists and Archaeologists*. London: Thames and Hudson.

2004 Towards a theory of material engagement. In E. DeMarrais, C. Gosden and C. Renfrew (eds.), *Rethinking Materiality: The Engagement of Mind with the Material World*. McDonald Institute Monographs, 23-31. Oxford: Oxbow.

Renfrew, C., and J.F. Cherry

1986 Preface. In C. Renfrew and J.F. Cherry (eds.), *Peer Polity Interaction and Socio-Political Change*, vii-viii. Cambridge: Cambridge University Press.

Renfrew, C., N. Brodie, C. Morris and C. Scarre (eds.)
2007 *Excavations at Phylakopi in Melos 1974-77*. British School of Athens. Supplementary Volume 42, 371-97. London: British School at Athens.

Reusch, H.
1956 *Die Zeichnerische Rekonstruktion des Frauenfrieses im Böotischen Theben.* Abhandlungen der Deutschen Akademie der Wissenschaften. Berlin: Akademie-Verlag.

Riede, F.
2006 *Chaîne opératoire, chaîne évolutionaire?* Putting technological sequences into an evolutionary perspective. In B. Gravina (ed.), *Technologies: Changing Matter, Changing Minds.* Archaeological Review from Cambridge 21.1: 50-75. Cambridge: Department of Archaeology.

Robins, G.
1994 *Proportion and Style in Ancient Egyptian Art*. Austin: Texas University Press.
2001 The use of the squared grid as a technical aid for artists in eighteenth dynasty painted Theban tombs. In W.V. Davies (ed.), *Colour and Painting in Ancient Egypt*, 60-62. London: British Museum Press.

Rodenwaldt, G.
1912 Die Fresken des Palastes. In G. Karo (ed.), *Tiryns. Die Ergebnisse der Ausgrabungen des Instituts*, Vol. II. Athens: Eleutheroudakis and Barth.
1921 *Der Fries des Megarons von Mykenai.* Halle: Max Niemeyerm.

Roe, P.G.
1995 Style, society, myth, and structure. In C. Carr and J.E. Neitzel (eds.), *Style, Society, and Person: Archaeological and Ethnological Perspectives*, 27-76. New York: Plenum Press.

Ronca, F.
1994 Protein determination in polychrome stone sculptures, stuccoes and gesso grounds. *Studies in Conservation* 39: 107-20.

Rosen, S.A.
1997 Craft specialisation and the rise of secondary urbanism: a view from the southern Levant. In W.E. Aufrecht, N.A. Mirau and S.W. Gauley (eds.), *Urbanism in Antiquity: From Mesopotamia to Crete*. Journal for the Study of the Old Testament Supplementary Series 244: 82-91.

Ross, J.C.
2005 Representation, reality and ideology. In S. Pollock and R. Bernbeck (eds.), *Archaeologies of the Middle East: Critical Perspectives*, 327-50. Oxford: Blackwell.

Roux, V.
1990 The psychological analysis of technical activities: a contribution to the study of craft specialisation. *Archaeological Review from Cambridge* 9.1: 142-53.

Rowlands, M.J.
1994 The politics of identity in archaeology. In G.C. Bond and A. Gilliam (eds.), *Social Construction of the Past: Representation as Power*. One World Archaeology 24: 129-43. London: Routledge.

Rutkowski, B.
1980 Religious elements in the Theran frescoes. In C. Doumas (ed.), *Thera and the Aegean World I. Papers Presented at the Second International Scientific Congress, Santorini, Greece, August* 1978, 661-66. London: Thera and the Aegean World.

Sackett, J.R.
1993 Style and ethnicity in archaeology: the case for isochrestism. In M.W. Conkey and C.A. Hastorf (eds.), *The Use of Style in Archaeology*, 32-43. Cambridge: Cambridge University Press.

Saitta, D.J.
1994 Agency, class, and archaeological interpretation. *Journal of Anthropological Archaeology* 13.3: 201-27.

Sasson, J.M.
1990 Artisans...artists: documentary perspectives from Mari. In A.C. Gunter (ed.), *Investigating Artistic Environments in the Ancient Near East*, 21-27.

Washington: Arthur M. Sackler Gallery, Smithsonian Museum.

Schiffer, M.B.

1987 *Formation Processes of the Archaeological Record.* Albuquerque: University of New Mexico Press.

2004 Studying technological change: a behavioural perspective. *World Archaeology* 36.4: 579-85.

Schiffer, M.B., and J.M. Skibo

1987 Theory and experiment in the study of technological change. *Current Anthropology* 28: 595-622.

1992 Theory and experimentation in the study of technological change. In M.B. Schiffer (ed.), *Technological Perspectives on Behavioural Change*, 44-76. Tucson-London: University of Arizona Press.

Schlanger, N.

1994 Mindful technology: unleashing the *chaîne opératoire* for an archaeology of mind. In C. Renfrew and E.B.W. Zubrow (eds.), *The Ancient Mind: Elements of Cognitive Archaeology*, 143-51. Cambridge: Cambridge University Press.

Schliemann, H.

1886 *The Prehistoric Palace of the Kings of Tiryns: The Results of the Latest Excavations.* New York: Charles Scribner's Sons.

Schoep, I., and C. Knappett

2004 Dual emergence: evolving heterarchy, exploding hierarchy. In J.C. Barrett and P. Halstead (eds.), *The Emergence of Civilisation Revisited.* Sheffield Studies in Aegean Archaeology 6: 21-37. Oxford: Oxbow.

Seeber, R.

2000 The technique of plaster preparation for the Minoan wall paintings at Tell el-Dab'a, Egypt—preliminary results. In S. Sherratt (ed.), *The Wall Paintings of Thera: Proceedings of the First International Symposium. Petros M. Nomikos Conference Centre, Thera, Hellas, 30 August-4 September 1997*, 91-102. Athens: Petros M. Nomikos and the Thera Foundation.

Seeher, J.

2002 *Hattusha Guide: A Day in the Hittite Capital.* Istanbul: Ege Yayınları.

Serpico, M., and R. White

2000a Oil, fat and wax. In P.T. Nicholson and I Shaw (eds.) *Ancient Egyptian Materials and Technology*, 390-429. Cambridge: Cambridge University Press.

2000b Resins, amber and bitumen. In P.T. Nicholson and I. Shaw (eds.) *Ancient Egyptian Materials and Technology*, 430-74. Cambridge: Cambridge University Press.

Settle, F. (ed.)

1997 *Handbook of Instrumental Techniques for Analytical Chemistry.* Upper Sadle River: Prentice Hall PTR.

Shaw, I.

2004 Identity and occupation: how did individuals define themselves and their work in the Egyptian New Kingdom? In J. Bourriau and J. Phillips (eds.), *Invention and Innovation: The Social Context of Technological Change 2: Egypt, the Aegean and the Near East, 1650–1150 BC. Proceedings of a Conference Held at the McDonald Institute for Archaeological Research, Cambridge, 4-6 September 2002*, 12-24. Oxford: Oxbow.

Shaw, J.W.

1973 *Minoan Architecture: Materials and Techniques.* Rome: Istituto Poligrafico del Stato.

Shaw, M.C.

1996 The bull leaping fresco from below the Ramp House at Mycenae: a study in iconography and artistic transmission. *Annual of the British School at Athens* 91: 167-90.

1997 Aegean sponsors and artists: reflections of their roles in the patterns of distribution of themes and representational conventions in the murals. In R. Laffineur and P.P. Betancourt (eds.), *TEXNH. Craftsmen, Craftswomen and Craftsmanship in the Aegean Bronze Age. Proceedings of the 6th International*

*Aegean Conference, Philadelphia, Temple University, 18–21 April 1996*. Aegaeum 16: 481-504. Liège-Austin: Université de Liège-University of Texas.

2000a Anatomy and execution of complex Minoan textile patterns in the Procession Fresco from Knossos. In A. Karetsou (ed.), *Kriti-Aigyptos. Polutismikoi Desmoi Trion Chilieton. Meletes*, 52-63. Athens: Ypourgeio Politismou-Archaiologiko Mouseio Herakleiou (in Greek).

2000b Sea voyages: the fleet fresco from Thera, and the Punt reliefs from Egypt. In S. Sherratt (ed.), *The Wall Paintings of Thera: Proceedings of the First International Symposium. Petros M. Nomikos Conference Centre, Thera, Hellas, 30 August–4 September 1997*, 267-82. Athens: Petros M. Nomikos and the Thera Foundation.

2003 Grids and other drafting devices in Minoan and other Aegean wall painting. A comparative analysis including Egypt. In K.P. Foster and R. Laffineur (eds.), *METRON. Measuring the Aegean Bronze Age. Proceedings of the 9th International Aegean Conference, New Haven, Yale University, 18–21 April 2002*. Aegaeum 24: 179-89. Liège-Austin: Université de Liège-University of Texas.

2004a The 'Priest-King' Fresco from Knossos: man, woman, priest, king or someone else? In A.P. Chapin (ed.), *Charis. Essays in Honor of Sara A. Immerwahr*. Hesperia Supplement 33: 65-84. Athens: American School of Classical Studies.

2004b The painted pavilion of the 'Caravanserai' at Knossos. In L. Morgan (ed.), *Aegean Wall Painting: A Tribute to Mark Cameron*. British School at Athens Studies 13: 91-111. London: British School at Athens.

Shelmerdine, C.W.

2001 The palatial Bronze Age of the southern and central Greek mainland. In T. Cullen (ed.), *Aegean Prehistory: A Review*. American Journal of Archaeology Supplement 1: 329-77. Boston: Archaeological Institute of America.

Shennan, S.

1989 Introduction: archaeological approaches to cultural identity. In S. Shennan (ed.), *Archaeological Approaches to Cultural Identity*. One World Archaeology 10: 1-32. London: Unwin Hyman.

Sherratt, A.

1997 *Economy and Society in Prehistoric Europe: Changing Perspectives*. Edinburgh: Edinburgh University Press.

1994 Comments on Ora Negbi, 'The Libyan Landscape from Thera. A Review of Aegean Enterprises Overseas in the Late Minoan IA Period'. *Journal of Mediterranean Archaeology* 7: 237-40.

Sherratt, S.

2001 Potemkin palaces and route-based economies. In S. Voutsaki and J. Killen (eds.), *Economy and Politics in the Mycenaean Palace States*. Cambridge Philological Society Supplementary Volume 27: 214-38. Cambridge: Cambridge Philological Society.

2003 The Mediterranean economy: 'globalization' at the end of the second millennium BCE. In W.G. Dever and S. Gitin (eds.), *Symbiosis, Symbolism, and the Power of the Past: Canaan, Ancient Israel, and Their Neighbors, from the Late Bronze Age through Roman Palaestina*, 37-62. Winona Lake, IN: Eisenbrauns.

Sherratt, S. (ed.)

2000 *The Wall Paintings of Thera: Proceedings of the First International Symposium. Petros M. Nomikos Conference Centre, Thera, Hellas, 30 August–4 September 1997*. Athens: Petros M. Nomikos and the Thera Foundation.

Sherratt, A., and S. Sherratt

2001 Technological change in the eastern Mediterranean Bronze Age: capital, resources and marketing. In A.J. Shortland (ed.), *The Social Context of Technological Change: Egypt and the Near East, 1650–1550 BC. Proceedings of a*

*Conference Held at St. Edmund Hall, Oxford, 12–14 September 2000*, 15-38. Oxford: Oxbow.

Shimada, I.
2005 Experimental archaeology. In H.D.G. Maschner and C. Chippindale (eds.), *Handbook of Archaeological Methods* 2: 603-42. Lanham/New York: Altamira Press.

Shortland, A.J.
2000 *Vitreous Material at Amarna: The Production of Glass and Faience in 18th Dynasty Egypt*. British Archaeological Reports, International Series 827. Oxford: Archaeopress.

Shortland, A.J. (ed.)
2001 *The Social Context of Technological Change: Egypt and the Near East, 1650–1550 BC. Proceedings of a Conference Held at St. Edmund Hall, Oxford, 12–14 Sept. 2000*. Oxford: Oxbow.

Skibo, J.M.
1992 *Pottery Function: A Use-Alteration Perspective*. New York/London: Plenum Press.
2000 Experimental archaeology. In L. Ellis (ed.), *Archaeological Method and Theory: An Encyclopedia*, 199–204. New York/London: Garland.

Skibo, J.M., and M.B. Schiffer
2001 Understanding artifact variability and change: a behavioural framework. In M.B. Schiffer (ed.), *Anthropological Perspectives on Technology*. Amerind Foundation New World Studies 5: 139-50. Albuquerque: University of New Mexico Press.

Sofaer Derevenski, J. (ed.)
2000 *Children and Material Culture*. London: Routledge.

Stark, M.T.
1991 Ceramic production and community specialization: a Kalinga ethnoarchaeological study. *World Archaeology* 23: 64-78.

Steel, L.
1998 The social impact of Mycenaean imported pottery in Cyprus. *Annual of the British School in Athens* 93: 285-96.

Steel, L.
2004 *Cyprus before History: From the Earliest Settlers to the End of the Bronze Age*. London: Duckworth.

Stein, G.J.
1996 Producers, patrons, and prestige: craft specialists and emergent elites in Mesopotamia from 5500–3100 BC. In B. Wailes (ed.), *Craft Specialisation and Social Evolution: In Memory of V. Gordon Childe*. University Museum Monograph 93: 25-38. Philadelpia: University Museum of Archaeology and Anthropology, University of Pennsylvania.
1999 Rethinking world-systems: power, distance, and diasporas in the dynamics of interregional interaction. In P.N. Kardulias (ed.), *World-Systems Theory in Practice: Leadership, Production, and Exchange*, 153-77. Oxford/New York: Rowman and Littlefield.

Stevenson Smith, W.
1998 *The Art and Architecture of Ancient Egypt*. Revised by W. Kelly Simpson. New Haven: Yale University Press. Pelican History of Art.

Stone, P.G., and P.G. Planel (eds.)
1999 *The Constructed Past: Experimental Archaeology, Education and the Public*. One World Archaeology 36. London-New York: Routledge.

Stos-Gale, Z.
2001 Minoan foreign relations and copper metallurgy in MM III-LM III Crete. In A.J. Shortland (ed.), *The Social Context of Technological Change: Egypt and the Near East, 1650–1550 BC. Proceedings of a Conference Held at St. Edmund Hall, Oxford, 12–14 September 2000*, 195-210. Oxford: Oxbow.

Strange, J.
1980 *Caphtor/Keftiu: A New Investigation*. Leiden: Brill.

Sutton, S.B.
1994 Settlement patterns, settlement perceptions: rethinking the Greek village. In P.N. Kardulias (ed.), *Beyond the Site: Regional Studies in the Aegean*

*Area*, 313-35. Lanham/New York/ London: University Press of America.

Swindler, M.H.
1929 *Ancient Painting: From Earliest Times to the Period of Christian Art*. New Haven: Yale University Press.

Talalay, L.E.
2005 The gendered sea: iconography, gender, and Mediterranean prehistory. In E. Blake and A.B. Knapp (eds.), *The Archaeology of Mediterranean Prehistory*, Blackwell Studies in Global Archaeology: 130-55. Oxford: Blackwell.

Taraqji, A.F.
1999 Nouvelles découvertes sur les relations avec l'Egypte à Tel Sakka et à Keswé, dans la région de Damas. *Bulletin de la Société Française d'Egyptologie* 144 (Mars): 27-43.

Taylor, J.
2000 Cultural depositional processes and post-depositional problems. In R. Francovich, H. Patterson and G. Barker (eds.), *Extracting Meaning from Ploughsoil Assemblages*. The Archaeology of Mediterranean Landscapes 5: 16–28. Oxford: Oxbow.

Televantou, C.A.
1992 Theran wall-painting: artistic tendencies and painters. In R. Laffineur and J.L. Crowley (eds.), *EIKΩN. Aegean Bronze Age Iconography: Shaping a Methodology. Proceedings of the 4th International Aegean Conference, University of Tasmania, Hobart, Australia, 6-9 April 1992*. Aegaeum 8: 145-59. Liège: Université de Liège.
2000 Aegean Bronze Age wall painting: the Theran workshop. In S. Sherratt (ed.), *The Wall Paintings of Thera: Proceedings of the First International Symposium. Petros M. Nomikos Conference Centre, Thera, Hellas, 30 August-4 September 1997*, 831-43. Athens: Petros M. Nomikos and the Thera Foundation.

Thaler, U.
2006 Constructing and reconstructing power. The palace of Pylos. In J. Maran, C. Juwig, H. Schwengel and U. Thaler (eds.), *Constructing Power: Architecture, Ideology and Social Practice*. Geschichte Forschung und Wissenschaft 19: 93–116. Hamburg: Lit Verlag.
2007 Ahhiyawa and Hatti: palatial perspectives. In S. Antoniadou and A. Pace (eds.), *Mediterranean Cross-Roads*, 291-324. Athens: Pierides Foundation.

Thompson, D.V.
1960 *The Craftsman's Handbook. "Il Libro dell'Arte" Cennino d'Andrea Cennini*. translated by D.V. Thompson Jr. New York: Dover.

Thyrza-Sparks, R.
2004 Canaan in Egypt: archaeological evidence for a social phenomenon. In J. Bourriau and J. Phillips (eds.), *Invention and Innovation: The Social Context of Technological Change 2: Egypt, the Aegean and the Near East, 1650–1150 BC. Proceedings of a Conference Held at the McDonald Institute for Archaeological Research, Cambridge, 4-6 September 2002*, 25-54. Oxford: Oxbow.

Tomas, H.
2003 Who were the readers of Linear A inscriptions? In A. Brysbaert, N. de Bruijn, E. Gibson, A. Michael and M. Monaghan (eds.), *SOMA 2002. Symposium on Mediterranean Archaeology*. British Archaeological Reports, International Series 1142: 129-41. Oxford: Archaeopress.

Tournavitou, I.
1995 *The 'Ivory Houses' at Mycenae*. British School at Athens, Supplement 24. London: British School at Athens.

Van de Mieroop, M.
1999 *Cuneiform Texts and the History of Writing: Approaching the Ancient World*. London: Routledge.

Vandenabeele, P., A. von Bohlen, L. Moens, R. Klockenkämper, F. Joukes and G. Dewispelaere
2000 Spectroscopic examination of two Egyptian masks: a combined method approach. *Analytical Letters* 33 (15): 3315-32.

Vandenabeele, P., and L. Moens
2003 Micro-Raman Spectroscopy of natural and synthetic indigo samples. *The Analyst* 128: 187-93.

Verhoeven, M.
2005 Ethnoarchaeology, analogy, and ancient society. In S. Pollock and R. Bernbeck (eds.), *Archaeologies of the Middle East. Critical Perspectives*, 251-70. Oxford: Blackwell.

Vickers, M., and D. Gill
1994 *Artful Crafts: Ancient Greek Silverware and Pottery*. Oxford: Clarendon Press.

Von Rüden, C.
2006 Die Wandmalerein von Tell Mishrife/Qatna im Kontext Interregionaler Kommunikation. Unpublished PhD Thesis, Albert Ludwig University of Freiburg, Germany.

Voutsaki, S.
2001 Economic control, power and prestige in the Mycenaean world: the archaeological evidence. In S. Voutsaki and T. Killen (eds.), *Economy and Politics in the Mycenaean Palace States. Proceedings of a Conference Held 1-3 July 1999 in the Faculty of Classics, Cambridge*. Cambridge Philological Society. Supplementary Volume 27: 195-213. Cambridge: Cambridge Philological Society.

Wace, A.J.B., and W. Lamb
1921–23 Excavations at Mycenae. § VIII. The palace. *Annual of the British School at Athens* 25: 147-282.

Wace, H., and C. Williams
1963 *Mycenae Guide*. Meriden: Meriden Gravure.

Wailes, B.
1996 V. Gordon Childe and the relations of production. In B. Wailes (ed.), *Craft Specialisation and Social Evolution: In Memory of V. Gordon Childe*. University Museum Monograph 93, 3-14. Philadelpia: University Museum of Archaeology and Anthropology, University of Pennsylvania.

Walls, N.
1996 Ugaritic documents from Ugarit. In A.B. Knapp (ed.), *Near Eastern and Aegean Texts from the 3rd to the 1st Millennium BC*. Sources for the History of Cyprus 2: 36-40. New York: Greece and Cyprus Research Center.

Warren, P.
1972 *Myrtos: An Early Bronze Age Settlement in Crete*. British School of Archaeology at Athens Supplementary Volume 7. London: British School of Archaeology at Athens-Thames and Hudson.
1995 Minoan Crete and Pharaonic Egypt. In W.V. Davies and L. Schofield (eds.), *Egypt, the Aegean and the Levant: Interconnections in the Second Millennium BC*, 1-18. London: British Museum Press.
2000 Crete and Egypt: the transmission of relationships. In A. Karetsou (ed.), *Kriti-Aigyptos: Polutismikoi Desmoi Trion Chilieton. Meletes*, 24-28. Athens: Ypourgeio Politismou-Archaiologiko Mouseio Herakleiou (in Greek).
2004 *The Early Minoan Tombs of Lebena, Southern Crete*. Paper presented at the Open Meeting of the British School at Athens, Greece.
2005 A model of iconographical transfer. The case of Crete and Egypt. In I. Bradfer-Burdet, B. Detournay and R. Laffineur (eds.), *Kris Technitis. L'Artisan Crétois. Recueil d'Articles en l'Honneur de Jean-Claude Poursat, Publié à l'Occasion des 40 Ans de la Découverte du Quartier Mu*. Aegaeum 26: 221-27. Liège-Austin: Université de Liège-University of Texas.

Watrous, L.V.
1987 The role of the Near East in the rise of the Cretan palaces. In R. Hägg and N. Marinatos (eds.), *The Function of the Minoan Palaces. Proceedings of the Fourth International Symposium at the Swedish Institute in Athens, 10–16 June 1984*, 65-70. Göteborg: P. Åström's Förlag.
2001 Crete from earliest prehistory through the Protopalatial period. In T. Cullen (ed.), *Aegean Prehistory: A Review*. American Journal of Archaeology Supplement 1: 157–223. Boston: Archaeological Institute of America.

Watson, P.J.
1999 Ethnographic analogy and ethnoarchaeology. In T. Kapitan (ed.), *Archaeology, History and Culture in Palestine and the Near East: Essays in Memory of Albert E. Glock*. ASOR Books 3: 47-65. Atlanta: Scholars Press.

Wattenmaker, P.
1998 Craft production and social identity in Northwest Mesopotamia. In C.L. Costin and R.P. Wright (eds.), *Craft and Social Identity*. Archaeological Papers of the American Anthropological Association 8: 47-55. Arlington, Virginia: American Anthropological Association.

Weatherhead, F.
2001 Recording and conservation of painted plaster from the early excavations at Amarna. In W.V. Davies (ed.), *Colour and Painting in Ancient Egypt*, 53-59. London: British Museum Press.
2007 *Amarna Palace Paintings*. London: Egypt Exploration Society.

Wedde, M.
2005 The Mycenaean galley in context: from fact to idée fixe. In R. Laffineur and E. Greco (eds.), *EMPORIA. Aegean in the Central and Eastern Mediterranean. Proceedings of the 10th International Aegean Conference, Athens, 14–18 April 2004*. Aegaeum 25: 29-38. Liège-Austin: Université de Liège-University of Texas.

Wehling, B., P. Vandenabeele, L. Moens, R. Klockenkämper, A. von Bohlen, G. Van Hooydonk and M. de Reu
1999 Investigation of pigments in medieval manuscripts by Micro Raman Spectroscopy and Total Reflection X-Ray Fluorescence Spectrometry. *Mikrochimica Acta* 150: 253-60.

Weingarten, J.
2000 The transformation of Egyptian Taweret into the Minoan Genius. In A. Karetsou (ed.), *Kriti-Aigyptos. Polutismikoi Desmoi Trion Chilieton. Meletes*, 114-19. Athens: Ypourgeio Politismou-Archaiologiko Mouseio Herakleiou (in Greek).

White, D.
2003 Multum in parvo: Bate's island on the NW coast of Egypt. In N.C. Stampolidis (ed.), *Sea Routes...From Sidon to Huelva. Interconnections in the Mediterranean. 16th-6th c.* BC, 71-82. Athens: Museum of Cycladic Art.

Whitelaw, T.M.
2005 A tale of three cities: chronology and Minoanisation at Phylakopi in Melos. In A. Dakouri-Hild and S. Sherratt (eds.), *AUTOCHTHON. Papers Presented to O.T.P.K. Dickinson on the Occasion of His Retirement. Institute of Classical Studies, University of London, 9 November 2005*. British Archaeological Reports, International Series 1432: 37-69. Oxford: Archaeopress.

Wiener, M.H.
1984 Crete and the Cyclades in LM I: the tale of the conical cups. In R. Hägg and N. Marinatos (eds.), *The Minoan Thalassocracy: Myth and Reality*, 17-26. Stockholm: Swedish Institute in Athens.
1990 The isles of Crete? The Minoan thalassocracy revisited. In D.A. Hardy, C.G. Doumas, I.A. Sakellarakis and P.M. Warren (eds.), *Thera and the Aegean World III-I: Archaeology. Proceedings of the 3rd International Congress, Santorini, Greece, 3-9 September 1989*, 128-60. London: Thera Foundation.

Wilson, D.
1994 Knossos before the palaces: an overview of the Early Bronze Age (EM I-III). In D. Evely, H. Hughes-Brock and N. Momigliano (eds.), *Knossos: A Labyrinth of History. Papers Presented in Honour of Sinclair Hood*, 23-44. Oxford: Oxbow.

Wingate, M.
1985 *Small-Scale Lime-Burning: A Practical Introduction*. London: Intermediate Technology.

Winter, I.J.
2000 Thera paintings and the ancient Near East: the private and public domains of wall decoration. In S. Sherratt (ed.),

*The Wall Paintings of Thera: Proceedings of the First International Symposium. Petros M. Nomikos Conference Centre, Thera, Hellas, 30 August-4 September 1997*, 745-62. Athens: Petros M. Nomikos and the Thera Foundation.

2003 'Surpassing work': mastery of materials and the value of skilled production in ancient Sumer. In T. Potts, M. Roaf and D. Stein (eds.), *Culture through Objects. Ancient Near Eastern Studies in Honour of P.R.S. Moorey*, 403-21. Griffith Institute Oxford: Cambridge University Press.

Woolley, C.L.

1955 *Alalakh: An Account of the Excavations at Tell Atchana in the Hatay, 1937–1949*. Reports of the Research Committee of the Society of Antiquaries of London, 18. Oxford: Oxford University Press.

Wright, J.C.

2006 The social production of space and the architectural reproduction of society in the Bronze Age Aegean during the second millennium BCE. In J. Maran, C. Juwig, H. Schwengel and U. Thaler (eds.) *Constructing Power: Architecture, Ideology and Social Practice*. Geschichte Forschung und Wissenschaft 19: 49-74. Hamburg: Lit Verlag.

Wright, R.H.

1986 The boundaries of technology and stylistic change. In W.D. Kingery (ed.), *Technology and Style*. Ceramics and Civilisation 2: 1-20. Columbus: American Ceramic Society.

Wright, R.P.

1998 Crafting social identity in Ur III southern Mesopotamia. In C.L. Costin and R.P. Wright (eds.), *Craft and Social Identity*. Archaeological Papers of the American Anthropological Association 8: 57-69. Arlington, Virginia: American Anthropological Association.

Wylie, A.

2000 Questions of evidence, legitimacy, and the (dis)unity of science. *American Antiquity* 65: 227-37.

Younger, J.

2005 Some similarities in Mycenaean palace plans. In A. Dakouri-Hild and S. Sherratt (eds.), *AUTOCHTHON. Papers Presented to O.T.P.K. Dickinson on the Occasion of His Retirement. Institute of Classical Studies, University of London, 9 November 2005*. British Archaeological Reports, International Series 1432: 185-90. Oxford: Archaeopress.

Zaccagnini, C.

1983 Patterns of mobility among ancient Near Eastern craftsmen. *Journal of Near Eastern Studies* 42 (4): 245-64.

1987 Aspects of ceremonial exchange in the Near East during the late second millennium BC. In M. Rowlands, M. Larsen and K. Kristiansen (eds.), *Centre and Periphery in the Ancient World*, 57-65. Cambridge: Cambridge University Press.

Zettler, R.L.

1996 Gordon Childe and the socioeconomic position of craft specialists in Early Mesopotamia. In B. Wailes (ed.), *Craft Specialisation and Social Evolution: in Memory of V. Gordon Childe*. University Museum Monograph 93: 17-23. Philadelpia: The University Museum of Archaeology and Anthropology, University of Pennsylvania.

## Websites

Centro d'Arte Dedalo: www.dedaloarte.org; last accessed: 11/02/2008.

Frieda Kahlo exhibition, London 2005: www.tate.org.uk/modern/exhibitions/kahlo/; last accessed: 11/02/2008.

Friedrich W.L. 2006. New Research in Science: Date of the Largest Volcanic Eruption in the Bronze Age Finally Pinpointed. (Media Release Dated 27/04/2006). http: //www.nat.au.dk/default.asp?id=11296&la=UK; last accessed: 11/02/2008.

Hattusha, Central Turkey: www.dainst.org/index_643_en.html; last accessed: 11/02/2008.

Lemba Experimental Village, Cyprus: www.arcl.ed.ac.uk/arch/lemba/homepage.html; last accessed: 09/02/2008.

Uluburun shipwreck webpage: http: //ina.tamu.edu/ub_main.htm with extensive bibliography; last accessed: 11/02/2008.

# Glossary

*Al fresco*
Painting technique whereby pigments, suspended in water, are applied onto a damp lime plaster wall. The pigments get incorporated by the drying and crystallization of the surface of the lime plaster which converts from calcium hydroxide back to calcium carbonate.

*Al secco*
Painting technique whereby pigments are mixed with a binder and are applied onto a dry surface of any kind.

*Arriccio* or *arricciato*
The base plaster layer or thicker plaster layer below the *intonaco*. In the case of the Aegean BA, *intonaci* were not very often used so one cannot really speak of an *arriccio* either, since the *arriccio* usually has a rougher mixture and texture than the *intonaco* which is not applicable here.

Backing support/substrate
The surface onto which the paint layer or plaster layer is applied. E.g. mud plaster may be the backing support for the lime plaster, the *intonaco* may be the backing support for the pigments, etc.

Binder/binding medium
Most often an organic sticky substance which is mixed with pigments to make them adhere to the backing support onto which they are painted.

Cake
A small mass of pure pigment powder that has formed a lump. It is made in advance of painting and needs to be wetted in order to be usable as a pigment.

Filler/aggregate
The material of organic and/or inorganic nature that is added to the plaster matrix to bulk it out and to help it dry faster.

*Giornata*
The amount of *al fresco* painting done in the time span of one day.

*Impasto*
The application of thick white 'paint' onto the painted surface, often as a way of adding final white detail. It also gives the impression of slight relief.

*Intonaco*
A very thin pure lime plaster layer (1-3 mm), applied onto the wetted but dried out thicker plaster layer (*arriccio*). This thin layer is the support for the *al fresco* painting in the Renaissance period and this layer shows the *giornata* if any.

*Malta*
Plaster layer with a strong red colour, consisting of lime plaster into which 5% v/wt red ochre pigment is mixed. Most often used in *sgraffito* decoration for colour-contrasting effects.

*Lucidato a mestola*
A specific sheen on a dry plaster surface obtained by trowelling and flattening the plaster very well while still wet.

Primary calcite
Tiny to small fragments (mean values for

Mainland sites: 26-63 μm) of crushed uncalcined limestone (or other lime source) that forms a filler in the lime plaster matrix.

*Sinopia*
The type of drawing and sketching done on the *arriccio*, usually in red, yellow or black, in preparation for the actual design that will appear on the *intonaco*. This is done as a sort of trial to see how much work may be done or fit in a one-day scheme.

Secondary calcite
Calcined limestone (or other lime source) that forms the lime plaster matrix, normally with a very small homogeneous grain size (around 5-10 μm).

*Sgraffito*
Meaning 'to scratch', the technique applied to ceramics and plaster whereby a punch (sharp tool) is used to scrape or scratch a design through an opaque layer in order to reveal the coat beneath, hence often revealing a contrasting colour.

To 'knock up'
Plasterers' term to describe the action of breaking down slaked lime that starts setting by throwing it with the trowel back into the mass. This action avoids having to add extra water to the mixture before it can be used.

# Index

compiled by Graham Shipley

www.ingramcontent.com/pod-product-compliance
Lightning Source LLC
LaVergne TN
LVHW070850210926
848482LV00016B/909

*9781781792537*